2

# Biotechnological Inventions and Patentability of Life

NEW DIRECTIONS IN PATENT LAW

**Series Editors:** Gustavo Ghidini, *University of Milan and LUISS Guido Carli University*, Emanuela Arezzo, *LUISS Guido Carli University and Catholic University of the Sacred Heart, Italy*, Rochelle Dreyfuss, *New York University School of Law* and Jerome H. Reichman, *Duke University, USA*

Patent law is a unitary system regulating different technological sectors, each with its own peculiarities and features. As science moves forward, new products and processes are created, hence new challenges arise for patent law to deal with. *New Directions in Patent Law* aims at promoting a constant monitoring and in-depth analysis of the emerging issues patent law is called to face. It also intends to spur the international academic debate by means of scholarly contributions offering a comparative view of how different patent systems approach and try to solve new questions posed by technological advancement. The books in this series will move beyond a mere legal approach, to engage in economic-based analysis of patent law as well as sociological and jurispolitical assessment of the impact that normative and jurisprudential evolutions might have on the development of science and technology.

Titles in the series include:

Biotechnology and Software Patent Law
A Comparative Review of New Developments
*Edited by Emanuela Arezzo and Gustavo Ghidini*

Biotechnological Inventions and Patentability of Life
The US and European Experience
*Andrea Stazi*

# Biotechnological Inventions and Patentability of Life

## The US and European Experience

Andrea Stazi

*Department of Human Sciences, European University of Rome, Italy*

NEW DIRECTIONS IN PATENT LAW

Edward Elgar PUBLISHING

Cheltenham, UK • Northampton, MA, USA

Published by
Edward Elgar Publishing Limited
The Lypiatts
15 Lansdown Road
Cheltenham
Glos GL50 2JA
UK

Edward Elgar Publishing, Inc.
William Pratt House
9 Dewey Court
Northampton
Massachusetts 01060
USA

A catalogue record for this book
is available from the British Library

Library of Congress Control Number: 2014957098

This book is available electronically in the **Elgar**online
Law subject collection
DOI 10.4337/9781784715908

MIX
Paper from responsible sources
FSC® C013056

ISBN 978 1 78471 589 2 (cased)
ISBN 978 1 78471 590 8 (eBook)

Typeset by Columns Design XML Ltd, Reading
Printed and bound in Great Britain by T.J. International Ltd, Padstow

# Contents

PART IV THE EVOLUTION OF THE DISCIPLINE AND PROBLEMATIC ISSUES IN EUROPE



PART V CONCLUSIONS: BIOTECHNOLOGICAL INVENTIONS AND THE PATENTABILITY OF LIVING BEINGS IN THE UNITED STATES AND EUROPEAN MODELS BETWEEN LEGAL ISSUES AND BIOETHICAL PROFILES – THE UTILITY OF THE COMPARATIVE METHOD, THE PERSPECTIVE OF THE LEGAL PROCESS AND RELATIONS WITH OTHER SCIENCES AND HUMANISTIC DISCIPLINES (IN PARTICULAR BIOETHICS)

# Acknowledgements

This book has been made possible within a Research Project of National Interest (PRIN) financed by the Italian Ministry of Education, University and Research, having as its object 'The legal regulation of the Information and Communication Technologies (ICT) as a tool for enhancement of inclusive, innovative and safe societies'.

The author would like to thank Professors Alberto Gambino, Gustavo Ghidini and Vincenzo Zeno-Zencovich for their valuable and insightful comments. However, the opinions expressed and any errors or omissions remain the responsibility of the author.

# PART I

# Patent protection of biotechnological inventions and the limits of patentability

## I.1 BIOTECHNOLOGY AND PROTECTION OF BIOTECHNOLOGICAL INVENTIONS

The concept of biotechnology refers to a wide range of techniques using living organisms. According to the accepted meaning, biotechnology can be defined as a group of techniques which use living organisms or their parts in order to create or modify products, to improve the characteristics of plants or animals, or to develop microorganisms or organisms destined for specific uses. In other words, biotechnology includes all techniques which use or cause organic changes in biological material, micro-organisms or animals, or cause changes in inorganic material, using biological methods.[1]

---

[1] According to the Convention on Biological Diversity (Rio de Janeiro, 5 June 1992): 'Biotechnology means any technological application that uses biological systems, living organisms, or derivates thereof, to make or modify products or processes for specific use'. With regard to the definition of the concept of biotechnology, in doctrine see, among others: A. Kornberg and T.A. Baker, *DNA Replication*, New York: University Science Books, 2005 (2nd edn.), *passim*; L.M. Houdebine, 'Biotecnologie', in *Nouvelle Enciclopedie de Bioéthique*, *sous la direction de* G. Hottois *et* J.N. Missa, Bruxelles: De Boeck, 2001, p. 145 et seq.; J.D. Watson and J. Tooze, *The DNA Story: a documentary history of gene cloning*, San Francisco/Oxford: W.H. Freeman, 1981, *passim;* H.F. Judson, *The Eighth Day of Creation: Makers of the Revolution in Biology*, New York: Cold Spring Harbor, 1996 (25th Anniversary Edition), *passim*; G.S.

A fundamental distinction within the subject exists between traditional biotechnology and so-called innovative or advanced biotechnology. The former consists in biological processes applied by man for centuries for the production of food and drink, cheese and wine, the cultivation of plants and the disposal of waste.[2] Innovative or advanced biotechnology, first developed in the 1970s, is based on the combined use of new biological techniques – such as genetic engineering, the culture of cells *in vitro*, the production of monoclonal antibodies, etc. – applied to various production sectors but brought together with organic material or biological organisms being the final or intermediate products.[3]

Today's biotechnology involves living organisms and their direct genetic modification. Until recently, the modification of living organisms could occur only through gradual selection. The genes which caused the modification of a particular organism could be chosen only from the full group of genes of the species to which the organism belonged.

In 1953 Watson and Crick designed a double helix model of DNA, which allowed for the theorization of the duplication mechanism of genetic material, providing a scientific explanation of the heredity phenomena (previously the subject of Mendel's observations), based fundamentally on subsequent study and intervention in the field of genetics.[4]

In particular, the technological innovations of the 1970s, which facilitated the development of the first recombinant DNA techniques aimed at creating new molecules of DNA through the unification of DNA

Stent and R. Calendar, *Molecular Genetics: An Introductory Narrative*, San Francisco: CBS Publishers & Distributors, 1986 (2nd edn.), *passim.*

[2] Until the second half of the 1800s, these techniques were limited to the activity of fermentation of microorganisms without the awareness of the related process. It was only after the discoveries of Pasteur that the fermentation processes began to use pure cultures of microorganisms, and as a consequence the industry of fermentation emerged.

[3] In this regard, see E.S. Grace, *Biotechnology Unzipped: Promises & Realities*, Washington DC: Joseph Henry Press, 1997, p. 1 et seq.

[4] Namely science, the branch of biology, which studies the genes, heredity and genetic variability of organisms. For a reconstruction of the developments recalled in the text, see D. Hartl and E. Jones, *Genetics: Analysis of Genes and Genomes*, Burlington (USA): Jones & Bartlett, 2008 (7th edn.), *passim*; R.C. King, W.D. Stansfield and P.K. Mulligan, *A Dictionary of Genetics*, Oxford: Oxford University Press, 2006 (7th edn.), *passim.*

fragments from different species, presented important opportunities for 'genetic engineering', an expression which is used to indicate artificially introduced modifications of the genetic information of a cell through the insertion of other genetic information into it.[5]

These techniques are based on the availability of scientific data regarding the structure of DNA.[6] Given that all living material contains the same type of DNA, the exchange of genetic material between organisms was deemed possible.

In 1973 Cohen and Boyer demonstrated that the DNA of different species could be assembled and inserted into a host organism. The process of assembling DNA is called recombinant DNA technology. Alternatively, the combination of genes of different organisms in order to introduce new properties to the host organism is called genetic engineering. The way in which the procedure is applied to animals is that of modifying genes through the insertion of altered DNA, which causes them to produce substances they normally do not.[7] The organisms whose genetic structure has been modified in this way are called transgenic organisms. These modified microorganisms, plants and animals are used as bioreactors[8] for the production of pharmaceutical substances. Insulin and the human growth hormone are examples of substances which have been produced by microorganisms for many years.

---

[5] On the theme see, among others: L. Yount, *Biotechnology and Genetic Engineering*, New York: Facts On File, 2008 (3rd edn.), *passim*; D.S.T. Nicholl, *An Introduction to Genetic Engineering*, Cambridge: Cambridge University Press, 2008 (3rd edn.), *passim*; J.D. Watson, *Recombinant DNA: Genes and Genomes: A Short Course*, San Francisco: W.H. Freeman, 2007, *passim*; S. Smiley, *Genetic Modification: Study Guide (Exploring the Issues)*, Cambridge: Independence Educational Publishers, 2005, *passim*; R.D. Hotchkiss, 'Portents for a Genetic Engineering' (1965) 56 *Journal of Heredity* 197 et seq.

[6] Deoxyribonucleic acid, the substance responsible for the hereditary characteristics revealed by Watson and Crick in 1953.

[7] The three principal methods of production of genetically modified animals are microinjection of DNA in the pronucleus, the retroviral invection and the constitution in vitro of embryonic cells then injected into the host cells. In this regard, see R. Moufang, 'Patentability of Genetic Inventions in Animals' (1989) 20 *IIC – International Review of Industrial Property and Copyright Law* 823 et seq.

[8] That is living containers for the production of medicines.

In 2003, the mapping of the human genome was completed,[9] which promises significant opportunities for the medical field and for the protection of health.[10]

---

[9] The human genome is constituted by the set of chromosomes contained in every cell, and contains the complete genetic heredity of a human organism, or, in other terms, the necessary genetic instructions for the development and management of every activity initiated by the organism. More correctly, the genome is the genetic material contained in a gamete, for which every somatic cell of a eukaryotic organism (typically diploid) such as humans, contain two complete genomes. Aploid organisms, like bacteria, have, instead, a single genome (usually a single circular DNA molecule). On this theme, see M. Ridley, *Genome*, New York: Harper Perennial, 2006; as well as the website of the National Human Genome Research Institute in the United States of America, accessed 12 December 2014 at http://www.genome.gov.

[10] The Human Genome Project was an international scientific research project, initiated in 1990 in 18 countries and coordinated by the United States of America, whose main objective was to determine the sequence of base pairs that make up DNA (in this regard, refer to Part II) and to identify and map all the genes of the human genome from the physical and functional point of view. The main objective of the Human Genome Project is to understand the function of genes belonging to the human race, but the project also studied various other non-human organisms. The genome of any individual (except that of monozygote twins and cloned organisms) is unique; thus, mapping the human genome means proceeding to the sequencing of the multiple variations of each gene. The first draft of the mapping of the genome was released in 2000, while the complete mapping was published in 2003 (except for some small residual intervals for which further research is being carried out, given that the methods of sequencing currently known do not allow their resolution). The researchers who carried out the basic sequencing were, on one hand, those of the Human Genome Project, with the conclusion of the sequencing being announced in June 2000 by the President of the United States Bill Clinton and the British Prime Minister Tony Blair; and on the other, those of a private company, Celera Genomics (founded and directed by the biologist Craig Venter), who announced the sequencing for the first time in April 2000. In February 2001 the results of both the studies were published, respectively in the magazines *Nature* and *Science*. The two sequences were not found to be significantly different from each other, and they were publicly available to the scientific community, meeting the demands made in this sense at an international level in the preceding months. What is more, following the sequencing of the nearly 3 billion base genes of the human genome is the discovery of single genes (which are held to be around 100,000 in all), or the parts of DNA responsible for the functioning of the human body, and if altered by illness (many of which, however, are monogenic, but are connected to a complex interaction between numerous genes and environmental factors). On this theme, besides what follows in Part II, see, among others: C. Kuppuswamy, *The International Legal Governance of the Human Genome*, London: Routledge,

With the advent of biotechnology, humanity is facing an extraordinary revolution, with profound implications for man and his relationship with other creatures.

Biotechnology, through its wide range of applications – relative, in particular to: (a) healthcare; (b) production and industrial processes; (c) the spheres of agriculture, breeding, veterinary and fish farming[11] – with the new awareness and opportunities which it provides, is having and will have an ever more significant impact on our way of life. Today, biotechnology seems able, for instance, to provide a solution for hunger in the world, to cure illnesses and improve the quality of life. Genetic and biological manipulation allows for diagnosis and gene therapy, as well as vaccines from plants and bio-medicines to create health treatments based on individual needs.[12]

In today's highly technological world, biotechnology is one of the most innovative and highly invested in industries for research, in the field of science. The main commercial applications of genetic engineering have been in the field of health care, agriculture and the environment. In the sector of health care, biotechnology has been used to produce medicines for the treatment of illnesses such as cystic fibrosis or various forms of cancer. With the help of biotechnology, diagnostic kits have also been produced. Currently, research is moving forward in regard to gene therapy,[13] which is aimed at correcting congenital pathologies. In agriculture, biotechnology uses recombinant DNA techniques in the process of

---

2009, *passim*; N.S. Mosier and M.R. Ladisch, *Modern Biotechnology: Connecting Innovations in Microbiology and Biochemistry to Engineering Fundamentals*, New York: Wiley & Sons, 2009, p. 385 et seq.; F. Lenzerini, 'Biotechnology, Human Dignity and the Human Genome', in F. Francioni and T. Scovazzi (eds), *Biotechnology and International Law*, Oxford/Portland (USA), Hart Publishing, 2006, p. 285 et seq.

[11] For an extensive analysis on the subject, see European Commission Joint Research Centre – Institute for Prospective Technological Studies, '*Consequences, Opportunities and Challenges of Modern Biotechnology for Europe ('Bio4EU study')*', 2007, accessed at: http://ipts.jrc.ec.europa.eu/publications/pub.cfm?id=1470.

[12] See B. Amani, *State Agency and the Patenting of Life in International Law*, Farnham (UK)/Burlington (USA): Ashgate Publishing, 2009, p. 20.

[13] Gene therapy consists in using nucleic acid as a therapeutic composition in medicine. The simplest method of use for gene therapy is to compensate for the abnormal expression of a gene. In this regard, see M.A. Croyle, 'Gene Therapy', in D.J.A. Crommelin, R.D. Sindelar and B. Meibohm (eds), *Pharmaceutical Biotechnology: Fundamentals and Applications*, New York: Informa Healthcare, 2007 (3rd edn.), p. 175 et seq.

raising animals or cultivating plants, to produce animals and plants with the required characteristics.[14] With regard to the environment, biotechnology has allowed for the production of transgenic microorganisms for the purpose of purifying the soil, water or air.

In order to protect the results of biotechnological research, legal regulations provide a system of protection based on the exclusive patent, which guarantees the inventor the right of exclusivity over the biotechnological finding.

In general, a biological patent is intended as that which is relative to an invention in the field of biology, the natural science regarding the study of life and living organisms. It may consist in a composition of matter, a method for obtaining or utilizing one or more of these, or a product which combines such elements. Contrarily, the hypothesis that a patent application may concern a naturally occurring biological substance in itself, regardless of whether it refers to a specific procedure or use associated with it, given that it has been sufficiently 'isolated' from its natural state, is one of the most controversial points of the debate on the subject.[15]

The genetic patent, according to the favored definition, is in reference to a specific and isolated genetic sequence, its chemical composition, the process used to obtain or use it, or a combination of these. Genetic patents fall under the broadest category of biological patents. What is more, the patentability of natural genetic sequences is not unanimous, and patents on genes have been allowed only in regard to isolated gene sequences with well-known purposes, and not for those naturally present in human beings or other living organisms.[16]

---

[14] To date, examples of transgenic plants include tomatoes, potatoes, sugar beets and tobacco. In this regard and more generally on the theme of protection for plant varieties and so-called genetically modified organisms (GMO), see, among others, O. Mills, *Biotechnological Inventions. Moral Restraints and Patent Law*, Farnham (UK)/Burlington (USA): Ashgate Publishing, 2010 (Revised Edition), p. 155 et seq.; C. Brega, 'La tutela giuridica delle novità vegetali in Europa: dalle origini alla direttiva 98/44/CE sulla protezione giuridica delle invenzioni biotecnologiche', in G. Ghidini and G. Cavani (eds), *Brevetti e biotecnologie*, Rome: Luiss University Press, 2007, p. 155 et seq.

[15] On this point refer, extensively, to the following Parts.

[16] '*Federal Register – Utility Examination Guidelines – Comments and Responses*', 5 January 2001; see also: Wikipedia.org, entries on *Biological patent* and *Gene patent*, respectively accessed 12 December 2014 at http://en.wikipedia.org/wiki/Biological_patent and http://en.wikipedia.org/wiki/Gene_patent (regarding the accuracy of the source, see J. Giles, 'Special Report: Internet Encyclopaedias Go Head to Head' (2005) 438 *Nature* 900 et seq.).

A human gene patent, finally, may be understood as that relative to a product or process which includes the single, specific human gene sequence. The sequence may be natural or synthetic, created in a laboratory through biotechnology (even if it is based on a natural human genetic sequence).[17]

In the United States of America, in 1980 the Supreme Court made a decisive contribution to the development of the biotechnology industry with its crucial decision in the case of *Diamond v. Chakrabarty*,[18] regarding the patentability of a microorganism not existing in nature and produced through genetic engineering. In particular, the Court ruled that 'anything under the sun made by man' is subject to patenting, identifying human intervention as a key element to distinguishing patentable inventions from non-patentable 'principles of nature and natural phenomena'.[19]

From the *Chakrabarty* ruling on, in the United States, and from the 1990s in Europe, thousands of patents related to inventions based on genetic material or information have been granted.[20] Some of these patents constitute important intellectual property assets for companies dedicated to the translation of biomedical research findings into diagnostic agents and life-saving therapies. They have played a key role in providing innovators with a sufficient period of exclusivity in order to recover the significant investments needed to develop and secure market interest for biotechnology products.[21]

---

[17] On this and in general regarding the definition of a genetic patent, see the extensive considerations of C. M. Holman, 'Gene Patents Under Fire: Weighing the Costs and Benefits', in E. Arezzo and G. Ghidini (eds), *Biotechnology and Software Patent Law: A Comparative Review of New Developments*, Cheltenham, UK and Northampton, MA, USA: Edward Elgar, 2011, p. 260 et seq.; E. Arezzo and G. Ghidini, 'The Impact of Human Gene Patents on Innovation and Access: A Survey of Human Gene Patent Litigation' (2007) 76 *UMKC Law Review* 295 et seq., especially at 307 et seq.; see also, for a more limited definition: K. Jensen and F. Murray, 'Intellectual Property Landscape of the Human Genome' (2004) 310 *Science* 239 et seq.

[18] *Diamond v. Chakrabarty*, 447 U.S. 303 (1980).

[19] In this regard, refer to what shall be said extensively, in Part III.

[20] See C.A. Fowler, 'Ending Genetic Monopolies: How the TRIPs Agreement's Failure to Exclude Gene Patents Thwarts Innovation and Hurts Consumers Worldwide' (2010) 25 *American University International Law Review* 1073 et seq.

[21] See J.H. Graham *et al.*, 'High Technology Entrepreneurs and the Patent System: Results of the 2008 Berkeley Patent Survey' (2010) 24 *Berkeley Technology Law Journal* 1255 et seq.; C.M. Holman, 'Learning from Litigation: What Can Lawsuits Teach Us about the Role of Human Gene Patents in Research and Innovation?' (2009) 18 *Kansas Journal of Law and Public Policy*

Today, the companies focused on the development of personalized medicine and on pharmacogenomics[22] – technology which is widely held to be central to future pharmaceutical development and healthcare – hold genetic patents as essential to securing the revenues necessary to introduce these products to the market.

In 2001, the Patent and Trademark Office of the United States published formal Guidelines,[23] in which it illustrated its position on the basis of which inventions based on genes are patentable, given that there is sufficient human intervention to satisfy the 'made by man' standard dictated by the *Chakrabarty* decision.[24]

Outside the United States of America, there has been greater objection to the patentability of genetic material. In Europe this objection has caused some countries to exclude or limit the patentability of genes existing in nature. In 1998, in any case, following an arduous procedure, the European Union adopted a directive on biotechnology inventions,[25] through which Member states were asked to allow the patentability of genes. Some Member states, however, further limited the sphere or

---

215 et seq.; K. Jensen and F. Murray, 'Intellectual Property Landscape of the Human Genome', cit., p. 239 et seq.

[22] Pharmacogenomics is the branch of biology which has developed from pharmacogenetics (regarding the study or clinical experimentation of genetic variations which result in different responses to medicines) and from the Human Genome Project, and can be defined as a science which is interested in the manner in which the knowledge of the human genome and its products (RNA and proteins) can be used in the discovery and development of new pharmaceuticals. This is based on the analysis of the entire genome of an individual in order to identify the genes that may be used as targets for new treatments, as well as individual genetic profiles from which the response to certain administered active ingredients may arise. See A. Squassina, M. Manchia, V.G. Manolopoulos, M. Artac, C. Lappa-Manakou, S. Karkabouna, K. Mitropoulos, M. Del Zompo and G.P. Patrinos, 'Realities and expectations of pharmacogenomics and personalized medicine: impact of translating genetic knowledge into clinical practice' (2010) 11 *Pharmacogenomics* 1149 et seq.; W. Kalow, U.A. Meyer and R.F. Tyndale (eds), *Pharmacogenomics*, New York: Taylor & Francis, 2005 (2nd edn).

[23] 'Utility Examination Guidelines', 66 *Fed. Reg.* 1092 (5 January 2001).

[24] In particular, the 'Guidelines' stated that the isolation of a DNA sequence existing in nature may lead to a product which is sufficiently different from the natural gene to render it patentable, citing a series of legal precedents which supported the patentability of biological molecules existing in nature, such as adrenaline (regarding the 'Guidelines' see, extensively, Part III).

[25] Directive 98/44/EC of the European Parliament and the Council of 6 July 1998 on the legal protection of biotechnological inventions.

applicability of these patents with respect to the American regulation. Nevertheless, all currently permit genetic patents.[26]

Despite these developments, patentability of genes remains a highly debated issue, even in the United States of America, and those opposed to it have been asking for some time that the possibility be brought to an end, or at least that the sphere of patents for innovations based on genetic sequences existing in nature be greatly limited.[27]

Biotechnology, further, also offers, besides the opportunities referred to above, new tools for information and operations regarding human life,[28] which raise fundamental questions regarding, on the one hand, the preclusive effects of genetic patents (in particular with regard to their impact on subsequent innovation and access to genetic diagnostic tests),[29] and on the other, the moral and bioethical profiles of the same.

Thus, despite the basic simplicity of the patent procedure, the complexity of today's technology, together with the pace at which it is evolving, leads to problems in patent law.

Further issues for inventors have arisen in regard to the development of patent legislation at the national and international levels. Today's inventor must face two problems that traditional mechanical inventors did not: the nature of the technology itself (which uses living material) and the deficiencies of international patent regulations.

The regulatory framework of the patent discipline has been conditioned, clearly, by the experience of man in forming and altering inanimate materials. Manufactured products were made from simple components. Patent protection was introduced in reference to the mechanical sector, concerning specific devices. Mechanical findings have certain basic characteristics, among which in particular is that of carrying

---

26 In this regard (referring for further details to Part IV), see *ex multis*: B.A. Brody, 'Intellectual Property and Biotechnology: The European Debate' (2007) 17 *Kennedy Institute of Ethics Journal* 69 et seq.

27 For a systematic framework of the debate see C.M. Holman, *The Impact of Human Gene Patents on Innovation and Access: A Survey of Human Gene Patent Litigation*, cit. in UKMC Law Review, 2007, accessed 5 January 2015 at http://papers.ssrn.com/sol3/papers.cfm?abstract_id=1090562, p. 295 et seq.

28 In this regard, see among others: Nuffield Council on Bioethics, 'Ethical Principles: Respect for Human Lives and the Human Body' (April 1995) *Human Tissue Ethical and Legal Issues* p. 39 et seq.

29 On this point see, for instance: Department of Health & Human Services (USA) – Secretary's Advisory Committee on Genetics, Health, and Society, *Gene Patents and Licensing Practices and Their Impact on Patient Access to Genetic Test*', April 2010, accessed 12 December 2013 at http://oba.od.nih.gov/oba/SACGHS/reports/SACGHS_patents_report_2010.pdf.

out generally one function and of not reproducing itself in subsequent generations. Thus, it was reasonable that every technical application could be characterized by a simple list of structural characteristics, without the need to require the inventor to supply specific indications of its function or to foresee rules regarding subsequent generations. What is more, in the traditional model, the sequence of steps in order to arrive at the invention could be clearly defined.[30]

The policy underlying the patent system, as was first recognized within the U.S. Constitution[31] and therefore in models of the 1800s, was based essentially on a 'quid pro quo' mechanism: strong protection through an exclusive right granted for a limited time – aimed at supplying the inventor with financial incentive for research and development of the invention – in exchange for the public disclosure of information regarding the same.[32]

Biotechnology, which concerns living matter, poses significant challenges for patent law,[33] given that biotechnological inventions do not fall so neatly under the classical model, as do those of mechanics. On the other hand, the reasons at the foundation of patent protection, which in particular are the incentive to innovation and disclosure of information regarding inventions, according to the prevalent reconstruction are also deemed to be essential to the matter of biotechnological research – which

---

30 In this regard see, among others: G. Ghidini, *Innovation, Competition and Consumer Welfare in Intellectual Property Law*, Cheltenham, UK and Northampton, MA, USA: Edward Elgar, 2010, p. 33 et seq.; S.A. Bent, R.L. Schwaab, D.G. Conlin and D. Jeffrey, *Intellectual Property Rights in Biotechnology Worldwide*, New York: Stockton Press, 1987, *passim*.

31 On the basis of Article 1, Section 8(8), according to which: 'The Congress shall have Power… To promote the Progress of Science and useful Arts, by securing for limited Times to Authors and Inventors the exclusive Right to their respective Writings and Discoveries'.

32 In this regard, in addition to referring extensively below, as well as the Conclusions, see, *ex multis*, O. Mills, *Biotechnological Inventions. Moral Restraints and Patent Law*, cit., p. 41.

33 For instance, according to some scholars, scientists and firms, the traditional assumption that the patent is granted as recognition for an invention appears to be reductive, due to the subtle line which exists between invention and discovery in the medical, pharmaceutical and agricultural fields. On this point see C.B. Onwuekwe, 'Plant Genetic Resources and the Associated Traditional Knowledge: Does the Distinction between Higher and Lower Life Forms Matter?', in J. Gibson (ed.), *Patenting Lives: Life Patents, Culture and Development*, Farnham (UK)/Burlington (USA): Ashgate Publishing, 2008, p. 139 et seq.

often requires significant investment and long time frames – justifying then the granting of patents in regard to their relative findings.[34]

Given that pre-existing legislation was not drafted with the particular characteristics of biotechnological inventions in mind, significant difficulties emerge in the application of patent law to such a constantly emerging and rapidly changing situation as that which is under discussion.[35] The conditions of novelty and originality are where much of the uncertainty lies with regard to the decision as to if and when biotechnological inventions are patentable. The condition of novelty requires that the invention not be available to others prior to the filing of the patent request. The issue is not if the subject of the request already exists, but if it is already known. The condition of originality requires, in addition, that the invention not be obvious to the average expert in the field.

There is a fundamental question that is posed regarding the patenting of biotechnological inventions concerning the interpretation of provisions of exclusion from patentability,[36] also in consideration of their close link with the moral/bioethical profiles[37] of biotechnology.

---

[34] See Mills, *Biotechnological Inventions. Moral Restraints and Patent Law*, cit., p. 14. With regard to the debate on the theme, see below, as well as in the Conclusions.

[35] See, among others: B. Sherman, 'Patent Law in a Time of Change: Non-obviousness and Biotechnology' (1990) 10 *Oxford Journal of Legal Studies* 278 et seq.

[36] Provided by article 53 of the European Convention on Patents, in regard to which refer below, to the subsequent paragraphs.

[37] In general, morality can be defined as the system of rules, values and ideals considered universally valid and important as guidelines for action (F.W.A. Brom, J.M.G. Vorstenbosch and E. Schroten, 'Public Policy and Transgenic Animals: Case-by-case Assessment as a Moral Learning Process', in P. Wheale, R. von Schomberg and P. Glasner (eds), *The Social Management of Genetic Engineering*, Farnham (UK)/Burlington (USA), Ashgate Publishing, 1998, p. 249 et seq.; see as well: Wikipedia.org, entry *Morality*, accessed 12 December 2014 at http://en.wikipedia.org/wiki/Morality). As relevant as this concept is in this case, it is held to be, on the other hand, too broad; as such it is necessary to identify limits and declinations of the same as far as biotechnology is concerned. In this sphere, then, it is opportune to refer to the concept of bioethics, a discipline which deals with moral issues in connection with biological and medical research; in particular, it is defined as 'the systematic study of the moral dimensions – including moral vision, decisions, conduct and policies – of the life sciences and health care, employing a variety of ethical methodologies in an interdisciplinary setting' (W.T. Reich (ed.), *Encyclopedia of Bioethics*, New York: MacMillan, 1995, 2nd edn, p. xxi). The coining of the term 'bioethics' is attributed to Fritz Jahr, who in 1927 spoke of 'imperative bioethics' in regard to

Though the science of biotechnology has been used for some time, in the United States of America and in Europe there is still confusion about the nature of the technology, its benefits and the limits which should be set in order to prevent related risks. In this regard, the most urgent and relevant issues of the debate concern the issues connected, in fact, to the risks that biotechnology poses, on the one hand, for the environment, and on the other, for human beings: the latter aspect, in light of the bioethical profiles it implies, constitutes the fundamental subject of analysis carried out hereinafter.

---

man's exploitation of flora and fauna. Concerning the relation between bioethics and religion, almost all believers of any religious faith share the idea that that which surrounds us originates from an act of creation, that is divine intervention, supernaturally. This initial setting, as a consequence, leads to issues involving bioethics, which themselves have an effect on some of the most significant issues of human life (birth, sexuality, death). In traditionally Catholic countries, the role of Catholic bioethics is relevant. Official Catholic bioethics – contained in the documents of the Church Magisterium, in the works of authors found to be doctrinally in line with them and in the scientific community to which it refers – is founded on the principles of the sanctity and unavailability of life, holding that a human person is not the creator or owner of his own life. According to Catholic bioethics, each human has the right to life, from the moment of conception until his natural death. In favor of this argument, besides referring extensively to Part II, see *ex multis,* Wikipedia.org, entry *Bioethics,* accessed 12 December 2014 at http://en.wikipedia.org/wiki/Bioethics; B. Bryan, 'Biotechnology, Bioethics and Liberalism: Problematizing Risk, Consent and Law' (2003) 11 *Health Law Journal* 119 et seq.; Committee of Inquiry into Human Fertilisation and Embryology (coord. M. Warnock), *Report of the Committee of Inquiry into Human Fertilisation and Embryology*, London: Her Majesty's Stationery Office, July 1984, accessed 12 December 2014 at http://www.hfea.gov.uk/docs/Warnock_Report_of_the_Committee_of_Inquiry_into_Human_Fertilisation_and_Embryology_1984.pdf; H. Jonas, *The Imperative of Responsibility: In Search of Ethics for the Technological Age*, Chicago: University of Chicago Press, 1984 (Original edition: *Das Prinzip Verantwortung: Versuch einer Ethik für die technologische Zivilisation*, Frankfurt am Main: Insel-Verlag, 1979), *passim*; G. Fornero, *Bioetica cattolica e bioetica laica*, Milan: Mondadori, 2009 (2nd edn); Congregation for the Doctrine of the (Catholic) Faith, *Instruction Dignitas Personae on certain bioethical questions*, Vatican City, 8 September 2008, accessed 12 December 2014 at http://www.vatican.va/roman_curia/congregations/cfaith/documents/rc_con_cfaith_doc_20081208_dignitas-personae_en.html; Catechism of the Catholic Church, 1997, accessed 12 December 2014 at http://www.vatican.va/archive/eng0015/_index.htm.

## I.2 THE PATENT SYSTEM IN THE UNITED STATES OF AMERICA AND IN EUROPE

The biotechnology industry is currently fundamentally important both in the United States of America and in Europe, given that biotechnology is now essential for many economic sectors and produces considerable benefits for consumers.[38]

The patent system is the key link between science and technology, on the one hand, and law on the other.[39] In the past, the junction between these disciplines led to significant tension, emphasized by the fact that certain industrial sectors (for instance foods or pharmaceuticals) received more limited protection than others. This tension was sharpened when companies began to carry out cross-border activities and to request patents outside of their countries of origin.

While scientific and technical profiles remained the same throughout the world, in particular in the United States and in Europe, the differences between patent legislation led to a wide variety of rules on the requirements, procedures and context of the protection. In particular, the various national regulations were influenced by factors such as: (a) different reasons which justify patent protection;[40] (b) the concept of patent law as economic policy;[41] and (c) the relation between patent law and the legal system of the interested country as a whole.[42]

---

[38] On this subject see, among others, G. Laurie, 'Biotechnology: Facing the Problems of Patent Law', in H.L. McQueen (ed.), *Innovation, Incentive and Reward*, Hume Papers on Public Policy vol. 5, Edinburgh: Edinburgh University Press, 1997, p. 46 et seq.

[39] See, in general, M. van Empel, *The Granting of European Patents, Introduction to the Convention on the Grant of European Patents, Munich, 5 October 1973*, Dordrecht (Netherlands)/Boston: Kluwer Academic Publishers Group, 1975.

[40] Traditionally there are four distinct theses: the thesis of 'natural right' or of 'intellectual property', that of 'compensation', that of 'incentive' and that of 'contract' or 'disclosure'. See O. Mills, *Biotechnological Inventions. Moral Restraints and Patent Law*, cit., p. 19. For a more wide framework of the various justifications underlying the patent system, see, among others, G. Ghidini, *Innovation, Competition and Consumer Welfare in Intellectual Property Law*, cit., p. 33 et seq.

[41] Of the four theses above, today the dominant one would seem to be that which emphasizes the function of patents as tools of economic policy.

[42] The enforcement of patent rights falls under the context of national rules in civil procedure. Procedures connected to property law are also subject to administrative and legal control of their respective national authorities.

In general, as is known, a patent is an intellectual property right.[43] It is, essentially, a negative right given by the state, who grants the owner the right to exclude third parties from the use or exploitation of the invention without her/his consent.[44]

---

[43] The fundamental discipline of intellectual property at an international level, as is well known, is provided in the 'Agreement on Trade Related Aspects of Intellectual Property Rights', signed in Marrakech on 15 April 1994, commonly known by the acronym TRIPS (accessed 12 December 2014 at http://www.wto.org/english/docs_e/legal_e/27-trips_01_e.htm). The TRIPS Agreement is an international treaty promoted by the World Trade Organization, in order to establish a standard for the protection of intellectual property. The agreement was formalized by the WTO at the end of the meeting held in Marrakech in 1994, at the conclusion of the so-called Uruguay Round talks. The agreement establishes the requirements that adhering countries must respect for the protection of intellectual property, in the context of patents, copyright, trademarks, protected geographical indications, industrial design, etc. TRIPS also establishes guidelines for the application of the regulations on the matter of intellectual property, for appeals and conflict resolution procedures. On the basis of the agreement, protection and application of intellectual property laws should contribute to the progress of technological innovation and to easing the transfer and disclosure of technological knowledge, in order to reciprocally benefit manufacturers and users of the relative knowledge; pursuant to art. 7, in particular: 'The protection and enforcement of intellectual property rights should contribute to the promotion of technological innovation and to the transfer and dissemination of technology, to the mutual advantage of producers and users of technological knowledge and in a manner conducive to social and economic welfare, and to a balance of rights and obligations'. The TRIPS Agreement represents an attempt to close the gap and differences in the way in which intellectual property rights are protected at a global level, in order to regulate them in a context of common international regulation. To this end, it establishes a minimum level of protection that each country must guarantee with respect to the intellectual property rights of WTO members. On this theme, see, *ex multi,*: N. Pires de Carvalho, *The TRIPS Regime of Patent Rights*, Alphen aan den Rijn (Netherlands): Kluwer Law International, 2010 (3rd edn); D. Gervais, *The TRIPs Agreement: Drafting History and Analysis*, London: Sweet and Maxwell, 2003 (2nd edn); D. Matthews, *Globalising Intellectual Property Rights: The TRIPs Agreement*, London: Routledge, 2002, *passim.*

[44] On economic profiles of the patent system, see in general, among others: V. Denicolò, 'Do Patents Over-compensate Innovators?' (2007) 22 *Economic Policy* 679 et seq.; S. Scotchmer, *Innovation and Incentives*, Cambridge (USA): MIT Press, 2004, *passim*; W.M. Landes and R.A. Posner, *The Economic Structure of Intellectual Property Law*, Cambridge (USA): Harvard University Press, 2003, p. 1 et seq.; V. Falce, *Profili pro-concorrenziali dell'istituto brevettuale*, Milan, Giuffrè, 2008 (also in regard to the philosophical foundation of the patent); D. Guellec and B. van Pottelsberghe de la Potterie, *The Economics of the*

The patent does not grant the property right of a specific tangible asset, which in the case of biotechnology would be for instance a specific sequence of DNA or protein.[45]

This confers a mere 'monopoly', more precisely a negative right of exclusion of third parties from the use of an invention relative to a product or process without the consent of the owner.

The patent exclusive, then, does not grant the owner the right to exploit the invention, but rather to prevent third parties from using the invention without her/his consent. The way in which this right is used, then, is usually entrusted to the control of the authority in charge of these matters.

The economic foundation of the patent system lies in the fact that it guarantees an incentive to innovation, representing a useful, at times indispensable, tool for the recovery of investments needed for the purposes of research and development of the invention.

In this sense, the main problematic issues within the patent system arise from the interests (seemingly conflicting) on the one hand of the innovators and/or investors, who want to recover their investments through protection obtained with the patent, and on the other of the other market operators (essentially competitors and consumers), who might be excessively damaged by the elimination of competition.[46]

Alongside such issues, but especially on the theme of biotechnological inventions of major significance, are the inherent profiles of risk of conflict between the granting of the exclusive and safeguarding the public interest, which may manifest in different hypotheses, from the

---

*European Patent System*, Oxford: Oxford University Press, 2007; for an heterodox point of view, see also M. Boldrin and D.K. Levine, 'The Case Against Intellectual Property' (2002) 92 *American Economic Review* 209 et seq.

[45] On this theme, among others, see P. Grubb and P. Thomsen, *Patents for Chemicals, Pharmaceuticals and Biotechnology*, Oxford/New York: Oxford University Press, 2010 (5th edn), p. 3 et seq.; O. Mills, *Biotechnological Inventions. Moral Restraints and Patent Law*, cit., p. 2 et seq.; L. Bently and B. Sherman, *Intellectual Property Law*, Oxford/New York: Oxford University Press, 2009 (3rd edn), p. 335 et seq.; C.M. Holman, *Gene Patents Under Fire: Weighing the Costs and Benefits*, cit., p. 267 et seq.; D.R. Koepsell, *Who Owns You? The Corporate Gold-Rush to Patent Your Genes*, New York: Wiley-Blackwell, 2009; D. Magnus, A. Caplan and G. McGee (eds), *Who Owns Life?*, Amherst, NY: Prometheus Books, 2002.

[46] See S.J.R. Bostyn, 'A European Perspective on the Ideal Scope of Protection and the Disclosure Requirement for Biotechnological Inventions in a Harmonised Patent System; The Quest for the Holy Grail?' (2002) 5 *Journal of World Intellectual Property* 1013.

conflict with fundamental rights, public order or principles of morality, to the obstruction to access of findings by other researchers at the expense of the freedom of research and subsequent innovation, or of patients to the detriment of the right to health.[47]

Where it is recognized, patent protection guarantees a right of exclusivity in order to answer the question of 'property': if a company cannot recover the costs of the invention because the necessary information is available to anyone at no cost, the level of innovation is greatly reduced. In other words, the granting of the patent right is aimed at eliminating the problem of free riding.[48]

Patent protection leads to positive externalities, consisting in opportunities for third parties to use the information indicated in the patent; which may make inventors reluctant to provide this information, or lead to a reduction of investment in research and development. The positive effects of these externalities consist, instead, in the creation of opportunities for the original innovator, who can benefit from the new information developed by competitors, or, in the case of dependent inventions, to recover a part of her/his investment through revenues from patent licenses.[49]

---

[47] In regard to these profiles refer extensively to the following Parts and the Conclusions.

[48] See K. Arrow, 'Economic Welfare and the Allocation of Resources for Invention', in R. Nelson (ed.), *Rate and Direction of Inventive Activity*, Princeton: Princeton University Press, 1962, p. 609 et seq. On the other hand, patent protection creates the so-called 'fishing problem': the possibility to obtain a patent (or the 'fish') pushes many individuals to invest their resources in order to obtain this result, but only one individual will succeed, while the investments of the others will be lost. Thus, the total social cost of the innovation is higher than that which is strictly necessary, giving rise to economic inefficiency. See S.M. Besen and L.J. Raskind, 'An Introduction to the Law and Economics of Intellectual Property' (1991) 5 *Journal of Economic Perspectives* 3 et seq.; F.M. Scherer and D. Ross, *Industrial Market Structure and Economic Performance*, Boston: Houghton Mifflin Company, 1990 (3rd edn), p. 624 et seq.; Y. Barzel, 'Optimal Timing of Innovations' (1968) 50 *Review of Economics and Statistics* 248 et seq.

[49] In this regard, among others, see S.J.R. Bostyn, *Patenting DNA sequences (polynucleotides) and scope of protection in the European Union: an evaluation – Background study for the European Commission within the framework of the Expert Group on Biotechnological Inventions*, Brussels: European Commission, 2004 (accessed 12 December 2014 at http://ec.europa.eu/internal_market/indprop/docs/invent/patentingdna_en.pdf); P.B. Joly, *Le rôle des externalités dans les systèmes d'innovation. Nouveaux regards sur le dilemme de la propriété intellectuelle* (1992) 43 *Revue économique* 785 et seq.

The objective of the patent system, both in the United States and in Europe, is to encourage innovation and the growth of new industries.[50] To these ends, the regulations grant the inventor rights over his invention in exchange for public 'disclosure' of the way it works. In this perspective, the interests of the inventor are balanced with those of the public. The patent system creates a monopoly in favour of private individuals, to whom appropriate limitations form a counterbalance. In the first place, a patent may be granted only for a finding that has the requirements of invention. What is more, the owner is granted exclusive rights regarding that specific invention. Finally, these exclusive rights, by their very nature, are temporary.

The patent right has a time limit, on the basis of the TRIPS[51] agreement (ratified by the United States and the European Union) and the European Patent Convention[52] of 20 years from the date of the application. The right is limited further by geographic terms, given that the patent is valid within the jurisdiction of the patent office from which it was issued.

In the United States, the Constitution provides the institution of a patent right granting Congress the power to promote progress of science and useful arts guaranteeing inventors, for a limited period of time, exclusive rights over their findings.[53] The provision simultaneously grants and limits power, in that Congress cannot ignore the grounds of the protection or override the limits imposed on it by constitutional dictates.[54]

In Europe, in order to guarantee states the possibility to compete at a global level, it was deemed necessary to grant the invention protection equivalent to that granted to other trading powers, such as, in particular,

---

50 In argument, see for all G. Ghidini, *Innovation, Competition and Consumer Welfare in Intellectual Property Law*, cit., p. 33 et seq.; J.M. Aubrey, 'A Justification of the Patent System', in J. Phillips (ed.), *Patents in Perspective*, Oxford: ESC Publishing, 1985, p. 1 et seq.

51 TRIPS Agreement, art. 33.

52 'Convention on the Grant of European Patents (European Patent Convention) of 5 October 1973', signed in Munich on 5 October 1973, implemented on 7 October 1977, and modified several times, art. 63.

53 United States Constitution, Article 1, Section 8(8). Concerning the U.S. regulations and connected problematic issues, see Part III.

54 In this regard, see Mills, *Biotechnological Inventions. Moral Restraints and Patent Law*, cit., p. 41 et seq.

the United States.[55] Under this context, powerful and harmonized patent protection has been introduced.

To this end, the Convention of Strasburg of 1963[56] and the European Convention on Patents of 1973[57] were originally adopted. The former was aimed in general at the harmonization of the requirements for substantial patent rights, with optional exceptions connected to bioethics. The latter aimed at the construction of a patent system which could sustain the economic structure of the EEC (at the time), with obligatory exceptions with regard to bioethics. However, the question of whether the EPC weakens patent rights or not in the name of bioethical instances remains unanswered. Similarly, on this point, Directive 98/44/EC on biotechnological inventions is problematic regarding the same issue.[58]

Both in the United States and in Europe, as provided, in fact, by the TRIPS Agreement,[59] in order to be protected by an exclusive patent, an invention must respond to three fundamental requirements, which are:

---

[55] What is more, the issue is extremely controversial. In this sense, besides ethical profiles concerning this situation, appropriate remarks in the literature are relevant, which highlight how excessive patent monopolies may damage the free development of the competitive market. In argument, beyond what shall be said in the following Chapters, see among others, S. Anderman and A. Ezrachi (eds), *Intellectual Property and Competition Law – New Frontiers*, Oxford: Oxford University Press, 2011, p. 1 et seq.; G. Ghidini, *Innovation, Competition and Consumer Welfare in Intellectual Property Law*, cit., p. 33 et seq. and 209 et seq.; J. Drexl (ed.), *Research Handbook on Intellectual Property and Competition Law*, Cheltenham, UK and Northampton, MA, USA, Edward Elgar, 2008, p. 1 et seq.; refer also to A. Stazi, *Intellectual Property Rights and Market Power in the European Union: The Fil Rouge of Consumer Welfare*, in *Comparazione e diritto civile*, June 2011 (accessed 12 December 2014 at http://www.comparazionedirittocivile.it/prova/files/stazi_intellectual.pdf), p. 1 et seq.

[56] 'Convention on the Unification of Certain Points of Substantive Law on Patents for Invention', signed by the Member states of the European Council in Strasburg on 27 November 1963, and implemented on 1 August 1980.

[57] 'European Patent Convention' (EPC), cited. The States parties to the Convention are different from the EU states: Switzerland, Liechtenstein, Turkey, Monaco, Iceland, Norway, The Republic of Macedonia, San Marino, Albania and Serbia are part of the Convention but do not belong to the EU. The Convention established the European Patent Office (EPO).

[58] In reference to the regulations and related problematic profiles, see Part IV.

[59] Art. 27 of the TRIPS Agreement explicitly lists the criteria for patentability, mentioning novelty, originality and industriality, and refers indirectly to the selection criteria of the finding, stating that 'patents shall be available for any inventions, whether products or processes, in all fields of technology, provided that they are new, involve an inventive step and are capable of industrial

- 'novelty';
- 'originality', or the inventive step (non-obviousness in the United States);
- 'industriality' (utility in the United States).

Further, the invention is required to provide 'sufficient description' (or disclosure) in the patent application.

One of the most significant questions asked at the base of patentability is the distinction between patentable inventions and non-patentable discoveries.

In the United States, the legislation does not give a clear definition of the concept of invention, nor does it make a distinction between an invention and a discovery.[60]

On the other hand, in case law a group of exceptions to patentability have been identified. In this sense, in the decision on the case *Diamond v. Diehr*[61] the Court

> has undoubtedly recognized limits to section 101 and every discovery is not embraced within the statutory terms. Excluded from such patent protection are laws of nature, natural phenomena, and abstract ideas.[62] An idea of itself is not patentable. A principle, in the abstract, is a fundamental truth; an original cause; a motive; these cannot be patented, as no one can claim in either of them an exclusive right.

Further, in the ruling on the case *Gottschalk v. Benson*[63] it has been established that:

> Phenomena of nature, though just discovered, mental processes, and abstract intellectual concepts are not patentable, as they are the basic tools of scientific and technological work. As we stated in Funk Bros. Seed Co. v. Kalo Co., 333 U.S. 127, 130, 'He who discovers a hitherto unknown phenomenon of nature has no claim to a monopoly of it which the law recognizes. If there is to be invention from such a discovery, it must come from the application of the law of nature to a new and useful end'.

---

application'; as far as concerns then the hypothesis of exclusion from patentability, see below the next paragraph.

60 See 35 USC section 101.

61 *Diamond v. Diehr*, 450 U.S. 175, 185 (U.S. Supreme Court 1981).

62 See: *Parker v. Flook*, 437 U.S. 584 (U.S. Supreme Court 1978); *Gottschalk v. Benson*, 409 U.S. 63, 67 (U.S. Supreme Court 1972); *Funk Brothers Seed Co. v. Kalo Inoculant Co.*, 333 U.S. 127, 76 U.S.P.Q. 280 (U.S. Supreme Court 1948).

63 *Gottschalk v. Benson*, 409 US, at 67.

The decision in *Parker v. Flook*,[64] again, added that:

> the rule that the discovery of a law of nature cannot be patented rests, not on the notion that natural phenomena are not processes, but rather on the more fundamental understanding that they are not the kind of 'discoveries' that the statute was enacted to protect. The obligation to determine what type of discovery is sought to be patented must precede the determination of whether that discovery is, in fact, new or obvious.

In the leading case *Diamond v. Chakrabarty*,[65] then, the Supreme Court has established the fundamental principle according to which 'everything under the sun made by man is patentable'. On the basis of U.S. case law, an invention is considered as such if it produces a useful, tangible and concrete result;[66] in other words, technical characteristics are not mentioned. This represents, without a doubt, a broad definition, but offers the advantage of clarity, and prevents the issues which arose in Europe in the attempt to determine the meaning of technical characteristics.[67]

The issue discussed up to this point is sometimes defined also as the 'product of nature doctrine'. On the basis of this doctrine, the products of nature as such are not patentable, while products derived from nature are. The principle has for some time been recognized in case law.[68] In the United States, for instance, in the ruling on the case of *Funk Brothers Seed Co. v. Kalo Inoculant Co.*, the Supreme Court held that the composite or a composition of unknown materials simply discovered in nature is not patentable.[69] In the European Union, similarly, Directive

---

64 *Parker v. Flook*, 437 US, at 593.

65 U.S. Supreme Court, *Diamond v. Chakrabarty*, 447 U.S. 303, 309, 65 L. Ed. 2d 144, 100 S. Ct. 2204 (1980) (which cites S. Rep. No. 82-1979, at 5 (1952); H.R. Rep. No. 82-1923, at 6 (1952)). Regarding the leading case, fundamental for the matter of biotechnology see Part III.

66 See, for instance: *State Street Bank & Trust Co. v. Signature Fin.Group, Inc.*, 149 F.3d 1368, 1374–75, 47 U.S.P.Q.2D (BNA) 1596, 1602 (CAFC 1998), U.S. 142 L. Ed. 2d 704, 119 S. Ct. 851(1999).

67 In regard to this point, see below.

68 See S.J.R. Bostyn, *Enabling Biotechnological Inventions in Europe and the United States. A study of the patentability of proteins and DNA sequences with special emphasis on the disclosure requirement*, Eposcript Series, no. 4, Munich: EPO, 2001, p. 69 et seq.

69 *Funk Brothers Seed Co. v. Kalo Inoculant Co.*, 333 U.S. 127, cited. In particular, Judge Douglas explained, expressing the opinion of the majority of the Court, that 'patents cannot issue for the discovery of the phenomena of nature. The qualities of these bacteria, like the heat of the sun, electricity, or the qualities of metals, are part of the storehouse of knowledge of all men. They are

98/44/EC on the protection of biotechnological inventions, in article 3(2), confirms the same concept, stating (positively) that: 'Biological matter that is isolated from its own natural environment or is produced through technical processes may be the subject of invention, even if it existed previously in its natural state'. In the same sense, the European Patent Office (EPO) 'Guidelines for Examination' clarify that:

> If a new property of a known material or article is found out, that is mere discovery and unpatentable because discovery as such has no technical effect and is therefore not an invention within the meaning of Art. 52(1). If, however, that property is put to practical use, then this constitutes an invention which may be patentable. … To find a previously unrecognized substance occurring in nature is also mere discovery and therefore unpatentable. However, if a substance found in nature can be shown to produce a technical effect, it may be patentable.[70]

In European law, the issue of the distinction between invention and discovery does not appear to be easily resolved, given that the legislation has yet to give a definition to each of the categories. This is not surprising, given that these concepts are subject to change. The EPO has traditionally held[71] that an invention must be technical in order to be considered patentable, regardless of the fact of whether or not it satisfies the requirements of patentability. The theme has long been considered hardly important, but with the developments in the field of biotechnology and software this position has become more and more problematic, especially considering that it is extremely difficult to define what

---

manifestations of laws of nature, free to all men and reserved exclusively to none'. (333 U.S., 130).

[70] 'Guidelines for Examination in the European Patent Office', Munich, EPO, 2010 (accessed 12 December 2014 at http://www.epo.org/law-practice/legal-texts/guidelines.html), C IV.2.3.1, according to which, for instance, 'a gene which is discovered to exist in nature may be patentable if a technical effect is revealed, e.g. its use in making a certain polypeptide or in gene therapy'. See as well, on this theme, the 'Joint Statement' with which, already in 1988, the USPTO, the EPO and the Japanese patent office had stated that: 'Purified natural products are not regarded under any of the three laws as products of nature or discoveries because they do not in fact exist in nature in an isolated form. Rather, they are regarded for patent purposes as biologically active substances or chemical compounds and eligible for patenting on the same basis as other chemical compounds' (cited in: Nuffield Council on Bioethics, 'The ethics of patenting DNA: a discussion paper', July 2002, accessed 12 December 2014 at http://www.nuffieldbioethics.org/patenting-dna, p. 26, note 9).

[71] On the basis of art. 52 of the EPC.

'technical' means. The case law of the EPO, on this point, has stated that this concept is intended to be a finding that has technical characteristics, or provides a technical contribution.[72] This definition, however, appears essentially tautological and therefore irresolvable.[73]

In consideration of the uncertainty regarding the concept of invention in European law, emphasis has been put on the definition of discovery. It has been established that it consists in the discovery of something already existing in nature, but that has not already been discovered. The distinction from invention lies in the fact that in order to make a discovery the inventive act is not required. A discovery, in other words, is the mere knowledge relative to something existing in nature, while invention implies the ability of a human being to use this knowledge in a technical way, the so-called 'technical information'.[74]

Once the finding has been determined to be an invention in the aforementioned sense, in order to be patentable it should also satisfy the requirements of novelty, originality, utility and of sufficient description.

The requirement of novelty requires that the invention is, in fact, new. It must not already be available to others, in any form of public disclosure, publication or use before the date of filing of the patent application. The entirety of this previous knowledge is known as 'state of the art',[75] from which to be new an invention has to differ.[76]

The reasoning behind the requirement of novelty lies in the fact that that which is already known to the public is not new, and that which is already in the public domain cannot be the subject of a private monopoly.

---

[72] See, for instance: Technical Board of Appeal, T 1173/97, *Computer program product/IBM*, 1 July 1998, EPO Official Journal, 1999, 609.

[73] See S.J.R. Bostyn, *Patenting DNA sequences (polynucleotides) and scope of protection in the European Union: an evaluation – Background study for the European Commission within the framework of the Expert Group on Biotechnological Inventions*, cit., p. 12.

[74] See again, S.J.R. Bostyn, *Patenting DNA sequences (polynucleotides) and scope of protection in the European Union: an evaluation – Background study for the European Commission within the framework of the Expert Group on Biotechnological Inventions*, cit., p. 14.

[75] Defined in the aforementioned 'Guidelines' of the EPO as 'everything made available to the public by means of a written or oral description, by use, or in any other way, before the date of filing of the European patent application' (EPO Guidelines, C IV.6).

[76] See art. 54 EPC; 35 USC 102. Further, in practice the application of the principle is not always clear, given that it is not easy to have full knowledge of that which is the existing state of the art in the field.

A specific controversial issue on the subject of novelty is made with regard to chemical substances based on substances which exist in nature. A substance can be new if it is produced in a purer form than that which occurs naturally. On the other hand, this is not sufficient in and of itself: in order to integrate the requirement of novelty, it is necessary to provide proof of the fact that the modification of the parameters of the process produces different results.[77] Basically, novelty exists if the chemical substance in a purer form is also different in form or structure from that which exists in the state of the art. The same principle is applied, *mutatis mutandis*, for DNA. What is more, in case law it has been shown that a substance with an identical structure to that which occurs in nature may be considered new if that which already exists is not readily available to the public.[78]

For the requisite of originality or the inventive step, it is difficult to establish, but it is without a doubt connected to the issue of obviousness. The invention must not be obvious to the average expert in the field regarding that which is the state of the art.[79] In other words, the invention must not merely follow logically that which is already known to be the state of the art.[80]

---

[77] See, for instance: T 0205/83, *Vinyl ester – crotonic acid copolymers/ HOECHST*, EPO Official Journal, 1985, 363.

[78] See EPO, T 0206/83, *Pyridine herbicides/ICI*, EPO Official Journal, 1987, 5, which specifies that 'a compound defined by its chemical structure can only be regarded as being disclosed in a particular document if it has been made available to the public in the sense of art. 54(2) EPC. … The requirement is fulfilled if a reproducible method is described in the same document'. Similarly, in the United States, it has been determined that a previous reference excludes the novelty of an invention which is subsequently the subject of a patent application if the reference describes or discloses the same invention in a way that makes it public domain. In this sense, see for instance: *In re Arkley*, 455 F.2d 586, 587 (CCPA 1972); *In re Brown*, 329 F.2d 1006, 1011 (CCPA 1964); *In re LeGrice*, 301 F.2d 929, 930 (CCPA 1962); in doctrine: S. Johnston, *Patent Protection for the Protein Products of Recombinant DNA* (1989) 4 *High Technology Law Journal* 257 et seq. In the sphere of DNA sequencing, the case law of the EPO determined that the mere fact that a sequence already existed in a library of DNA does not eliminate novelty, given that the sequence is not readily available to the public (See T 0301/87, *Alpha-interferons/BIOGEN*, EPO Official Journal, 1990, 335).

[79] Art. 56 EPC; 35 USC 103.

[80] According to the cited 'Guidelines' of the EPO: 'The term "obvious" means that which does not go beyond the normal progress of technology but merely follows plainly or logically from the prior art, i.e. something which does

In the United States, the test to be carried out in order to analyze the inventive step or obviousness was indicated by the Supreme Court in the leading case *Graham v. John Deere Co.*, stating that 'under section 103, the scope and content of the prior art are to be determined; differences between the prior art and the claims at issue are to be ascertained; and the level of ordinary skill in the pertinent art resolved. Against this background, the obviousness or the non-obviousness of the subject matter is determined'.[81]

In Europe, the EPO uses the so-called 'problem-and-solution approach'. At the basis of this approach, the nearest state of the art and the technical effect resulting from it is first examined. Then, the finding under discussion with its relative technical effect is identified. Finally, the objective technical problem to be resolved is defined, in order to pass from the nearest state of the art previous to the invention: if the solution of this technical problem is not obvious, the requirement of originality is deemed to be satisfied.[82]

The requirement of industriality, which up until a short time ago was not a particularly relevant theme in patent law, requires that the invention be subject to use in any industrial sector, including agriculture.[83]

---

not involve the exercise of any skill or ability beyond that to be expected of the person skilled in the art' (C IV-11.4).

[81] U.S. Supreme Court, *Graham v. John Deere Co. of Kansas City*, 148 USPQ 459, 467 (1966).

[82] See EPO 'Guidelines', CIV-11.5; Technical Board of Appeal, *Metal Refining/BASF*, 13 October 1982, EPO Official Journal, 1983, 13; T 0296/93, *Hepatitis B virus antigen production/BIOGEN*, 28 July 1994, EPO Official Journal, 1995, 627, where it is specified (7.4.4) that 'the fact that other persons (or teams) were also working on the same project might suggest that it was "obvious to try" or that it was "an interesting area to explore", but it does not necessarily imply that there was "a reasonable expectation of success". "A reasonable expectation of success", which should not be confused with the understandable "hope to succeed", implies the ability of the skilled person to reasonably predict, on the basis of the existing knowledge before the starting of a research project, a successful conclusion to the said project within acceptable time limits. The more unexplored a technical field of research is, the more difficult is the making of predictions about its successful conclusion and, consequently, the lower the expectation of success'. In doctrine, see S.J.R. Bostyn, *Enabling Biotechnological Inventions in Europe and the United States. A study of the patentability of proteins and DNA sequences with special emphasis on the disclosure requirement*, cit., p. 74.

[83] Art. 57 EPC; 35 USC 101. On the basis of the EPO 'Guidelines': "Industry" should be understood in its broad sense as including any physical activity of "technical character" … , i.e. an activity which belongs to the useful

The reasoning behind this requirement is that the invention which cannot be applied to industry does not provide benefits to society, and therefore is not worthy of patent protection. With the development of technological progress, the requirement of utility has taken on increased importance. In the field of biotechnology, the inventions concern DNA sequences without a known function. This has created doubt as to the interpretation of the utility requirement. The issue to resolve is whether, in order to satisfy this requirement, determining that the invention may potentially be used for specific purposes is sufficient or if the identification of at least one specific use is required.

In the United States, in 2001 the USPTO published a new version of the 'Utility Guidelines', according to which 'a claimed invention must have a specific and substantial utility. This requirement excludes "throw-away", "insubstantial", or "non-specific" utilities ...', specifying, in particular, that 'an invention has a well-established utility (1) if a person of ordinary skill in the art would immediately appreciate why the invention is useful based on the characteristics of the invention (e.g., properties or applications of a product or process), and (2) the utility is specific, substantial, and credible'.[84]

Similarly, in the recent case law of the EPO, a restrictive interpretation of utility has been provided.[85]

Such careful application of the requirement of utility, for instance, regarding the inventions concerning human stem cells, may hinder the granting of excessively broad patents, which block further technological developments, as well as the patentability in and of itself of fundamental research tools, supplying in this way an important 'filter' in order to

---

or practical arts as distinct from the aesthetic arts; it does not necessarily imply the use of a machine or the manufacture of an article and could cover e.g. a process for dispersing fog or for converting energy from one form to another' (IV 5.1). It is, in particular, specified that: 'Methods of testing generally should be regarded as inventions susceptible of industrial application and therefore patentable if the test is applicable to the improvement or control of a product, apparatus or process which is itself susceptible of industrial application. In particular, the utilization of test animals for test purposes in industry, e.g. for testing industrial products (for instance for ascertaining the absence of pyrogenetic or allergic effects) or phenomena (for instance for determining water or air pollution) would be patentable' (IV 5.2).

[84] 'USPTO Guidelines for Examination of Applications for Compliance with the Utility Requirement' (2001) 66 Federal Register, 5 January, § 1092 et seq.

[85] EPO Opposition Division, *ICOS/SmithKline Beecham and Duphar International Research*, 20 June 2001, EPO Official Journal, 2002, 293.

prevent monopolization of entire fields of research through excessively broad and unspecific patents.[86]

The application of requirements for patentability may indirectly influence the sphere of patent extension. This must be broad enough in order to compensate for the costs of the invention. If the field of the invention is not broad enough, minor improvements may be made by third parties without violating the patent, at the expense of the possibility to recover investments made for the achievement of the invention itself. If the field is too limited, the exclusive right may have a preclusive effect on technological development.

A limited field has, from an economic point of view, the advantage of leaving more room for subsequent innovation and competition, after the original innovation. On the other hand, increased competition may also lead to higher social costs, in the sense that it may lead, for instance, to the duplication of entry costs, inefficient production, etc.[87]

A broad field has the advantage of supplying better protection to the original creator with regard to minor improvements and from *tout court* second generation creators. This leads the original creator to disclose the invention, with the related positive effects in terms of the reduction of social costs of innovation. On the other hand, a broad field inhibits new entries and subsequent creators, reducing competition. When the field of exclusive patents is broad, there may be a tendency to disinvest, given

---

[86] On this point, see P.L.C. Torremans, 'Patentability of Human Stem Cell or Synthetic Biology Based Inventions', in E. Arezzo and G. Ghidini, *Biotechnology and Software Patent Law: A Comparative Review of New Developments*, cit., p. 288 et seq.; R.J. Aerts, 'Biotechnological Patents in Europe – Functions of Recombinant DNA and Expressed Protein and Satisfaction of the Industrial Applicability Requirement' (2008) 39 *IIC* 283 et seq.; K. Bergman, G. Graff, 'The Global Stem Cell Patent Landscape: Implications for Efficient Technology Transfer and Commercial Development' (2007) 25 *Nature Biotechnology* 419 et seq. In this regard, clearly, besides economic situations which pose fundamental ethical questions, regarding which one may refer below to the subsequent paragraphs.

[87] In virtue of the fact that the field is limited, and that more operators are attracted to entry to the market with competitive products, there is less profit for the original creator. This may lead to a reduction in the incentive to innovation, or to a tendency to keep it secret; this secret may lead to duplication of investment in research and development, given that more people will be occupied in 'reinventing the wheel', in the absence of knowledge available to the public. On this theme, see V. Denicolò, 'Patent Races and Optimal Patent Breadth and Length' (1996) XLIV *The Journal of Industrial Economics* 249 et seq.; H.F. Chang, 'Patent Scope, Antitrust Policy, and Cumulative Innovation' (1995) 26 *Rand Journal of Economics* 34 et seq.

that the broadness of the field makes it less interesting for potential competitors to enter the market or invest in improvements or in other innovations based on the original invention. The consequence of this may be a preclusion of technological development.[88]

Thus, according to the preferable reconstruction, it is better for the system to be more open, though it may create the inefficiencies referenced above, given that such a system may allow for the preservation of the opportunity for technological progress.[89]

In addition to the requirements of novelty, originality and utility, both in the United States and in Europe, there are further limits to patentability, among which that concerning the 'sufficient description' of the invention in the patent application is particularly relevant. Based on this principle, the invention must be described in a sufficiently clear manner to allow the average expert in the field to put it into practice.[90]

The reasoning is that the owner of the patent obtains a temporary monopoly on the invention in exchange for the fact of making the way in which it functions known, which is called disclosure.[91]

---

[88] See, among others, H.F. Chang, 'Patent Scope, Antitrust Policy, and Cumulative Innovation', cit., p. 52; S. Scotchmer, 'Standing on the Shoulders of Giants: Cumulative Research and the Patent Law' (1991) 5 *Journal of Economic Perspectives* 29 et seq.

[89] See S.J.R. Bostyn, *A European Perspective on the Ideal Scope of Protection and the Disclosure Requirement for Biotechnological Inventions in a Harmonised Patent System; The Quest for the Holy Grail?*, cit., p. 1016.

[90] Art. 83 EPC; 35 USC 112.

[91] See EPO Technical Board of Appeal, T 0169/83, *Wallelement/ Vereinigte Metallwerke*, decision 3.2.1 of 25 March 1985, EPO Official Journal, 1985, 193; U.S. Supreme Court, *Bonito Boats, Inc. v. Thunder Craft Boats, Inc.*, 489 U.S. 141, 109 S.Ct. 971, 9 USPQ2d 1847 (1989), which reveals, in particular, how: 'From their inception, the federal patent laws have embodied a careful balance between the need to promote innovation and the recognition that imitation and refinement through imitation are both necessary to invention itself and the very lifeblood of a competitive economy. … The federal patent system thus embodies a carefully crafted bargain for encouraging the creation and disclosure of new, useful, and not obvious advances in technology and design in return for the exclusive right to practice the invention for a period of years. [The inventor] may keep his invention secret and reap its fruits indefinitely. In consideration of its disclosure and the consequent benefit to the community, the patent is granted. An exclusive enjoyment is guaranteed him for seventeen years, but upon expiration of that period, the knowledge of the invention inures to the people, who are thus enabled without restriction to practice it and profit by its use'. See also U.S. Supreme Court, *Grant v. Raymond*, 31 U.S. (6 Pet.) 218 (1832).

The sufficient description of the claims contained within the patent application is significant to the definition of the field of patent monopoly.[92] The patent may be granted only if and in as much as the description indicates the method for reproducing that which the inventor claims.[93] Adequate disclosure of the invention, then, is the quid pro quo of the granting of a patent to the owner for temporary monopoly over his invention.

---

[92] What is more, from the systematic point of view it is useful to distinguish the context of the invention from the context of protection, given that the latter concept is used in relation to the phase following the granting of the patent, in particular in the context of a request for violation of the patent and any application of the equivalency doctrine. See also, for reference to European and United States case law on the subject, S.J.R. Bostyn, *Patenting DNA sequences (polynucleotides) and scope of protection in the European Union: an evaluation – Background study for the European Commission within the framework of the Expert Group on Biotechnological Inventions*, cit., p. 28 ss., and S.J.R. Bostyn, *Enabling Biotechnological Inventions in Europe and the United States. A study of the patentability of proteins and DNA sequences with special emphasis on the disclosure requirement*, cit., p. 145 et seq.; C.M. Holman, *Gene Patents Under Fire: Weighing the Costs and Benefits*, cit., p. 268; J.M. Mueller, *Patent Law*, New York: Aspen Publishers, 2009 (3rd edn), p. 332 (which, citing Judge S. Rich, recalls that '*the name of the game is the claim*'); M.J. Adelman, R.R. Rader, J.R. Thomas, H.C. Wegner, *Cases and Materials on Patent Law*, St. Paul, Minnesota: West Group, 1998, p. 567 et seq. (where it is specified that: 'The enablement requirement serves to delimit the boundaries of patent protection by ensuring that the scope of a patent claim accords with the extent of the inventor's technical contribution'); D.S. Chisum, C.A. Nard, H.F. Schwartz, P. Newman, F.S. Kieff, *Principles of Patent Law*, New York: Foundation Press, 1998, p. 163 et seq.; J.H. Barton, 'Patent Scope in Biotechnology' (1995) 26 *IIC* 605 et seq.

[93] In reference to the sphere of extension of patent exclusivity, in the ruling of the House of Lords regarding the *Biogen* case Lord Hoffmann stated that 'if the invention discloses a principle capable of general application, the claims may be in correspondingly general terms. … On the other hand, if the claims include a number of discrete methods or products, the patentee must enable the invention to be performed in respect of each of them. Thus if the patentee has hit upon a new product which has a beneficial effect but cannot demonstrate that there is a common principle by which that effect will be shared by other products of the same class, he will be entitled to a patent for that product but not for that class, even though some may subsequently turn out to have the same beneficial effect. … On the other hand, if he has disclosed a beneficial property which is common for the class, he will be entitled to a patent for all products of that class even though he has not himself made more than one or two of them' (*Biogen Inc. v. Medeva plc*, decision of 31 October 1996, [1997] RPC 1, 48-49). In the same sense, it seems to move in the same direction of the EPO case law; see for instance: Technical Board of Appeal, T 0694/92, *Modifying plant cells/*

In this regard, as far as, in particular, the issue of accessibility to the public of biological material is concerned, the tool which prevails in international practice consists in the depositing of samples of the specific organism claimed in the patent application with the designated authorities delegated for this purpose, based on the Treaty of Budapest of 1977.[94]

## I.3 UNPATENTABLE INNOVATIONS: THE PROVISIONS OF THE TRIPS AGREEMENT AND DIFFERENCES BETWEEN PATENT MODELS

The TRIPS Agreement, in article 8, para. 1, considers[95] the possibility that the States parties to it, when forming or modifying their legislative or regulatory provisions, may adopt the measures necessary to ensure the protection of public health and nutrition, and to promote the public interest in spheres which are fundamentally important for their social-economic and technological development, as long as those measures are compatible with the provisions of the Agreement.

On the basis of article 27, para. 1, of the same Agreement, the attainment of patents and enjoyment of relative rights are not subject to discriminations on the basis of the place of invention, the technological sector or whether products are imported or of local production.

The provisions regarding non-discrimination among the various fields of technology subject to patent protection have particularly relevant effects for the development of biotechnology. As a consequence, on one hand, the use of biological materials of plant, animal or human origin in and of itself does not exclude patentability of the finding (where they satisfy the conditions laid down in that Article); on the other, ethical or environmental limits to patentability of biotechnology, though legitimized

---

*MYCOGEN*, ruling 3.3.4, 8 May 1996, EPO Official Journal, 1997, 408. In doctrine, see Bostyn, *Enabling Biotechnological Inventions in Europe and the United States. A study of the patentability of proteins and DNA sequences with special emphasis on the disclosure requirement*, cit., p. 190 et seq.; R.P. Merges, *Patent Law and Policy*, Virginia: The Michie Company, 1992, p. 515 et seq.

[94] 'Treaty on the International Recognition of the Deposit of Microrganisms for the Purposes of Patent Procedure', Budapest, 28 April 1977. The term 'microorganisms', which the Treaty provides no definition for, has been in practice interpreted in an extensive sense, including not only bacteria, fungi, algae and yeasts, but also plant, animal and human cells, seeds, DNA or RNA samples and oncogens.

[95] In the first part relative to general provisions and fundamental principles.

by the Agreement itself,[96] may be introduced into national regulations and applicative practices only as exceptions which are defined and limited by a general rule, with the result that they must be interpreted in a restrictive sense.[97]

Article 27, para. 2, provides that the States parties to the Agreement may exclude patentability for inventions whose commercial use in their territory must be inhibited for reasons of public order or morality, as well as to protect human, animal or plant life or health or to prevent serious environmental damage, as long as the exclusion is not dictated solely by the fact that its use is prohibited by their legislation. It is advisable, therefore, that the States reconsider their pre-existing legislation, should it be in conflict with this provision.[98]

In the context of regulations on intellectual property, the clauses which safeguard public order[99] and morality[100] are traditionally considered a sort of formula of style destined to remain mostly unimplemented. These profiles fall under the competence of national political authorities (and not technical offices such as that of patents), which may intervene through implementation or marketing prohibitions regarding the inventions in question. The argument is based on the rules, including those of the TRIPS Agreement,[101] which confer rights to owners of exclusive

---

[96] Pursuant to the provisions referenced in this paragraph.

[97] See, for a broad analysis of the classification of the principle of non-discrimination in the sphere of the TRIPS Agreement, T. Sommer, 'The Scope of Gene Patent Protection and the TRIPs Agreement – An Exclusively Nondiscriminatory Approach?', (2007) 38 *IIC* 30 et seq.

[98] In this way, a system of decentralized and distributed policy is put in place, which has been defined in doctrine as 'neo-federalist', with a broad sphere of discretion left to the States parties to the Agreement. See, among others, G.B. Dinwoodie and R.C. Dreyfuss, 'International Intellectual Property Law and the Public Domain of Science' (2004) 7 *Journal of International Economic Law* 435 et seq.

[99] The concept of public order is understood as the totality of fundamental regulations regarding ethical and political principles at the base of a specific legal system, with respect to which it is not possible to vary without putting its very existence in jeopardy. See C. Kuppuswamy, *The International Legal Governance of the Human Genome*, cit., p. 137; Gervais, *The TRIPs Agreement: Drafting History and Analysis*, cit., p. 222.

[100] Or rather, the totality of conventions and ethical values constituting the morals of a social group of a specific time and place. With regard to the concepts of morals and morality, besides referring to what has been said above, please see Wikipedia.org, entry *Morality*, cit.

[101] TRIPS Agreement, art. 28, on the basis of which: '(a) where the subject matter of a patent is a product, to prevent third parties not having the owner's

patents, related only to the exclusion of use of the protected invention by anyone, and not to the creation and commercial use of the invention itself. According to this perspective, the patent must be considered a neutral tool with regard to the discipline based on principles connected to fundamental rights, morals, public order, environmental protection, and so on, derived from the distribution of the products or patent processes. Such a discipline may take precedence over the granting of a patent and, vice versa, nothing prevents, even where it is not granted, products and processes based on controversial technology from being put into circulation.[102]

In any case, the development of biotechnology has led to a broad debate on the conformity of biotechnological inventions to public order and morality, forcing different patent offices to respond to the legality of the patenting of such inventions. In particular, the EPO, starting with the *Harvard Onco-Mouse* decision,[103] considered the concept of public order in a broad sense, as the totality of principles which in a specific legal system has the purpose of protecting basic values of the collective. This approach of the EPO, then, appears to be in contrast with the theory of irrelevance of the aforementioned fundamental principles on the matter of patentability of biotechnological inventions.

Once again, in article 27, para. 3, the States parties to the Agreement are allowed (but not required) to exclude from patentability: (a) diagnostic, treatment and surgical methods for the treatment of man or animals; (b) plants and animals, except for microorganisms, and essentially biological processes for the production of plants or animals, except for processes which are not biological or microbiological. However, the

---

consent from the acts of: making, using, offering for sale, selling, or importing for these purposes that product; (b) where the subject matter of a patent is a process, to prevent third parties not having the owner's consent from the act of using the process, and from the acts of: using, offering for sale, selling, or importing for these purposes at least the product obtained directly by that process'.

[102] In this regard, among others, see G. Van Overwalle, 'Biotechnology and patents: global standards, European approaches and national accents', in D. Wüger and T. Cottier (eds), *Genetic Engineering and the World Trade System*, Cambridge: Cambridge University Press, 2008, p. 77 et seq.

[103] A genetically modified mouse from the University of Harvard modified in order to make it predisposed to contracting cancer, and in this way to serve as a test subject for the study of the illness. See Examining Division, *Onco-Mouse/ HARVARD*, 14 July 1989, EPO Official Journal, 1989, 451.

States are required to provide protection for plant varieties either through patents or an effective *sui generis* system or some combination of the two.

In regard to this provision, with specific reference to (b), it has been shown that the situation does not refer specifically to biotechnology, but rather is more broadly applicable to all research and/or industrial activities which involve the use of plants and animals. In particular, regarding the patenting of products the regulation allows for prohibitions regarding plants and animals considered as a whole, with the exception of microorganisms, while the processes not subject to protection are those which are essentially biological for the production of plants or animals, except also those which are microbiological and non-biological processes. Such broad power is still limited in part by the regulation which provides the obligation of States to introduce a tool for protection of the so-called plant varieties.[104]

What is more, the prevalent interpretation of article 27, para. 3(b), implies that a large number of biotechnological inventions related to the animal and plant world result in patentable effects.

Concerning process patents, the prohibition of exclusion from patentability of microbiological processes, contrary to that which is provided for those which are essentially biological, act in order that techniques of genetic engineering are almost all patentable, given that they use and create microbiological material. The context of patentability of genetic engineering is broadened, on the one hand, from the extensive definition of the material concept of microbiological material, also for the difficulty at a scientific level of a clear distinction between microbiology and biology; on the other, the practice of patent offices to deem that the existence of human intervention for the development and completion of processes of genetic engineering does not permit them to be held as essentially biological.

With regard to product patents, the same considerations in substance are valid. In particular, there is a clear consensus that DNA, beyond the fact that it is living matter responsible for the transmission of hereditary characteristics, is in any case a chemical substance, or a particular microbiological material, and therefore patentable. Animal or plant genes, in turn, may be patentable in that they are fragments of a chemical substance. This position finds support in practice, both in the decisions of

---

[104] The reason for the provision of obligation of protection for specific taxonomic classifications of plants is traceable back to the 'International Convention for the Protection of New Varieties of Plants', so-called UPOV Convention, adopted in Paris on 2 December 1961 and last modified in 1991.

patent offices and as far as the filing of microorganisms on the basis of the Treaty of Budapest are concerned.

The TRIPS Agreement, therefore, does not provide a unified standard which outlines the limits of the concept of invention: this has led various States parties to the Agreement to adopt their own specific definitions (except with respect to the requirements of patentability cited from article 27, para. 1).[105]

In particular, the Agreement does not take material existing in nature into consideration, and it does not include genetic material among the exceptions of patentability.[106] Thus, single States parties to the Agreement, characterized by different levels of social and economic development, must face complex issues of patentability on a case-by-case basis.[107]

The generic terminology used by the TRIPS Agreement – which does not specifically mention biotechnology in any provision[108] – has led to

---

105 See C.A. Fowler, 'Ending Genetic Monopolies: How the TRIPs Agreement's Failure to Exclude Gene Patents Thwarts Innovation and Hurts Consumers Worldwide', cit., p. 1073 et seq.; D. Matthews, 'Globalising intellectual property rights: the TRIPs Agreement', cit., p. 46 et seq.

106 See, among others: G. Dinwoodie and R. Dreyfuss, 'International Intellectual Property Law and the Public Domain of Science', cit., p. 435 et seq.; M. Khor, *Intellectual Property, Biodiversity and Sustainable Development: Resolving the Difficult Issues*, London/New York: Zed Books; Penang, Malaysia: Third World Network, 2002, p. 70 et seq.

107 In this regard, see World Health Organization, 'Genetics, Genomics and the Patenting of Dna: Review of Potential Implications for Health in Developing Countries', 2005, accessed 12 December 2014 at http://www.who.int/genomics/FullReport.pdf, p. 21 et seq.

108 Vice versa, the agreement projects preceding the so-called 'Dunkel Draft' ('Draft Final Act Embodying the Results of the Uruguay Round of Multilateral Trade Negotiations', doc. MTN.TNC/W/FA of 20 December 1991, or the 'Dunkel Draft,' from the name of the CEO of GATT at that time, a document which the definitive text of the TRIPS Agreement has remained almost entirely unchanged), considered an option, introduced upon the request of developing countries, according to which, on the subject of biotechnological inventions, the limits of patentability established at a national level and beyond would have been admissible with respect to those provided by the Agreement for the generality of inventions. The failure to include this clause in the final text is due to the opposition of industrialized countries, which held that it might hinder the development of the emerging biotechnology sector. On the other hand, it has been highlighted how the fact that during negotiations biotechnological patents were considered by developing countries as phenomena to be avoided or in any case limited, or rather in terms of exemptions and exceptions to the forms of protection guaranteed by the Agreement, has impeded a specific reflection on the

many differences in national patent legislation regarding biotechnology and genetic material.[109] The States parties to the Agreement have adopted a variety of approaches in the interpretation and implementation of their respective patent laws on the matter, in an attempt to conform to international law while maintaining specific national disciplines.[110]

---

issues connected to the theme of patentability of biotechnological inventions, which would have in theory been able to lead to solutions which took into consideration their particularity, for instance through a provision of further requirements for granting patents in this field, or the promotion of a sort of specific collateral agreement, etc. (for a detailed diachronic reconstruction of the approval process and of the inspired reasoning behind the TRIPS Agreement and in particular the provisions referenced, see among others: N. Pires de Carvalho, *The TRIPS Regime of Patent Rights*, cit., especially p. 245 et seq.).

[109] Regardless, it is clear that the TRIPs Agreement holds significant importance for the discipline of biotechnology. The objectives that it indicates as fundamental for the protection of intellectual property, in fact, consists essentially in the promotion of technological innovation and in the transfer and distribution of technology (see art. 7, cit.). The Agreement is based on the assumption that only adequate and effective protection of intellectual property can lead to technological transfer in favour of less advanced countries. In the sphere of biotechnology, for instance, this means the actual recognition and respect of countries for a patent regarding recombinant DNA techniques, which make agricultural plants resistant to drought, representing the conditions necessary to make this technology available to the country itself. The perspective adopted in the Agreement, therefore, is fundamental, and significant resistance from some countries arises – in particular from those which are less developed – with regard to the implementation in their legal systems of regulations which are completely compliant with the Agreement. Such a perspective, in fact, implies that (at least in this moment in history) the protection of intellectual property in less advanced countries works almost completely in favour of companies of industrialized countries, given that the former do not have the adequate local industry to benefit from it. On this theme, see, among others, G. Ghidini, *Innovation, Competition and Consumer Welfare in Intellectual Property Law*, cit., p. 247 et seq.; C.M. Correa, *Intellectual Property and Competition Law – Exploring Some Issues of Relevance to Developing Countries*, ICTSD, Issue Paper No. 21, 2007, accessed at http://ictsd.net/downloads/2008/06/corea_oct07.pdf.

[110] In this sense, the main issue both in the United States and in Europe regarding inventions based on genetic material is in relation to the doubt that only relative processes are patentable or also the genes themselves. In this regard, besides that which will be said hereafter, see S. Soini *et al.*, 'Patenting and Licensing in Genetic Testing: Ethical, Legal and Social Issues' (2008) 16 *European Journal of Human Genetics* S10 et seq.; World Health Organization, 'Genetics, Genomics and the Patenting of Dna: Review of Potential Implications for Health in Developing Countries', cit., p. 26 et seq.

In general, unlike the United States of America, many European States provide a list of material which is not considered to be subject to patent protection. Though certain materials may constitute the substrata for an invention, they are in any case considered unpatentable inventions pursuant to patent legislation.

As far as patent requirements are concerned, both in the United States and European models, as previously mentioned, the reasoning for the exclusion lies overall in the fact that, as they are not inventions which are applicable at an industrial level, they would not produce benefits for society, and therefore it is not worthwhile to grant the applicant the status of monopoly for its use.

With specific reference to patentable material, then, in Europe, inventions which are essentially biological and those contrary to public order or morality are also excluded from patent protection. In any case, it should be noted that the single European states differ in their approach to patentable subject matter and application of the relevant discipline.

Given that the U.S. and European patent systems have basically the same objectives, one could argue that they operate in substantially the same way. In reality, there are significant differences between the two systems.

The first fundamental difference lies in the fact that in the United States the origin of intellectual property is in the Constitution, in that the Constitution grants Congress the power to adopt regulations relative to patents.[111] On the contrary, in European countries the origin of intellectual property is national and – except for the principle of intellectual property protection introduced at article 17, para. 2 of the 'Charter of Fundamental Rights of the European Union'[112] – there is not a directly binding legislation or centralized mechanism of enforcement.[113]

In the United States, patents are granted by the federal government, through the United States Patent Office, a centralized institution. In the

---

[111] United States Constitution, Article I, Section 8, Clause 8, so called 'Copyright Clause', grants Congress the power: 'To promote the Progress of Science and useful Arts, by securing for limited Times to Authors and Inventors the exclusive Right to their respective Writings and Discoveries'.

[112] Charter of Fundamental Rights of the European Union, proclaimed on 7 December 2000 in Nice, and afterwards, in an adapted version, on 12 December 2007 in Strasbourg; The Charter actually, given the Treaty of Lisbon, has the same legal value as the Treaties (see below, especially in Parts III and V).

[113] At least until the entry into force of the 'European Union patent', approved by the European Parliament on 11 December 2012, which would constitute a single patent right valid for the entire territory of the EU (see in Part IV).

European Union, patents are granted by the European Patent Office, which is not an EU institution, but rather functions on the basis of the European Patent Convention.

United States public policy basically considers patents a tool for the creation of new industries. Besides this, the European legal system regulates them, both by promoting and limiting them, for the adequate development of the internal market.[114]

In the United States, requirements for patentability and identification of patentable material are derived overall from the work of the case law. In Europe, they are provided at a regulatory level within the European Patent Convention, as well as, of significance here, in Directive 98/44/EC.

In general, in the United States, the limits of patentable material appear to be well defined,[115] given the legislation drafted in positive terms and the 'activism' of case law in its interpretation.

In Europe, both in the European Patent Convention and Directive 98/44/EC, the provisions relative to unpatentable materials are drafted in negative and complex terms, with the relative legal uncertainty for aspiring patent owners.

In the United States, there is a clear distinction between the granting of the patent and use of the invention. On the contrary, in Europe it is held that where use is contrary to public order or morality the patent may not be granted.

The origin of the policy and the history of patent regulation in the United States indicate how in that system there is not a patent regime based on morality. Though the courts may have sometimes denied patents for inventions which were considered immoral, in fact these cases fall under two possible hypotheses: inventions used to defraud buyers and machines used for gambling. As far as the Patent Office is concerned, the only case relevant to a denied patent on the basis of moral considerations

---

[114] In this regard, see *ex multis*: S.W. Halpern, C.A. Nard and K.L. Port, *Fundamentals of US Intellectual Property Law. Copyright, Patent, Trademark*, Alphen aan den Rijn (Netherlands): Kluwer Law International, 2010 (3rd edn), p. 1 et seq.; T. Cook, *EU Intellectual Property Law*, Oxford/New York: Oxford University Press, 2010, p. 1 et seq.; C. Seville, *EU Intellectual Property Law and Policy*, Cheltenham, UK and Northampton, MA, USA, 2009, p. 1 et seq.; H.C. Hansen, *US Intellectual Property Law and Policy*, Cheltenham, UK and Northampton, MA, USA, 2006, p. 1 et seq.

[115] Even with the lack of which will be spoken of hereafter.

is that which was requested in 1997 by Stuart Newman and Jeremy Rifkin for the creation of a human-animal chimera.[116]

On the other hand, in Europe the history of patent legislation indicates that morality plays a broader role, even with all the issues it presents (concepts, limitations, time frame, etc.).[117] Both the European Patent Convention and Directive 98/44/EC provide exclusion from patentability for reasons of public order or morality.[118]

With reference to the matter of biotechnology, United States case law has interpreted the criteria of general patent legislation as suitable to allow for protection of biotechnological inventions, regulating patentability of the forms of life according to the general protection regime.

On the contrary, the European Patent Convention, as the Strasburg Convention, prohibited the patenting of plant[119] or animal varieties, a tendency continued in Directive 98/44/EC, which, further, provided broad exclusions, specifying the requirements of patentability for biotechnological inventions, with the result of a dual framework of patent protection (general and biotechnological).[120]

## I.4 THE ISSUE OF THE ROLE OF MORAL EVALUATION IN PATENT RIGHTS

The debate on legal, social and moral issues connected with today's biotechnology gives rise to very different points of view – and underlying interests – not only between individuals within the patent system, in particular scientists, lawyers and economists, but also in the general collective, who ask who determines the methods of use of technology and who will reap the benefits.

---

[116] In this regard, refer to Part III.

[117] See R.P. Merges, 'Intellectual Property in Higher Life Forms: The Patent System and Controversial Technologies' (1988) 47 *Maryland Law Review* 1051 et seq.

[118] See A. R. Chapman, 'The Ethics of Patenting Human Embryonic Stem Cells' (2009) 19 *Kennedy Institute of Ethics Journal* 261 et seq.; R.K. Seide and C. Stephens, 'Ethical Issues and Application of Patent Laws in Biotechnology', in D. Magnus, A.L. Caplan and G. McGee (eds), *Who Owns Life?*, cit., p. 59 et seq.

[119] Previously protected by the UPOV Convention and today on the basis of EC Regulation no. 2100/94.

[120] On these aspects, besides referring below, see O. Mills, *Biotechnological Inventions. Moral Restraints and Patent Law*, cit., p. 2 et seq. and 155 et seq.

The fundamental question, in this regard, is if public control over these aspects must be carried out in some way through the process of a moral evaluation in the context of the patent system.

The controversy regarding the role of patents in research and development of biotechnology products, together with the impact that they may have on this activity, seems to suggest that patent law may represent the 'competent court' for the debate on biotechnology.

However, to limit the debate to patent rights may be misguided. On one hand, the 'moral issue' on the matter essentially concerns the act of technological development, and as such its analysis within the patent system is problematic, given that the patent right concerns the protection of technology once it exists, therefore 'applied research'[121] and not 'pure research'.[122] On the other hand, however, the opposition to a patent on moral grounds obviously amounts in essence to preventing the activity altogether, through withdrawal of the incentive; this indicates that the patent right is a highly significant component of the discipline, at least directly, of biotechnological development.

In any case, there seems to be little evidence to indicate that the 'moral' provisions of patent law are meant to regulate the matter in and of itself. These provisions, first of all, do not have a *de iure* legal basis. Secondly, it is not clear from case law that a *de facto* regulatory approach exists. Rather, moral considerations are mostly applied in order to highlight concerns regarding the determination of that which is or is not socially acceptable, and as such must be faced.

Patent law traditionally remained far from moral arguments, especially in the United States. In Europe, however, moral profiles entered the patent 'ring' directly upon the creation of art. 53 of the European Patent Convention. Essentially inactive until the advent of biotechnology, the rule provides, among other things, that patents not be granted for inventions whose use would be contrary to public order or morality. In this way, issues were entered into the patent system which, until the development of biotechnology, were not considered particularly relevant to it. Thus, today, in the European framework, patent law is an area in which moral issues take on greater importance.[123]

---

121 That is, aimed at the development of industrial applications.

122 Freed from practical aims, and in any case results. In this regard, in addition to referencing that which will be said in the following Chapters, see in particular, G. Ghidini, *Innovation, Competition and Consumer Welfare in Intellectual Property Law*, cit., p. 37 et seq.

123 In argument, see L. Bently and B. Sherman, 'The Question of Patenting Life', in L. Bently and S. Maniatis (eds), *Intellectual Property and Ethics*

The main aim of the patent system, as is stated both in the U.S. and European systems, is to encourage investment and innovation through the protection of inventions.

There is no doubt that economic policy is fundamental to patent law, which contributes to the economic regulation of the market, through the attribution of 'monopoly islands' in a 'sea of competition', which ultimately stimulates the attribution of these temporary monopolies, favoring innovation, investment and therefore exchange.[124]

Conversely, the aim of 'moral policy', if one can be identified, does not appear to be reached adequately through patent law.[125] The doubt which is raised, then, is if patent, designed to pursue mainly economic interests, may effectively intervene to protect moral claims.[126]

A policy which protects moral values may suggest the adoption of excessively limited behaviour with respect to patentability. Thus, it is necessary first to understand what the aims of a moral policy are. Once these have been identified, it is necessary to determine what is the relative 'specific weight' that the economic and moral profiles may have on the patent discipline. At this point, the issue may be brought up as to whether and perhaps in what way it may be possible to achieve the purpose of each policy through their incorporation into patent law.

In this sense, on the one hand, the basic concept of the instances of the participants in the debate, particularly scientists, economists and lawyers,

---

*(Perspectives on Intellectual Property, vol. IV)*, London: Sweet & Maxwell, 1998, p. 109 et seq.

124 On this point, see S. Anderman and A. Ezrachi (eds), *Intellectual Property and Competition Law – New Frontiers*, cit., p. 1 et seq.; Ghidini, *Innovation, Competition and Consumer Welfare in Intellectual Property Law*, cit., p. 209 et seq.

125 See G. Laurie, 'Biotechnology: Facing the Problems of Patent Law, in H.L. McQueen (ed.), Innovation, Incentive and Reward', cit., p. 61.

126 In this sense, for instance, it has been stated that when patentability of biotechnological inventions is discussed under an ethical profile, one must ask about the ethical compatibility of the exclusive regime, as that of professional production – and only this. This implies that biotechnological innovation one is dealing with has already gone beyond the ethical range of applied research, experimentation, professional and business methods of production of goods and services. In short: the issue of patentability of biotechnology under an ethical profile is translated, with remains, into that of ethical compatibility of business production of biotechnological inventions in a regime of exclusivity. Thus, innovation ethics should not be confused with patent ethics (see P. Spada, *Liceità dell'invenzione brevettabile ed esorcismo dell'innovazione*, in *Rivista di diritto privato*, 2000, p. 6 et seq.).

is that a strong patent system is an essential prerequisite for economic development.

The reasoning of the main economic issue is that according to which it is correct to reward the inventor for the time, ability and effort he has put into the creation of an invention. The inventor must be given the possibility to recover her/his investment, without being forced by competitors to lower the price, and therefore never finding her/himself in a position to recover the money invested. On the basis of this argument, exclusivity is justified due to the fact that it is a necessary element to stimulate research and development.

Another argument made in order to justify the patent system is that in which patents help to stimulate innovation and industrial development through the dissemination of technical knowledge. The patent system, operating on the basis of the principle of disclosure, ensures that the details of the invention are made available to the public. Researchers are then free to build upon this knowledge, and in such a way to further contribute to the advancement of technological progress. This argument works also inversely. If technology is excluded from patentability, there is no incentive to invest in research. In a similar situation, collectivity may be denied the knowledge and advantages that a particular technology may offer.

On the other hand, not all accept the assumption that patents have a positive effect, and it is shown how it is not clear up to what point research and development may be damaged following the withdrawal of incentive. In the first place, the patent does not protect the inventor who first has the idea for the invention.[127] It is the first one to present the application for the patent, rather than the first who invents it, who is given priority.[128] Moreover, it is held that from the moment in which the inventions are ready to be developed the industries which have financed them up to that point would inevitably proceed to their development, therefore there is no need to create artificial incentives.[129] Further, in the field of biotechnological innovation, legal systems now require lengthy experimentation of new products, due to which the period of protection

---

[127] See W.R. Cornish, *Intellectual Property: Patents, Copyright, Trade Marks and Allied Rights*, London: Sweet & Maxwell, 1999 (4th edn), p. 79.

[128] The United States traditionally operated on a 'first to invent' system; the Patent Reform Act 2009, however, introduced a 'first to file' system similar to that of the EPC.

[129] As such, W.R. Cornish, *Intellectual Property: Patents, Copyright, Trade Marks and Allied Rights*, cit., p. 80.

on the market is reduced.[130] For these reasons, some doubt the effective role of the patent system on encouraging the use of inventions. This regards in particular, considering their peculiarity, biotechnological inventions.

In addition, the granting of patents is opposed by many outside the system,[131] on the more fundamental basis of traditional arguments; this is, in particular, the reason why, contrary to the past, today the issue of the role of moral evaluations is a debated theme in patent law.

Those opposed to inventions obtained through genetic engineering hold that the recognition of exclusive rights to use such an invention is simply inappropriate. The consideration at the basis of this position is that according to which intellectual property rights may not and should not take precedence over the fundamental rights of man, such as the rights of autonomy, human dignity and the right to live a full and productive life.[132] In this view, the patent system would put in jeopardy especially the rights for personal identity, the development of the individual, food and health.

Moreover, those opposed to the patent system argue that the high level of protection provided by existing national and international regimes moves the delicate balance between public and private in favour of private interests, resulting in a sort of 'marketing of life'.[133]

---

130 For instance, in the pharmaceutical industry often clinical trials prior to the release of a product on the market last 11 or 12 years, as a result, reducing their period of protection. On the importance of patent incentive to encourage innovation in the pharmaceutical and biotechnology sectors, see among others, J.D. Wright, 'Implications of Recent Patent Law Changes on Biotechnology Research and Biotechnology Industry' (1997) 1 *Virginia Journal of Law and Technology* 1 et seq.

131 Such as religious movements or organizations such as Greenpeace or Friends of the Earth.

132 With reference to the relation between the discipline of biotechnology and human rights, see among others, C. Kuppuswamy, *The International Legal Governance of the Human Genome*, cit.; B.R. Schaller, *Understanding Bioethics and the Law. The Promise and Perils of the Brave New World of Biotechnology*, Westport (USA)/London: Praeger, 2008; F. Francioni (ed.), *Biotechnologies and International Human Rights*, Oxford/Portland (USA), Hart Publishing, 2007; F. Francioni and T. Scovazzi (eds), *Biotechnologies and International Law*, cit., *passim*.

133 See A. Oyewunmi, 'The Right to Development, African Countries and the Patenting of Living Organisms: A Human Rights Dilemma', in J. Gibson (ed.), *Patenting Lives: Life Patents, Culture and Development*, Farnham (UK)/ Burlington (USA): Ashgate Publishing, 2008, p. 53 et seq.; F. Vandenabeele, *Patentability of Living Organisms: Legal and Ethical Aspects of the Question*,

Biotechnological patents encounter opposition also on the basis of the opinion according to which allowing the patentability of a new human or animal trait is equated with 'condoning' the marketing of life, which would be morally unacceptable.[134] In this sense, the Humane Society of America defined genetic engineering as 'a step backwards in the evolving recognition of the significance of animal life, the sanctity of being and the interconnectedness of all life'.[135]

The position is rooted in the concepts of property and possession, and inappropriateness of the application of them, in particular, over human DNA; while there are not strong objections, for instance, regarding the possession of steers by a breeder and the possibility he has in selling or butchering them. The argument is based on the consideration of the existence of a 'slippery slope': once patents have been granted for animals, it would only be a matter of time before they were granted for humans. In particular, the argument assumes that it is impossible to make a distinction of principle between human and animal life, and that the patentability of animals would lead to the consideration of human life as a commodity, over which patents may be granted.

On the other hand, assumptions made on the basis of such an argument are not fully convincing. The assumption that patentability of animals leads to similar treatment of human life and patentability of it, with a sort of process of 'downgrading' of humans, could be seen on the contrary as a process of 'upgrading' for animals. While the worries related to marketing of life undoubtedly need to be faced, the issue is not resolvable with the aforementioned *a priori* comparison. Rather, it should be entrusted to the appropriate evaluation of specialized authorities in charge of such activities.

In this sense, a particularly relevant profile concerns the risk of the so-called 'tragedy of anticommons', which may result from the sum of patentability of essential information for biotechnological research and inadequate practices of licensing for the relative patents. A situation such as this would create significant obstacles for the work of researchers,

---

Montreal: McGill University, 2005, p. 69 et seq.; J. Rifkin, *The Biotech Century: Harnessing the Gene and Remaking the World*, New York: J.P. Tarcher/Putnam, 1998, p. 37 et seq.

[134] See, among others: Y. Joly, 'Genetic Research Tools: Recent Trends and Future Outlook', in E.R. Gold and B.M. Knoppers (eds), *Biotechnology IP & Ethics*, Markham, Canada: LexisNexis, 2009, p. 39 et seq. (47 et seq.); D.B. Resnik, *The Morality of Human Gene Patents*, cit., p. 54 et seq.

[135] In this regard, see M. Paver, 'All Animals Are Patentable, But Some Are More Patentable Than Others' (1992) 9 *Patent World* 9 et seq.

who would be discouraged by the prohibitive costs of obtaining the necessary licences to use a product or process fundamentally necessary for the development of their research.[136]

In any case, it appears clear that patent law is characterized mainly as a tool for economic policy. It provides an incentive to invest and innovate. What is more, through the tool of disclosure, it provides a centralized system of information gathering in order for other inventors to gather ideas or develop new inventions.

On this basis, supporters of patentability of living organisms argue that a mechanism of incentive is essential to guarantee the production of benefits resulting from biotechnological research, including in particular those for health treatment, agriculture and environmental protection.

The arguments against biotechnology in and of itself, again, state that the creation itself of relative inventions is problematic. The key objection appears to be that biotechnology, and in particular genetic engineering, is conceptually wrong in and of itself, even where benefits outweigh the damage caused. The reasons are essentially the following: (a) biotechnology is an attempt to 'play God';[137] (b) genes represent a 'common patrimony of humanity', 'common interest of humanity', or 'public asset',[138] and, according to a more radical formulation, should be passed on from generation to generation without human intervention; and (c) genes naturally exist in organisms and should not be interfered

---

[136] In this regard see, *ex multis*, M. Heller, 'The Tragedy of Anticommons' (1998) 111 *Harvard Law Review* 621 et seq.; M. Heller and R.S. Eisenberg, 'Can Patent Deter Innovation? The Anticommons in Biomedical Research' (1998) 280 *Science* 698 et seq.; G. Ghidini, *Innovation, Competition and Consumer Welfare in Intellectual Property Law*, cit., p. 37 et seq.; R. Gold and B.M. Knoppers, *Biotechnology IP & Ethics*, cit., p. 52. For further consideration on the theme, refer to Parts III and V.

[137] See, among others, M. Bratton (ed.), *God, Ethics & the Human Genome*, London: Church House Publishing, 2009, *passim*; J. Boyle, 'Enclosing the Genome: What Squabbles Over Genetic Patents Could Teach Us', in F. Scott Kieff (ed.), *Perspective on Properties of the Human Genome Project*, San Diego: Elsevier Academic Press, 2003, p. 97 et seq.; J. Rifkin, *The Biotech Century: Harnessing the Gene and Remaking the World*, cit., p. 60 et seq.; M.J. Hanson, 'Religious Voices in Biotechnology. The Case of Gene Patenting' (1997) 27 *Hastings Center Law Report* 1 et seq.

[138] See, *ex multis*, R. Gold and B.M. Knoppers, *Biotechnology IP & Ethics*, cit., p. 47 et seq. (as well as the numerous references cited herein); B.M. Knoppers and Y. Joly, 'Our Social Genome?' (2007) 25 *Trends in Biotechnology* 284 et seq.

with.[139] The argument following such considerations is that which states that it does not consider how much men progress in biotechnology, because they should simply not be using it at all.

Those opposed to this argument reveal that this approach appears in reality to take a position, rather than justify it. Considering that any interference in natural processes may be described as 'playing God', it is stated that the mere fact that genetic engineering allows for control over life is not an acceptable moral objection. Unless the control used by human beings over nature in virtue of genetic engineering can be distinguished from the control they exercise routinely over nature in other ways, genetic engineering should be considered morally equivalent to the other human intrusions on nature; among these, clearly, are the actions which benefit life, for instance, the use of medicines (also obtained thanks to genetic engineering).[140]

With regard to human interest, then, in reality it seems to be served through the incentive to scientific and technological research which allows benefits to be obtained from inventions resulting from it, without it inhibiting the establishment, where appropriate, of any limits of patentability.[141]

What is more, the fact that genetic engineering is a risky practice, the consequences of which are widely unknown, is not held to be a legitimate moral argument against biotechnology. For new technology, in the first stages of development, the actual results are unknown. The problem of security is not exclusive to the sector of genetic engineering, but must be faced with specific reference to the given context.[142]

On the other hand, it should be noted that there is a moral, as well as legal, obligation of individuals to be responsible for the benefits granted them by legislature. In this view, aside from the consideration according to which the patent system is not the appropriate place to evaluate moral

---

139 See S. Sterckx, 'European Patent Law and Biotechnological Inventions', in S. Sterckx (ed.), *Biotechnology, Patents and Morality*, Farnham (UK)/ Burlington (USA): Ashgate Publishing, 1997, p. 1 et seq.

140 On this point, see F. Vandenabeele, *Patentability of Living Organisms: Legal and Ethical Aspects of the Question*, cit., p. 68; B. Hoffmaster, 'The Ethics of Patenting Higher Life Forms' (1988) 4 *Intellectual Property Journal* 1 et seq. (especially p. 4).

141 See World Health Organization, 'Genetics, genomics and the patenting of DNA. Review of potential implications for health in developing countries', Geneva, 2005, accessed 12 December 2014 at http://www.who.int/genomics/ FullReport.pdf, p. 31 et seq.

142 According to B. Hoffmaster, 'The Ethics of Patenting Higher Life Forms', cit., p. 1 et seq.

profiles, it is frequently used explicitly to exclude certain inventions from patentability, in virtue, in fact, of moral considerations. Patent regimes have often functioned as 'moral filters', allowing a certain form of innovation to enter the market and blocking others.[143] The limitations to patentability have allowed the courts and patent offices to pass value judgments on the issue of social interest.[144]

A further argument on the moral order of things against biotechnology is in regard, then, to the experimentation on animals for purposes of genetic engineering, which is held to be unacceptable given that it inflicts pain and suffering on animals for aims which in comparison appear to be of minor importance.[145] The difficulty in sustaining such an argument lies in its being based on the absolutist approach which highlights only one value, that of protecting animals. A discipline which takes morals into account requires a continuous balancing of conflicting values.[146]

Further, the difficulty in identifying the difference between genetic engineering and other technology has been shown to be such that many moral questions have arisen in regard to the former. This, even more in virtue of the fact that also in Europe the European Patent Office, like the U.S. Patent Office, in consenting to the patentability of life forms, has apparently reconciled science and profit. The essential point concerns the examination of whether a higher moral standard can be justified on the basis of the consideration for which biotechnology deals with life, and the negative consequences of it which could be catastrophic. At most, every science, including genetic engineering, may demonstrate advantages and disadvantages at a moral level. The issue, then, is that of identifying where the line should be drawn, which consequence society will see as a tolerable evil produced by genetic engineering.

In the context of the patent system, then, the function of morality appears to identify relevant issues for the determination of what is and is

---

143 See A.J. Wells, 'Patenting New Life Forms: An Ecological Perspective' (1994) 3 *European Intellectual Property Review* 111 et seq.

144 Think, for instance, of the case of prohibition of patentability of nuclear weapon technology in the United States (42 USC s.218(a) 1982).

145 In this regard, see again B. Hoffmaster, *The Ethics of Patenting Higher Life Forms*, cit., p. 8.

146 See R. Ford, 'The Morality of Biotech Patents: Differing Legal Obligations in Europe?' (1997) 6 *European Intellectual Property Review* 315 et seq.

not acceptable for society. At this point, specialized authorities responsible for this issue can, through adequate regulations, exercise the relative control.[147]

In this sense, with particular reference to microorganisms, the worry is that modified organisms may spread in the environment escaping from the lab where they were produced, without scientists knowing what effects they may have on future generations.

With regard to plants, the worry is substantially the same, or rather that modified plants may spread in the environment, with related risks for ecological balance.[148] Those in opposition to genetic engineering, on the other hand, are not convinced by the consideration according to which, once it has been established that modified plants are not damaging, there are not negative effects for the environment. In the case of plants, then, there is a further objection, which is that genetic modification will give rise to a reduction in biological diversity, or biodiversity, a concept that indicates the variability of living organisms of every type, including those on land, in the sea and ecological complexes and water ecosystems of which they are a part.[149] In this regard, on the one hand, it is stated that in the long term only plants useful for the interest of man will remain and biodiversity will be reduced in an unacceptable way. On the other hand, it is affirmed, however, that biodiversity may be enriched by the introduction of external genes (from non-plants or other plants) to existing plants.

As far as animals are concerned, in the first place, arguments against genetic modification are similar to those raised for plants. The concern, also in this case, is regarding the danger of destroying the ecosystem due

---

147 See O. Mills, *Biotechnological Inventions. Moral Restraints and Patent Law*, cit., p. 15 et seq.

148 Though biotechnology has been relatively successful in transforming plants for agriculture, concerns remain for profiles of security and sustainability. What is more, the more recent engineering techniques of precision seem able to relieve also some of the concerns regarding biosecurity. In this regard, see C.N. Stewart and D.W. Ow, 'The Future of Plant Biotechnology', in C.N. Stewart (ed.), *Plant Biotechnology and Genetics: Principles, Techniques, and Applications*, New York: Wiley & Sons, 2008, p. 357 et seq.

149 Biodiversity detected within species, between species and ecosystems. On this theme, see the Convention on Biological Diversity, opened and signed in Rio de Janeiro on 5 June, 1992 and implemented on 29 December, 1993, art. 2. In doctrine, regarding the role of patents in order to achieve the objective of the Convention, see C. Lawson, 'Patents and Biological Diversity Conservation, Destruction and Decline? Exploiting Genetic Resources in Queensland under the Biodiscovery Act 2004' (2006) 28 *European Intellectual Property Review* 418 et seq.

to the release of transgenic animals into the environment. The argument of the reduction of biodiversity, then, is invoked also in relation to animals, stating that only those useful to man would remain. On one hand, the statement is reinforced by the observation according to which the characteristics of animals are modified in a way that will benefit man, but rarely the animals themselves. On the other, the objection is that also in this case biodiversity may be enriched as a result of the introduction of foreign genes (of non-animals or other animals) into existing animals.

Once again, in consideration of the fact that the animals are considered to fall under the category of higher forms of life (with humans), a further objection concerns the methods and consequences of experimentation on animals. In particular, it is asked whether these experiments are necessary, if it is acceptable to carry them out on animals for different purposes other than their health and well-being, and if and in which case the use of animals as bio-reactors should be allowed.

Beyond these questions, the fundamental theme appears to be that of evaluation of whether the damage done to animals takes precedence or not over the benefit to human beings.[150] In addition, the main doubts regard the negative impact that transgenic animals may have on their own species. The concern is based on the assumption that the transfer of genes between species crosses the natural barriers existing between them, violating their integrity.[151]

In any case, evidently the more complex situations presented by biotechnology arise when considering genetic modification of human beings. The threat posed by today's biotechnology lies in the possibility that they may alter human nature in an immoral and perhaps unrecoverable way.

The main moral issues raised by the application of biotechnology to human life concern, in particular,[152] the doubts regarding the possibility

---

150 See G. Annas, 'Of Monkeys, Man and Oysters' (1987) 17 *Hastings Center Report* 20 et seq.

151 With regard to arguments for and against the patentability of animals, see U.S. Congress – Office of Technology Assessment, '*New Developments in Biotechnology: Patenting Life – Special Report*', OTA-BA-370, Washington DC, U.S. Government Printing Office, April 1989, accessed 12 December 2014 at ota-cdn.fas.org/ota/reports/8924.pdf.

152 In this regard, besides what will be said more extensively below, see among others, E. Arezzo and G. Ghidini, *Biotechnology and Software Patent Law: A Comparative Review of New Developments*, cit., especially p. 221 et seq.; D.R. Koepsell, *Who Owns You? The Corporate Gold-Rush to Patent Your Genes*, cit.; D. Magnus, A.L. Caplan and G. McGee (eds), *Who Owns Life?*, cit., *passim*; Gold and Knoppers, *Biotechnology IP & Ethics*, cit., *passim*; Bratton, *God Ethics*

for humans to possess their own genetic material, held by many as being common property of humanity and as such belonging to society as a whole; the risks of violation of human dignity which interventions on the human genome may involve; the criticism of genetic therapy, or rather the production of healthy couples of defective genes, as a tool which may surpass the morally acceptable limits of human intervention on life; and the condemnation of practices furthest from the moral standards of our society, such as human cloning or eugenics.

*& the Human Genome*, cit., *passim*; B. Looney, 'Should Genes Be Patented? The Gene Patenting Controversy: Legal, Ethical and Policy Foundations of an International Agreement' (1994) 26 *Law and Policy in International Business* 231 et seq.

# PART II

# Interventions on human genetic material: Legal and bioethical issues

## II.1 BIOTECHNOLOGY, GENETICS, BIOETHICS AND BIO-LAW: DOCTRINAL DEBATE AND FUNDAMENTAL PRINCIPLES

The development of biotechnology and the progress of genetic science, with the technical opportunities they create, have increased the power of human beings over themselves and over nature immensely. In this sense, problems arising from the application of new genetic techniques on man are particularly significant. The threat that may arise from an uncontrolled and destructive use of the increased technical power with regard to man and life provokes fundamental questions of an ethical nature. With respect to technical-mechanical errors, which allow for the reversibility of the effects, technical-genetic errors are unsettling, given that they are irreversible.[1]

All of these reasons have contributed to the rise of bioethics, through which it is possible to continue the analysis of biotechnological applications which affect the fundamental value of man, life, or better, the entire biosphere where human life takes place, with important possibilities for progress, but also with the risk of tragic errors or abuses.

[1] Mechanical errors 'involve the object directly; biogenetic ones spread out from it' (H. Jonas, *Technik, Medizin und Ethik – Zur Praxis des Prinzips Verantwortung*, Frankfurt am Main: Suhrkamp, 1987, p. 216; see also p. 163 et seq.).

Bioethics consists, according to the so-called classic definition, in the systematic study of moral dimensions – including the moral vision, decisions, conduct and policy – of life sciences and the protection of health, through the use of a variety of ethical methods according to an interdisciplinary setting.[2] This is part of general ethics and is characterized as applied ethics.[3]

Bio-law, then, indicates the part of legal reflection which deals with distinguishing that which is just and unjust in the context of new possibilities of manipulation (even if it is not therapeutic) of human life.[4]

The practical applications of new technology must be affected by the ethical premises of a theoretical nature; they respond to a fundamental inquiry, or rather if it is legal to do everything that can be done to man.[5] The answers to this inquiry are significantly divergent, in line with the various roots of culture, religion and philosophy of modern thought.

A first position, which could be defined as scientific-radical, holds that everything that is an expression of freedom is legal.[6] Here, any limits are purely of an instrumental nature, or rather, to verify the adequacy of the

---

[2] This definition was accepted by the *Encyclopedia of Bioethics* (edited by W.T. Reich, New York: Macmillan, 1995, 2nd edn; 1st edn: New York: The Free Press), p. xxi. The origin of the term is attributed by various authors to V.R. Potter, 'Bioethics: The Science of Survival' (1970) 14 *Perspectives in Biology and Medicine* 120 et seq., who, starting with the conviction according to which the future existence of the human species is not only not guaranteed, but seriously threatened, viewed bioethics as the 'science of survival'; see as well, *amplius*, V.R. Potter, *Bioethics: Bridge to the Future*, Englewood Cliffs (USA): Prentice-Hall, 1971.

[3] See A. Kaufmann, 'Riflessioni giuridiche e filosofiche su biotecnologia e bioetica alla soglia del terzo millennio', in *Rivista di diritto civile*, 1988, p. 208.

[4] See A.R. Jonsen, *The Birth of Bioethics*, New York/Oxford: Oxford University Press, 1998; H. Kuhse and P. Singer (eds), *A Companion to Bioethics*, Oxford: Wiley-Blackwell, 1998; H.T. Engelhardt Jr., *The Foundations of Bioethics*, New York/Oxford: Oxford University Press, 1996 (2nd edn); C. Byk, 'From Ethics to Law, from Law to Ethics' (1997) 48 *International Digesto of Health Legislation* 252 et seq.

[5] See W.C. Zimmerli, 'Dürfen wir, was wir Können? Zum Verhältnis von Recht und Moral in der Gentechnologie', in R. Flöhl (ed.), *Genforschung – Fluch oder Segen? Gentechnologie. Chancen und Risiken*, Munich: Schweitzer, 1985, p. 59 et seq.

[6] In this regard, among others, J. Huxley, 'The evolutionary vision', in S. Tax and C. Callender (eds), '*The Evolution after Darwin, III – Issues in Evolution*, Chicago: University of Chicago Press, 1960, p. 249 et seq.

objectives pursued. On behalf of more moderate supporters, it is recognized that freedom, including that of scientific research, is limited by the freedom of others.[7]

A second position, which may be defined as reductionist-evolutionist, claims that, as from a biological aspect life forms are sorted by an evolutionary and interrelated process and by adaptation between the organism and the environment, so is the history of culture distinguished by a continuous evolutionary process of scientific and cultural adaptation. Thus, scientific progress in the field of biology should supply ethical indications, and not vice versa: ethics should limit itself to recognizing and adapting to the changing conditions of life.[8]

A third position, pragmatist-efficient, is inspired by utilitarianism and considers mainly the results, on the basis of a cost and benefit analysis.[9] The utility of new technology, for instance to defeat illnesses which are not yet curable or to bring well-being to populations in developing countries through agricultural or zoology techniques, is considered to be fundamental criteria for legality, and utility is seen essentially as economic efficiency. In this sense, the existence of non-monetary values is described as mere psycho-social factors,[10] or even as taboos,[11] nuisance

---

[7] See M. Horkheimer, *Sozialphilosophisce Studien*, Munich: Fischer, 1972, p. 47 et seq. In this view, biologists and geneticists could not withdraw from precepts and prohibitions of ethics and law simply by calling for freedom of science and research, because there are limits inherent in the values of the body, life, self-determination and human dignity, and are not subject to these limits solely from the moment of technological application, but from their basic investigation (Kaufmann, 'Riflessioni giuridiche e filosofiche su biotecnologia e bioetica alla soglia del terzo millennio', cit., p. 211; see also H. Hofmann, 'Biotechnik, Gentherapie, Genmanipulation – Wissenschaft im rechtsfreien Raum?' (1986) *Juristenzeitung* 253 et seq.).

[8] See R.G. Edwards (the biologist who delivered the first in vitro baby), 'The ethical, scientific and medical implications of human conception *in vitro*', in C. Chagas (ed.), *Modern Biological Experimentation*, Vatican City: Pontificia Academia Scientiarum, 1984, p. 193 et seq.

[9] This position is traceable to the sector of economic analysis of law. In argument, see for all, R.A. Posner, *Economic Analysis of Law*, Boston: Aspen Publishers, 1977.

[10] See G. Calabresi, *Costo degli incidenti e responsabilità civile*, Milan: Giuffré, 1975, p. 330; S. Rodotà, *Il problema della responsabilità civile*, Milan: Giuffré, 1964, p. 170.

[11] G. Calabresi and A.D. Melamed, 'Property Rules, Liability Rules and Inalienability: One View of the Cathedral' (1972) 85 *Harvard Law Review* 1111 et seq.

factors for the achievement of more rational and efficient objectives.[12] On this point, it has been shown how the extension of this setting to the biomedical field, taking it to the most extreme consequences, could lead to a fight against hereditary illnesses through the elimination of foetuses who are carriers.[13] Along this line, open experimentation on embryos in the first phases of development has been proposed.[14]

A fourth position, a personal approach, recognizes the value and dignity of the human being and claims the inviolability of physical life.[15] In this context, the reference to natural law as the established order and superior to positive law no longer has a central role, given today's lack of consensus about the existence of a legal regulation beyond positive law.[16] Even a generic reference to values does not appear in and of itself able to constitute an adequate basis for shared regulatory solutions. For instance, in countries like Germany and Italy, reference is often made to human dignity,[17] which is the fundamental principle upheld at a constitutional

---

[12] See G. Calabresi, *Costo degli incidenti e responsabilità civile*, p. 339. *Contra*: I. Englard, 'The System Builders: A Critical Appraisal of Modern Tort Theory' (1980) *Journal of Legal Studies* 27 et seq.

[13] In this manner: E. Sgreccia and M.L. Di Pietro, 'Manipolazioni genetiche e procreazione artificiale: orientamenti giuridici e considerazioni etiche' in *Diritto di Famiglia e delle Persone*, 1987, p. 1356 et seq. (1357).

[14] See the 'Report of the Committee of inquiry into human fertilisation and embryology' (the so-called 'Warnock Report'), presented to the British parliament in July 1984, which proposed the authorization of experiments on human embryos who were residual from in vitro fertilization, following obtainment of a specific license, within the second week of life (Report, no. 44), a proposal which was carried over in the Human Fertilisation and Embryology Act of 1990.

[15] See E. Sgreccia and M.L. Di Pietro, 'Manipolazioni genetiche e procreazione artificiale: orientamenti giuridici e considerazioni etiche', cit., p. 1358 et seq.

[16] On this theme, see in general, among others, D. Mayyer-Maly and R.M. Simons, *Das Naturrechtsdecken, heute und morgen – Gedenkschrift für René Marcic*, Berlin: Duncker & Humblot, 1983, *passim.*

[17] See, among others, H. Seesing, K. Buttner and K. Bèuttner, '*Technologischer Fortschritt und menschliches Leben; Die Menschenwürde als Maβtab der Rechtspolitik. Gentechnologie – Chancen und Risiken*, Frankfurt: J. Schweitzer Verlag, 1987, *passim*; G. Püttner and K. Brühl, 'Fortpflanzungsmedizin, Gentechnologie und Verfassung' (1987) *Juristenzeitung* 529 et seq.; E. Benda, 'Erprobung der Menschenwürde am Beispiel der Humangenetik', in *Aus Politik und Zeitgeschichte, Das Parlament*, suppl. 3/1985, p. 18 et seq.; A. Baldassarre, *Diritti della persona e valori costituzionali*, Turin: Giappichelli, 1997, p. 1 et seq. as well as p. 151 et seq.; F. Bartolomei, *La dignità umana come concetto e valore costituzionale*, Turin: Giappichelli, 1987, in particular p. 35 et seq. and 105 et seq.

level;[18] in this sense, the *Bundesverfassungsgericht* claimed that the dignity of human life must be recognized from the moment of conception, drawing from this precise restriction consequences regarding the admissibility of pregnancy interruption.[19] On the other hand, the concept of human dignity in and of itself is considered by many as insufficient to form the aforementioned common regulatory base, given that it is the subject of many differences of opinion and uncertainty with respect to its ability to function as a 'constitutional emergency break', as it does not overcome ideological, cultural and moral pluralism present in society.[20]

However, it is clear that there are certain fundamental ideas regarding human dignity, over which there is almost complete general agreement, in particular: (a) the prohibition of its degradation, which for instance seems to be considered violated in the case of the production of chimeras or man-animal hybrids; and (b) the guarantee of minimum individual protection, which for instance, seems to be absent where cloning of human beings is permitted. Even if on these points, however, human dignity is not a self-evident formula, but rather a formula which assumes sufficiently unique axiological meaning where more or less clear general agreement is made in its regard.[21]

In today's reality, globalization of technology and scientific research seem to have given rise to a scenario of inadequate development of national bioethical and bio-law solutions with respect to the new and complex issues raised by the continuous evolution of genetic technology. Concretely, the significant distinction between regulatory axes of some countries which have adopted specific regulations on the subject and of others in which such disciplines are absent, encourages the transfer of practices of which the legality and morality with regard to the latter is uncertain.[22]

---

[18] *Grundgesetz*, art. 1, § 1; Italian Constitution, articles 2, 3 and 41, paragraph 2.

[19] Ruling of 25 February 1975, in *Entscheidungen der Bundesverfassungsgericht*, 39 Band, p. 1 et seq., and in (1975) *Juristenzeitung* 205 et seq., with a note by Kriele.

[20] In this manner: F. Mastropaolo, 'Valori fondamentali e persona umana nell'evoluzione del diritto privato', (1987) *Iustitia*, in particular p. 214 et seq.

[21] See A. Kaufmann, 'Riflessioni giuridiche e filosofiche su biotecnologia e bioetica alla soglia del terzo millennio', cit., p. 214.

[22] Which, further, sometimes purposely do not adopt disciplines, in order to increase their level of development through the importation of genetic technology. On this point, among others, see F. Lenzerini, 'Biotechnology, Human Dignity and the Human Genome', in F. Francioni and T. Scovazzi (eds),

Thus, the values upon which bio-law should be founded seem, on one hand, to be traceable to the fundamental principles of positive law shared in the context of reference, and on the other hand, to originate only from the processes of an adequately broad and shared cultural dialogue.

In this sense, from historical experience some behavioural, ethical and legal principles emerge over which various authors have attempted to structure and position bioethics.[23] Among these, the principle of equality is particularly relevant, which includes equal treatment, but also the categorical Kantian imperative and the so-called 'golden rule' in its positive and negative meaning[24] (which contains in and of itself the

---

*Biotechnology and International Law*, Oxford/Portland (USA): Hart Publishing, 2006, p. 292 et seq.

[23] See, among others, A. Kaufmann, 'Riflessioni giuridiche e filosofiche su biotecnologia e bioetica alla soglia del terzo millennio', cit., p. 215, and A. Kaufmann, 'Der entefesselte Prometheus; Fragen der Humangenetik und der Fortpflanzungstechnologien aus rechtlicher Sicht', in R. Flöhl, *Genforschung – Fluch oder Segen? Gentechnologie. Chancen und Risiken*, cit., p. 275 et seq.; J. Reiter, 'Ethische Implikationen des Genforschung' (1984) 202 *Stimmen der Zeit* 435 et seq.

[24] The golden rule, both in its negative form ('do not do to others what you would not want done to you'), and in the positive ('do unto others as you would have done unto you'), is in substance the expression of the Christian principle 'love your neighbor as you love yourself'; St Thomas Aquinas, though not expressing it in exactly the same way, explicitly references it (I-II, 94, 4 to 1) and claims that all moral principles and norms are implicitly included under its sphere (I-II, 91, 1 to 2; 100, 2 and 3); see also J. Finnis, *Aquinas*, Oxford: Oxford University Press, 1998, p. 138; J. Wattles, *The Golden Rule*, Oxford: Oxford University Press, 1996. Beyond the dependence on ethical-religious perspectives, there is a thesis of convergence between what is said by reason and what is taught by faith; in this regard, see F. D'Agostino, 'La "regola aurea" e la logica della secolarizzazione', in L. Lombardi Vallauri and G. Dilcher (eds), *Cristianesimo, secolarizzazione e diritto moderno*, vol. II, Milan: Giuffré, 1981, p. 941 et seq. In the same ethical sense of the golden rule, but philosophically, there is the categorical imperative of Kant, which in regard to biotechnology appears relevant to the formula: 'Act so that you treat humanity, both in your own person as in the person of any other, always at the same time as an end and never merely as a means.' (I. Kant, *Grounding for the Metaphysics of Morals*, Indianapolis: Hackett, 1993, 3rd edn, p. 30). Criticism of this position and the attempt to form a fundamental ethical formula for humanity, with respect either to the golden rule or to the categorical imperative of Kant, have been authoritatively described by Hans Kelsen (H. Kelsen, *Das Problem der Gerechtigkeit*, Wien: Franz Deuticke, 1960. This criticism, however, appears unfounded, given that it is based on a misunderstanding. The golden rule leads to contradictory results (as Kelsen states) only when it is understood that you must desire for others that which you

principle of *non nocere*). The freedom to act is limited and may not be allowed if it violates the liberty (life, health) of another, nor is it possible to accept discrimination among men should some be subject to prejudice to the benefit of others,[25] given that this would violate the principle of equality. Therefore, tolerance of a high risk activity from which an irreparable event may occur with regard to the life or health of man, even a single man, is inadmissible: in this view, the philosopher Hans Jonas wrote that bioethics must be an 'ethic of responsibility in the long term', leading back to the categorical imperative 'act in such a way that the effects of your actions are compatible with the permanence of authentic human life on Earth'.[26]

---

desire for yourself, without qualifying this desire in a properly legal sense. What is more, one cannot transform his own personal preferences into the preferences of others (think e.g. of the masochist) and, consequently, hold himself obliged in this sense. On the other hand, the meaning of the golden rule is, rather, the following: the good that I would like to be done to me, or that which I desire as good for me, as my well-being, must also be desired as a good to others, which is as good for all or good in and of itself. In this sense, St. Augustine, in his comment to the Gospel of Matthew 7:12, noted that in the traditional form of the golden rule the term 'good' must be understood, or rather, that it is necessary to do that which is good with respect to others, and that is implicit in reference to will, which is the faculty of good (Augustine, *De Sermone in Monte*, II, sec. 74). On this theme, see as well, G. Bognetti, 'The Concept of Human Dignity in European And U.S. Constitutionalism', in G. Nolte (ed.), *European and U.S. Constitutionalism*, Cambridge: Cambridge University Press, 2005, p. 85 et seq., which highlights how the categorical imperative of Kant, of the Jewish-Christian tradition and of the golden rule are considered to be at the origin of today's concept of human dignity.

[25] The World Medical Organization, in the 'Declaration of Helsinki' ('Recommendations guiding physicians in biomedical research involving human subjects', published in (1996) 313 *British Medical Journal*, December, 1448 et seq.), approved in 1964, and subsequently modified, solemnly proclaimed, among other things that: 'Every biomedical research project involving human subjects should be preceded by careful assessment of predictable risks in comparison with foreseeable benefits to the subject or to others. Concern for the interests of the subject must always prevail over the interests of science and society' (Declaration cit., § I.5).

[26] H. Jonas, *Das Prinzip Verantwortung, Versuch einer Ethik für die technologische Zivilisation,* Frankfurt: Inselverlag, 1985, p. 36. See also W. Kern, 'Ethische Normen für die Zukunftsgestaltung' (1986) 204 *Stimmen der Zeit* 594 et seq.; J. Reiter, 'Ethische Implikationen des Genforschung', cit., p. 440.

## II.2 BIOTECHNOLOGY AND HUMAN LIFE: TECHNIQUES OF INTERVENTION AND TREATMENT OPTIONS

Genetics has brought to light for some time the presence and function of chromosomes in the cells of living organisms. 'Chromosomes' are constituted by a double helix filament of DNA and by proteins around which this filament is wound.

'Genes' are enclosed within chromosomes. Every gene has its own important information or instruction on which a specific bodily structure or function depends. Each gene is a polymer constituted by a sequence of 'triplets', or rather by the different combination of three molecules of the four which constitute the elementary basis of information: adenine (abbreviation A), cytosine (C), guanine (G) and thymine (T).

Each of the sequences of the gene is extremely complex, being formed by a large number of triplets, which together determine a specific instruction to induce the gene to produce a specific substance (a protein for instance[27]) or to carry out a certain function.

In turn, within a cell there are a large number of genes: in every cell there are all the genes of the organism, even if only some are active.[28]

In the two helixes of filament of the chromosome, the A, T, G, C bases are paired up, two by two.[29] This sequence of paired bases, interlinked in every chromosome and extremely numerous, constitute 'DNA' (deoxyribonucleic acid).[30]

---

[27] Proteins are polymers of 20 small molecules called amino acids. Each protein is characterized by a specific sequence of amino acids. Genes synthesize, or produce, proteins. Each gene codifies for a specific protein: the sequence of amino acids of a protein X is determined by the sequence of bases present in a segment of DNA called 'gene for protein x'. The triplet which makes up the gene, then, corresponds to the sequence of amino acids of the protein which it codifies.

[28] Though all genes in the body are present in all the cells, in every cell only certain genes are active, or synthesizing certain proteins; the others, though they are physically present, are inactive.

[29] The complementary bases A-T and G-C are coupled for the function of hydrogen links located in the center of the molecule.

[30] Deoxyribonucleic acid (DNA) is a nucleic acid which contains the genetic information necessary for the biosynthesis of RNA and proteins, indispensable molecules for the development and correct functioning of most living organisms.

The chromosomes transport their informative content from cell to cell and from parents to children, mediated by 'RNA' (ribonucleic acid)[31] molecules, which are at the basis of DNA replication, in a constructive and reproductive process also for new cells, according to the specific cellular structure of different tissues and organs.[32]

The sum of all chromosomes of a cell constitutes the 'genome'. The human genome counts around three billion pairs of bases contained in 46 chromosomes, which are held to contain in total around 100,000 genes.[33]

The knowledge of (virtually) the entire human genome sequence opens important new possibilities for medicine and health treatments. The analysis of the content of the genome and the ways in which its components interact may allow scientists to discover the genetic basis for health and the pathology of genetic illnesses. As a consequence, it may allow medical science to improve existing diagnostic tools significantly, to develop an advanced approach to preventive treatment of genetic illnesses and to provide new and efficient medicines for these illnesses.

The treatment needs of every single human, in particular, may be treated on an individual basis, taking into consideration the peculiarity of

---

[31] Given that DNA, in eukaryotic organisms, is located mostly in the chromosomes found in the nucleus of the cell, while protein synthesis is carried out prevalently in the cytoplasm (gelatinous substance within which there are various organelles that carry out the functions of the cell, the most complex of which is the nucleus), the genetic information of the DNA must be transferred to an intermediate molecule, which then moves in the cytoplasm, where it determines the order of the amino acids: this 'intermediary' is the ribonucleic acid (RNA). In particular, there are three types of RNA: (a) messenger RNA (mRNA), functionally defined as molecule which is capable of transferring genetic information from the DNA to the sites where protein synthesis is carried out, codifying a specific protein: the different messenger RNA in a cell have base sequences complementary to specific segments of DNA (genes) capable of codifying specific proteins; (b) ribosomal RNA (rRNA), synthesized in the nucleus which joined to proteins forms ribosomes, which is where protein synthesis occurs in the cell; and (c) transfer RNA (tRNA), which recognizes the codons on the sequence of messenger RNA and binds to the corresponding amino acids, transferring them onto the ribosomes. For each of the 20 amino acids present in nature, there is at least one specific tRNA able to recognize and bind to it.

[32] See F. Mastropaolo, entry 'Ingegneria genetica', in *Digesto delle discipline privatistiche. Sezione civile*, IX, Turin: Utet, 1993, p. 427 et seq.

[33] J.C. Venter *et al.*, 'The Sequence of the Human Genome' (2001) 291 *Science* 1304 et seq. With regard to the mapping of the human genome and the *Human Genome Project*, see above, in Part I.

her/his genetic code. In the short term, the capacity to diagnose individuals at risk for genetic diseases may improve. In the medium term, it may be possible to know the entire sequence of events which provoke a certain disease, with the consequent possibility to develop new and efficient medicine, even 'personalized' on the basis of individual characteristics. In the long term, when all aspects of the human genome will actually be known, it may be possible to discover the genetic composition of each human organ or tissue; as a consequence, it would be possible to produce organs to be used as perfectly compatible 'replacement parts', given that they would be produced through the use of a cell belonging to the person in need, resolving in this way the current problem of organ scarcity or transplant rejection. Finally, the mapping of the genome may be of great significance for the development of gene therapy, based on the use of genes as a therapeutic tool, in particular for those diseases which are visible only when they have reached an advanced stage (such as cancer or degenerative diseases).[34]

As far as new opportunities offered by biotechnology regarding the research of human genetic material is concerned, some basic concepts should be clarified.

In order to produce a specific substance, for instance protein X, with interesting properties from a biomedical point of view, it is necessary to complete a research project to purify the protein and characterize it as a product.[35]

For the production of protein X, it is possible to use isolation, also-called purification or cloning, of the gene which codifies the protein itself. Every molecule in a biological composition may be theoretically purified or separated from its initial complex mixture, through a series of manipulative interventions.

Following 1970, methods for the extraction of DNA, or rather, genes, from cells were developed. The duplication of the hereditary material in DNA in order to transmit it to progeny is allowed by a specific sequence of DNA, called 'origin of replication'. A sequence which allows the origin of replication, if introduced into a microorganism, for instance a bacteria, may give rise to the replication of any molecule of DNA bound to it, independently from where it comes from.

---

[34] See F. Lenzerini, 'Biotechnology, Human Dignity and the Human Genome', cit., p. 287. Regarding genetic therapy, refer to what shall be said more extensively below.

[35] In most cases, in fact, the obstacle to the project is the extremely limited quantity of substances present at its natural source.

Human DNA may be fragmented, every fragment can be bound to an origin of replication and introduced individually into a bacteria cell. In this way, an entire population of bacteria in which every single cell contains a different fragment of human DNA can be generated, which will duplicate with every generation. Given that bacteria replicate quickly, in a few hours millions of bacteria can be obtained from a single one: all of these bacteria together, identical to the parent cell, is called a clone. Given that each of them contains an identical copy of the fragment of human DNA originally introduced into the parent cell, it is said that the fragment of DNA has been cloned.[36] In order to separate the clone of interest from the others, a suitable method is needed for its identification.

The entire sequence of amino acids of the protein whose gene that needs to be isolated is rarely known. Often, however, through conventional biochemical techniques, it is possible to determine the sequence of some fragments. From the sequence of the fragment, it is possible to deduce a certain number of DNA sequences which may be able to codify it, among which will be the gene sequence to be isolated. On the other hand, the gene sequence (base composition) and the sequence of the connected protein (composition of amino acids) are not aligned in correspondence with one another. Most genes in the base sequence which codify for the protein are not continuous, but are sometimes interrupted by non-codifying sequences; the codifying parts are called 'exons', the non-codifying parts are called 'introns'. Thus, even where it is possible to identify a clone which contains the gene of the protein of interest, it is not possible to distinguish the exons from the introns, and therefore it will not be possible to know if the gene is complete.

One method to avoid the problem is to not use the library of genes described, but rather to build and use a library of 'cDNA' (complementary DNA).[37] In cells in which a gene is active, as was said, copies of the gene are present in a form called mRNA (messenger RNA). Only the exonic parts are copied in the messenger RNA, which constitutes the 'useful' part of the gene. Given that mRNA cannot, as such, be cloned, it is necessary first to convert it into DNA, called in fact in this case cDNA.

---

[36] Despite the significant dimensions of human DNA, it can be fragmented into around a million fragments, which can be introduced in as many bacteria cells, which will then generate a million clones containing then, in fragments, all of human DNA. The clones together constitute a library of human DNA.

[37] Complementary DNA (cDNA) is synthesized double helix DNA from a sample of mature messenger RNA. In order to produce cDNA the two helixes are synthesized in two steps: the first helix is produced using mRNA as a mold, while the second is synthesized using the first helix produced.

At this point, as long as the appropriate source from which to extract messenger RNA is known, it is possible to create a library of cDNA which reproduces copies of the gene in question which contain only the exonic parts. From this, then, it is possible to conceptually deduce, through the so-called gene expression,[38] the sequence of amino acids of the corresponding protein, which then may be produced in the desired quantity.[39]

Genetic engineering, or recombinant DNA technology, allows for the creation of new combinations of DNA,[40] and is the most important advanced biotechnology.

In the past, traditional selection techniques permitted modification of genetic makeup of plants and animals of similar species through selective breeding.

Today, through the transfer of genes or DNA on which the technology of recombinant DNA is based, it is possible to modify the genetic properties of a living organism in a well-defined and rapid way.

Recombinant DNA technology consists, essentially, in the isolation and modification, reproduction and insertion of single genes of DNA into the chromosomes of cells which may also be different from those from which it was extracted.

A molecule of recombinant DNA transferred to a host cell, as hinted at, is cloned, or replicated, during the process of cellular division. In this way, genes may be passed from one species of the genome to a different species. Today, the transfer and replication of a fragment of DNA through a host is a technique often used for the selection and study of genes.

---

38 Or rather in fact the process through which information contained in a gene, constituted by DNA, is converted into a functional macromolecule (typically a protein).

39 In particular, in order for the cloned gene to be used for the production of protein, it is necessary always to substantially modify it, removing the natural structures it has and that were necessary and sufficient for their activation in the artificial system chosen. In this way, the cloned gene may be introduced in cultivated microorganisms or cells (in origin derived from a living organism), in which it is possible to be activated and produce the protein of interest. Sometimes, then, besides the amino acidic function to form the active protein it is necessary to add another molecule to it, through the action of enzymes present only in some types of cells.

40 The first molecule of recombinant DNA was created by Paul Berg, a biochemist of the University of Stanford, who in 1972 was able to manipulate the genetic makeup combining fragments of DNA of different origins and creating then a new combination of DNA.

The transfer of cytoplasm of a cell of a DNA fragment coming from the outside is obtained through the actions of enzymes,[41] capable of cutting the molecule of DNA in specific places. In order for the DNA fragment to replicate, it is necessary to use the so-called cloning vectors of the same (plasmids, bacteriophages, viruses, and so on), which must first be purified.

The technologies of genetic engineering, then, may be divided, on the basis of their aims, in the following way: (a) sequencing of genes, whose aim is the study of the structure of single genes, and therefore the position of the four molecules which compose the triplet; (b) gene mapping, which is aimed at locating genes, of which the products and effects on chromosomes is known; (c) isolation of genes, which is carried out by special sensors on the DNA fragments of interest, obtained through aforementioned enzymes; (d) transfer of genes, which is aimed at the insertion of genes in cells or tissues, different from those of origin;[42] and (e) cloning of genes, which aims at rapidly reproducing single genes, through the use of cloning vectors, in order to study and use them.[43]

Alongside techniques of genetic engineering, then,[44] there are cloning techniques. These consist in the removal of chromosomes or nuclei of the ovum, through microsurgical manipulations or methods of cellular fusion, and in their substitution with a nucleus donor, from a cell (somatic or follicular) of the same female from whom the ovum was taken, or from a cell of a different individual, male or female, or using a blastomere from an embryo. These techniques, which allow for the obtainment even of 'children with one parent', are distinguished from techniques of genetic engineering since they do not modify or combine genes.

---

41 So-called 'endonuclease restriction'.

42 The process of introducing exogenic biological material into eukaryotic cells is called 'transfection'. The process of transfection may be carried out: (a) *in vitro*, on targeted cells in long term cell cultures; (b) *ex vivo,* on isolated cells of an organism and transferred onto a culture medium; or (c) *in vivo*, directly on the cell of an organism.

43 See F. Mastropaolo, 'Ingegneria genetica', cit., p. 429.

44 Besides the techniques of artificial insemination, intended to result in the artificial meeting of two gametes, which are not analyzed in this book in consideration of their relevance which is quite remote from patent protection profiles.

The development of biotechnology which considers intervention on the genetic makeup of humans, in particular, increases significantly possibilities of diagnosis and treatment. Biotechnology allows for the diagnosis of diseases caused by genetic mutations and even to verify if the foetus is a carrier of genetic defects.[45]

What is more, they allow for the use of gene therapy, or rather the transfer and expression of foreign genetic material within the patient's cells with a cellular function which has a genetic alteration, in order to restore normal function with permanent effects.[46]

As a general rule, there are two types of gene therapy: (a) gene therapy carried out on germinal (or reproductive, eggs and spermatozoa) cells is intended to reintroduce the lacking function or suppress that which is impaired, with the effect of changing some characteristics of the individual which is then transmitted to descendants; or (b) gene therapy carried out on somatic cells, or on cells making up tissue, allows for the introduction of functional genes into somatic cells with impaired function due to defective genes; in this case the therapeutic effect is limited to the person receiving the treatment (and is not transmitted to descendants). Gene therapy of somatic cells, in turn, is divided into two groups: *ex vivo*[47] and *in vivo*.[48]

Both types of therapy are still in their experimental phases, even if there have been significant results in an extremely limited number of

---

[45] Genetic tests consist in the analysis of specific genes, their product or function, as well as every other type of investigation of DNA or chromosomes, aimed at identifying or excluding modifications (of DNA) similarly associated with genetic pathologies, and which may have various applications (diagnostic, pre-clinical, prenatal and postnatal diagnosis of genetic diseases, prediction of genetic predisposition, forensic investigation, genetic screening, etc.).

[46] Contrary to traditional therapy, which usually achieves limited results over time. In gene therapy, specific vectors transport the segment of DNA containing the gene to be introduced onto specific cells of the organism. The sequences transported by the vector are transcribed and translated by the cell, in order to finally reach synthesis of the therapeutic gene product.

[47] This is the type used first and consists in the withdrawal of somatic cells from the person of interest, which afterwards are put in culture in the laboratory. During this time, the desired gene is inserted into the cells, through specific vectors, and subsequently is reimplanted into the body of the subject. This procedure is longer and more costly than the other, but is more effective. Currently it is the most used procedure, but is reserved only for those cases in which it is possible to extract cells, put them in culture and reinsert them into the organism.

[48] This type is carried out in all cases in which the cells cannot be put in culture, or extracted and re-implanted, like for most internal organs; on the other

cases. Therapy based on the transfer of genes of other organisms into germinal line cells, raising strong doubts on ethics, has been applied only in animals and plants. At a scientific level, there is not, in fact, enough of a basis to predict the effects of such therapy on the development of the individual and his descendants. At the level of bioethics, germinal gene therapy impairs the right of every human being to be born with an unchanged genetic makeup, coinciding with eugenic practices, which are unacceptable for human beings.[49] Instead, it has been agreed that therapy on somatic cells may also be carried out on humans.[50]

Gene therapy is suitable, overall, for the treatment of damage resulting from spontaneous mutations which are verified in relation to one or more genes, or the treatment of chromosomal abnormalities arising during the process of maturation of germinal cells.[51]

Significant development, in particular, has been made in the experimentation of therapy based on the use of stem cells (embryonic, fetal, amniotic and adult), unspecialized primitive cells with the singular ability to be transformed into various other types of cells of the body. They could be used, in particular, for the reparation of damaged organs, in substitution of transplants, to prevent organ rejection, as well as, through the use of specialized cells derived from stem cells, in cell therapy of many illnesses.[52] The use of stem cells from embryos or foetuses, however, raises fundamental questions of bioethics.[53]

---

hand, currently, its application is more difficult. In this case the desired gene is directly inserted into the organism through a suitable vector, locally or systemically.

[49] See F. D'Agostino and L. Palazzani, *Bioetica. Nozioni Fondamentali*, Brescia: La Scuola, 2007, p. 101 et seq.

[50] See D.J. Watt and G.E. Jones, 'Skeletal Muscle Stem Cells: Function and Potential Role in Therapy', in C.S. Potten (ed.), *Stem Cells*, London: Academic Press, 1997, p. 75 et seq.; D.W. Emery and G. Stamatoyannopoulos, 'Stem Cell Gene Therapy for the β–chain Hemoglobinopathies', therein, p. 94 et seq.; J.A. Nolta and D.B. Kohn, 'Haematopoietic Stem Cells for Gene Therapy', therein, p. 447 et seq.

[51] In this regard: N.S. Roy, S. Wang, L. Jiang *et al.*, '*In Vitro* Neurogenesis by Progenitor Cells Isolated from the Adult Hippocampus' (2000) *Nature Medicine* 271 et seq., n. 6.

[52] In particular, stem cells from bone marrow are used, easily withdrawn, transferable *in vitro*, re-transferable into the original organism as if it were an autologous transplant, and therefore capable of multiplying in order to replace the diseased cells. The product obtained in this manner, then, taken into circulation and formed from the same modified cells, will reach the cells of other organs and tissues carrying out the desired correction.

[53] In regard to which, see below.

Finally, tissue engineering is an innovative technology useful for the *in vitro* reconstruction of tissues used for therapeutic treatments. They consist in living cells, isolated for instance from the patient awaiting treatment, from a donor or, rarely, from an animal, made to proliferate through the seeding of specific structures and the cultivation of specific bioreactors, in order to reconstruct specific tissues.[54]

In general, for interventions on human beings which have already been born, gene therapy requires no more attention than current techniques used for any other experimental therapy treatment.[55] Therefore, gene therapy, besides being applied on the condition that there is detailed information on the gene and its function, must be justified by the seriousness of the illness, the lack of an effective alternative therapy and with lasting effects, and by the low occurrence of undesirable side effects. What is more, particular care must be taken when obtaining the informed consent of the patient,[56] offering him the possibility to resort to traditional therapies, which may be less effective but sometimes using more established procedures. The experimental nature of the therapy, on the other hand, may justify refusal to practice it by the doctor, even if it has been specifically requested by the patient.[57]

Gene therapies on embryos in utero present particular difficulties. In this case, besides the aforementioned necessary requirements in general, with specific information regarding the advantages, the contraindications and the authorization of legitimate subjects, an accurate evaluation of the related risks to the patients (the unborn child and the mother)[58] from such

---

[54] Think, for instance, of the recent discovery of stem cells, isolated from human lung tissue, which seem able to rapidly reconstruct lung alveoli, allowing for significantly effective new treatments for people with damaged lungs (see A. Coghlan, 'Stem-cell Find Breathes New Life into Lung Repair', in *New Scientist*, n. 2837, November 2011, accessed 16 December 2014 at http://www.newscientist.com/article/dn21102-stemcell-find-breathes-new-life-into-lung-repair.html).

[55] See E. Sgreccia and M.L. Di Pietro, 'Manipolazioni genetiche e procreazione artificiale: orientamenti giuridici e considerazioni etiche', cit., p. 1382.

[56] If the patient is capable of acting; if instead she/he is not, it is necessary to obtain the consent of her/his legal representatives (on this point see also below).

[57] See F. D'Agostino and L. Palazzani, *Bioetica. Nozioni Fondamentali*, cit., p. 102.

[58] In this regard, the Congregation for the Doctrine of the (Catholic) Faith, in the *Instruction on respect for human life in its origin and on the dignity of procreation – Replies to certain questions of the day* of *Instruction Dignitas Personae on certain bioethical questions* (Vatican City, 8 September 2008, accessed 16 December 2014 at http://www.vatican.va/roman_curia/congregations/cfaith/

a therapy, as was said of an experimental nature, is required. In theory, there may be less difficulty where action is taken on a zygote *in vitro*, or on the mono-cellular state of the individual,[59] originating from out of body fertilization. Further, on one hand, technical difficulties and undesirable results do not seem to be completely avoidable, on the other, a general method of prevention of hereditary illnesses, through a general use of *in vitro* fertilization and zygote diagnosis, appears to be – for the enormous distortion between advantages and bioethical, legal, social and biological contraindications (not to mention the technical difficulty) – 'more of a nightmare than a chance'.[60]

Developments in genetic therapy on humans, in any case, require adequate experimentation. In cases in which already tried methods are lacking and in situations in which known therapies are not effective, it may be necessary to apply new therapies, which have not yet been sufficiently tried, directly on the ill individual. Therapeutic experimentation is legal on conditions common to every health treatment, should ordinary treatments be ineffective. Thus, it is necessary to obtain the patient's consent. In this regard, the problem is that consent may be considered valid when, although there is a risk of permanent reduction of the physical integrity (which in and of itself would render consent invalid), therapeutic experimentation is justified by the usefulness to the ill individual and when it was preceded by adequate experimentation on animals. On the state of need and the reasonable probability of a cure for the ill individual, or at least to offer treatment which may relieve suffering, experimental therapies may be authorized when the patient is not able to express his own consent or, in the case of incapacity, minors or unborn children, when that of their respective legal representatives is lacking.[61]

---

documents/rc_con_cfaith_doc_19870222_respect-for-human-life_en.html), stated that: 'As with all medical interventions on patients, one must uphold as licit procedures carried out on the human embryo which respect the life and integrity of the embryo and do not involve disproportionate risks for it but are directed towards its healing, the improvement of its condition of health, or its individual survival' (Instruction cited, I, 3).

[59] The zygote is a single cell which is obtained from the fertilization from the fusion of two specialized cells, the gametes (male and female).

[60] F. Mastropaolo, 'Ingegneria genetica', cit., p. 445; *contra*, R.G. Edwards, '*In Vitro* Fertilization and Embryo Replacement: Opening Lecture', in *Annals of the New York Academy of Sciences*, Vol. 442, New York: Academy of Sciences, 1985, p. 1 et seq.

[61] See again F. Mastropaolo, 'Ingegneria genetica', cit., p. 446–7.

## II.3 FUNDAMENTAL LEGAL AND BIOETHICAL ISSUES ARISING FROM INTERVENTION ON HUMAN GENETIC PROPERTY

As a consequence of the various forms of intervention on the human genetic makeup previously discussed, there is a series of fundamental legal and bioethical questions, among which, in this case, the following are particularly relevant, concerning: (i) the conditions and limits of research on the human genome; (ii) the definition of the so-called 'statute of the embryo'; (iii) the conditions and limits on stem cell research; and (iv) the prohibition or admissibility of reproductive or therapeutic cloning.

### II.3.1 The Conditions and Limits of Research on the Human Genome

The development of scientific research concerning the human genome[62] offers important new possibilities in the field of biomedical science. In particular, it provides tools for the discovery of genetic alterations which render some individuals and families more predisposed than others to genetic illness. This raises high expectations regarding the possibility that treatments be developed even specifically for a single person, for the prevention and/or treatment of such illnesses.[63]

However, the development of biomedical activities connected to the human genome raises also a group of controversial questions of an ethical and legal nature, such as for instance those related to the genetic identity and privacy of each person, to the nature of informed consent which should be guaranteed to those who are subject to a genetic test, to the access to information resulting from this test, to the ownership and possible use of information following any 'patent claims', as well as to the possibility that the aforementioned activities may lead to some form of genetic discrimination. There are many further examples that could be made. Among the most worrying from an ethical point of view is human cloning, which seems to be an option which is no longer far from the realm of possibility, or the repugnant option of the combination of human genes with those belonging to primates, in order to produce beings

[62] Following the completion of mapping the human genome, regarding which refer to Part I.

[63] In this regard, *ex multis*, see K. Davies, *The $1,000 Genome: The Revolution in DNA Sequencing and the New Era of Personalized Medicine*, New York: Free Press, 2010, and K. Davies, *Cracking the Genome: Inside the Race to Unlock Human DNA*, Baltimore: John Hopkins University Press, 2001, *passim*.

intelligent enough to carry out human activities, but who, given their presumed non-human status, could legally be exploited in conditions of near slavery.[64]

At the beginning of the Human Genome Project, the aversion of such questions was attempted through the request that all participants observe guiding criteria and binding procedures. The project was designed based on a group of principles, which in 1997 were inserted into the Universal Declaration on the Human Genome and Human Rights of UNESCO.[65] The principles which were adopted, in particular, are the following: (a) recognition that the human genome is an integral part of a 'common heritage of mankind';[66] (b) recognition of the fundamental rights and freedoms of man and the compliance with relative international regulations; and (c) respect for the values, integrity, traditions and culture of the subject of genetic research.

On the basis of these guidelines, specific indications were given regarding the rights of people involved in the study of the genome, on the limits of research and the conditions under which to carry it out.

In particular, in order to start genetic experimentation the informed consent of the person involved was deemed necessary. If she/he is not able to give it, the research may only be carried out for her/his direct benefit, in compliance with the law, or in exceptional cases the benefit of others, if the risks are minimal and the potential beneficiaries are in the same genetic conditions and of the same age, as long as the research is compatible with the human rights of the individual. It was decided that it

---

[64] See among others, C. Kuppuswamy, *The International Legal Governance of the Human Genome*, London: Routledge, 2009, p. 1 et seq.; M. Bratton (ed.), *God, Ethics & the Human Genome*, London: Church House Publishing, 2009, p. 1 et seq.; M. Rimmer, *Intellectual Property and Biotechnology: Biological Inventions*, Cheltenham, UK and Northampton, MA, USA Edward Elgar, 2008, p. 138 et seq.; B.A. Brody, 'Intellectual Property and Biotechnology: The U.S. Internal Experience – Part II' (2006) 16 *Kennedy Institute of Ethics Journal* 105 et seq.; Lenzerini, *Biotechnology, Human Dignity and the Human Genome*, cit., p. 285 et seq.; F. Vandenabeele, *Patentability of Living Organisms: Legal and Ethical Aspects of the Question*, Montreal: McGill University, 2005, p. 79 et seq.

[65] 'Universal Declaration on the Human Genome and Human Rights' ('Universal Declaration'), adopted by the General Conference of UNESCO on November 11, 1997 and implemented at the General Assembly of the United Nations in 1998.

[66] 'The human genome underlies the fundamental unity of all members of the human family, as well as the recognition of their inherent dignity and diversity. In a symbolic sense, it is the heritage of humanity' (Universal Declaration, art. 1).

is necessary to respect the right of everyone to decide whether they would or would not be informed of the result of the genetic test and the related consequences for their health. The prohibition of discrimination based on genetic characteristics of the person and the obligation to maintain the confidentiality of genetic data belonging to an identifiable person used for research or other purposes were established.[67] The prohibition of activities contrary to the dignity of man, such as the cloning of humans or others to be established by states and international organizations was reiterated.[68] It was also provided that the benefits resulting from advances in biology, genetics and medicine, concerning the human genome, be made available to the public and especially to the scientific community, so that it can pursue research as it is deemed appropriate.[69] Finally, states were given the task of adopting measures necessary to favour the exercise of freedom of research on the genome and in order to face the relevant ethical, legal and social questions, including through the creation of specific ethics committees.[70]

Similar provisions are contemplated, more briefly, in the European Convention on Human Rights and Biomedicine of the Council of Europe,[71] in which[72] it is stated in particular that: 'An intervention seeking to modify the human genome may only be undertaken for preventive, diagnostic or therapeutic purposes and only if its aim is not to introduce any modification in the genome of any descendants.'[73]

As far as the tension between genome research[74] and intellectual property is concerned, it may be identified in various provisions of the Declaration of UNESCO and European regulations. Due to the moral and legal motives, restrictions on the possibility to obtain patent rights over the human genome were introduced. On the basis of article 4 of the Universal Declaration on the Human Genome and Human Rights of

---

67 Universal Declaration, arts 5–9.

68 Universal Declaration, art. 11.

69 Universal Declaration, art. 12.

70 Universal Declaration, arts 14–25.

71 European Council, 'Convention for the protection of Human Rights and dignity of the human being with regard to the application of biology and medicine: Convention on Human Rights and Biomedicine', CETS No. 164, Oviedo, 4 April 1997, entered into force on 1 December 1999.

72 After being provided in general, in art. 2, according to which: 'The interests and welfare of the human being shall prevail over the sole interest of society or science'.

73 'Convention on Human Rights and Biomedicine', art. 13.

74 Genomics is the branch of molecular biology which deals with the study of the genome of living organisms.

UNESCO, 'the human genome in its natural state shall not give rise to financial gains'. Article 21 of the European Convention on Human Rights and Biomedicine[75] declares, similarly, that 'the human body and its parts shall not, as such, give rise to financial gains'.

Clear evidence of the uncertainty the aforementioned tension creates is due to the circumstances of the partial DNA sequences, the so-called Expressed Sequence Tags (EST).[76] Though in the last few years numerous patents were granted for them,[77] there is still a significant difference of opinion regarding the issue of whether they satisfy the requirements of patentability or not, given that they represent part of the genes falling under the sphere of public domain.[78] In any case, on one hand, the growing number of patents granted to ESTs shows the importance of their patent protection for the economy and industry;[79] on the other hand, it seems appropriate to observe that ESTs[80] are not part of the genome 'in its natural state', nor are they part of the human body 'as such', but are obtained through specific laboratory procedures.

Another profile in which tension is felt between intellectual property and genome research is that which relates to the status of the human genome as 'common heritage of mankind' (CHM).[81] Some have claimed,

---

75 Oviedo, 4 April 1997, cit.

76 Expressed Sequence Tags (ESTs) are short fragments of DNA transcribed and sequenced by a longer cDNA sequence. Basically, they represent part of the expressed genes and may be used as a tool for biotechnological research, for instance as a sensor in order to identify which protein is expressed in a cell, or to discover new genes or to determine their sequence and position in the genome. Single Nucleotide Polymorphism (SNP), instead, are DNA fragments which, inserted into broader gene sequences, determine the codification of a protein in the latter which is different or contains new properties. This characteristic feature leads to the belief that, contrary to ESTs, SNPs have a specific use. With regard to ESTs and SNPs, see, among others, J.H. Barton, 'United States Law of Genomic Patents' (2002) 33 *IIC* 783 et seq.

77 In this regard, refer to the following Parts.

78 See D. Nicol, 'On the Legality of Gene Patents' (2005) 29 *Melbourne University Law Review* 809 et seq.; J. Murray, 'Owning Genes: Disputes Involving DNA Sequence Patents (1999) 75 *Chicago-Kent Law Review* 231 et seq.

79 On the theme of patentability of ESTs, see, among others, A. Oser, 'Patenting (Partial) Gene Sequences Taking Particular Account of the EST Issue' (1999) 30 *IIC* 1 et seq.

80 Like SNPs.

81 In argument, see Nuffield Council on Bioethics, '*The Ethics of Patenting DNA*', accessed 16 December 2014 at http://nuffieldbioethics.org/wp-content/uploads/2014/07/Patenting-DNA-Chapter-3-Patenting-DNA.pdf, p. 21 et seq.; in

in this regard, that the patent system is not applicable to the human genome in virtue of this status.[82]

When, in 1993, the International Bioethics Committee of UNESCO began to examine the notion of CHM as a principle to be introduced to the discipline of the human genome, the concept was developed with an adequate structure concerning its origin and function with regard to different common spaces and natural resources. Subsequently, however, the attention shifted from the understanding of the system and the function of the concept as a whole to the elements regarding ownership. In this regard, while ownership rights on parts of the body were clearly excluded from the beginning, at the start those of patent protection of products or processes which use DNA were not, and protection in this sense was deemed possible.

In the final wording in 1997 of the Universal Declaration on the Human Genome and Human Rights, in any case, the human genome is considered the 'common heritage of mankind', as noted, 'in a symbolic sense', with the intention of minimizing the meaning of the notion as a guiding principle. This is due to the prevalent intergovernmental expert perspective according to which a CHM framework would lead to the reduction in motivation of innovation.

The notion of a 'common heritage of mankind', according to the most proper reconstruction, essentially concerns the equal use of some resources and the respect of peoples' right to development.[83] This is not

---

doctrine: P.N. Ossorio, 'The Human Genome as Common Heritage: Common Sense or Legal Nonsense?' in (2007) 35(3) *Journal of Law, Medicine & Ethics* 425 et seq.; J. Frakes, 'The Common Heritage of Mankind Principle and the Deep Seabed, Outer Space, and Antarctica: Will Developed and Developing Nations Reach a Compromise?' (2003) 21 *Wisconsin International Law Journal* 409 et seq.; K. Baslar, *The Concept of the Common Heritage of Mankind in International Law*, Dordrecht (Netherlands): Martinus Nijhoff Publishers, 1997; G.M. Danilenko, 'The Concept of the "Common Heritage of Mankind" in International Law', in *XIII Annals of Air and Space Law*, 1988, p. 247 et seq.; C.C. Joyner, 'Legal Implications of the Concept of the Common Heritage of Mankind' (1986) 35 *International and Comparative Law Quarterly* 190 et seq.

[82] In this sense, see L.B. Andrews and J. Paradise, 'Gene Patents: The Need for Bioethics Scrutiny and Legal Change' (2005) 5 *Yale Journal of Health Policy, Law & Ethics* 403 et seq.

[83] The concept, in fact, originates from the 'Treaty on Principles Governing the Activities of States in the Exploration and Use of Outer Space, including the Moon and Other Celestial Bodies' (27 January 1967, accessed 16 December 2014 at http://www.oosa.unvienna.org/oosa/SpaceLaw/outerspt.html), and appears hereafter, *inter alia*, in the 'Agreement Governing the Activities of States

contrary to the granting of exclusive rights and considers the possibility of alienation of resources which are the subject of those rights. At the same time, it does not allow for its appropriation and excludes the function of the exclusive right with respect to some common elements of all men and therefore not appropriable 'by their very nature'. A similar diagram of operation of the CHM system shows that the concept of exclusivity is not central to it, but rather the equal use of certain resources is. Thus, it appears incorrect to claim that the aforementioned system denies the possibility to become the owner of patents on genes.[84]

Finally, as far as regulatory limitations that influence research on the human genome are concerned, there are specific exceptions to patentability which were introduced to the European patent system on the basis of moral concerns regarding human rights and public health.[85]

### II.3.2 The Definition of the So-called 'Statute of the Embryo'

The recent possibilities in the manipulations of the initial phase of human life made available by biomedical progress have started a bioethical debate on the so-called 'statute of the human embryo'.

New technology allows, for instance, for the production of embryos in vitro through assisted fertilization, the freezing of embryos in order to preserve them (so-called cryopreservation), prenatal diagnosis and interventions with regard to any pathology, the transfer of embryos into the uterus of the biological or surrogate mother, the cloning of embryos, the extraction of embryonic stem cells, and so on. Such practices raise urgent questions, which require in fact clarification regarding the legal status of the unborn child. In the context of the current debate on the matter,[86] two

---

on the Moon and Other Celestial Bodies' (5 December 1979, accessed 16 December 2014 at http://www.unoosa.org/oosa/SpaceLaw/moon.html).

[84] On this theme, see in particular: C. Kuppuswamy, *The International Legal Governance of the Human Genome*, cit., p. 140 et seq.

[85] In this regard, refer to Part IV. For a general framework, see L.B. Andrews and J. Paradise, 'Gene Patents: The Need for Bioethics Scrutiny and Legal Change', cit., p. 403 et seq.; F. Abbott, T. Cottier and F. Gurry, *The International Intellectual Property System: Commentary and Materials, Part I*, The Hague: Kluwer Law International, 2003, p. 503 et seq.

[86] In Roman law, the issue appeared to be solved more easily, in that those who were in utero were considered in almost all civil law as if they had already come into the world, on the basis of a noteworthy passage of Giuliano: '*Qui in utero sunt, in toto paene iure civili intelleguntur in rerum natura esse*' ['Those who are in the womb, almost in the whole law of the state are understood to be in the nature of things'] (D.1,5,26 Iul. 69 dig.); in this regard, *ex multis*, see

main lines of thought can be distinguished, one that denies intrinsic dignity of the unborn child and one that instead supports it.

The denial of dignity to the unborn child is derived from a scientific-reductionist and empirical viewpoint, on the basis of which the human embryo is held to be a mere cell belonging to the human species, which casually forms and is multiplied into a group of cells which then differentiate into tissues and organs. In this way, the human embryo is simply considered to be an object, which is organic matter with a more or less probable development, without attributes of anthropological (personal identity) characteristics, or rather moral or legal characteristics (human dignity and legal subjectivity). Thus, the embryo is not yet an individual, but will become one in a subsequent stage of development, identified, depending on the author, in syngamy,[87] in implantation in the uterus, in the formation of the primitive streak (14th day following conception),[88] in the formation of the central nervous system, in the formation of the cerebral cortex or in the exercise of reason.[89]

In the United States, in particular, the theory contrary to the existence of a person prior to birth, has been sustained with authority, both in doctrine[90]

---

E. Bianchi, *Per un'indagine sul principio "conceptus pro iam nato habetur" (Fondamenti arcaici e classici)*, Milan: Giuffrè, 2009, p. 289 et seq.; P. Ferretti, *In rerum natura esse in rebus humanis nondum esse. L'identità del concepito nel pensiero giurisprudenziale classico*, Milan: Giuffrè, 2008, p. 132, who recalls the hypothesis of interpolation advanced by a part of modern doctrine.

[87] This is the moment of complete fusion of all chromosomes of male and female germinal cells (gametes) into a zygote.

[88] The moment in which the group of cells (some called 'pre-embryonic') lose their totipotency, or rather the property for which every cell may be isolated from the others and lead to the development of a complete individual (such as for twin pregnancies), as well as, if they are packed with the others, uniting in a single organism.

[89] In this view, the fact of the material existence of biological life of a human embryo does not lead to any obligation on the axiological or regulatory level; the division provided by Hume (see D. Hume, *A Treatise of Human Nature*, III, part I.1) among facts and values/rights, between being and should be, prevents the ontological consideration of the unborn child as an individual with rights. See, in this regard, E. Sgreccia, *Manuale di bioetica, Volume 1 – Fondamenti ed etica biomedica*, Milan: Vita e Pensiero, 2007 (4th ed.), p. 550 et seq.; F. D'Agostino and L. Palazzani, *Bioetica. Nozioni Fondamentali*, cit., p. 75 et seq.

[90] See Engelhardt Jr., *The Foundation of Bioethics*, cit., p. 135 et seq., who in particular claims that there is no sense in speaking of autonomy of a foetus, just as infants and those with serious mental disabilities, who have never been rational.

and in case law.[91] Here, emphasis is not placed on the concept of legal capacity, but rather on the principle of 'autonomy', which identifies people 'in a strict sense' as individuals capable of autonomy. From this framework, then, one may claim that unborn children have never been legally recognized as people in the fullest sense.

A similarly utilitarian[92] or rational[93] approach is based on the common origin of functionality, or the identification of personal subjectivity with the detection of external conditions or current performance of certain functions.

The functionalist theories have been criticized given the presence of a sensory, conscious or intellectual function, or the conditions for their use, assumes the existence of an individual; the existence of the individual which renders the performance of certain functions possible, and not the functions themselves, constitutes subjectivity.[94]

On the basis of the line of thought that, instead, supports the intrinsic dignity of the unborn child, the biological observation of the human body in initial phases of development shows the need to overcome the scientific-reductionalist and utilitarian reconstruction, in favour of interpretive meta-biological hypothesis, in that the human body, from its state as a single-celled zygote, or according to others from the spermatozoa's

---

91 See Court of Appeals for the Fourth Circuit, decision *Mary Scott Doe, et al., v. Barack H. Obama, President of the United States, et al.*, 21 January 2011, 180 L. Ed. 2d 226, 131 S.Ct. 2938, in which it was stated that the frozen embryo has no standing to assert constitutional rights; U.S. Supreme Court, decision *Roe v. Wade*, 22 January 1973, 410 U.S. 113, which affirmed the constitutional right to abortion as an expression of the more general *right of privacy*. In regard to this decision, see extensively below and in Part III.

92 Utilitarians deem it legal to suppress non-sentient embryos, but hold also that embryos, once they have developed sensibility, may equally be destroyed, on condition of using painless techniques.

93 Rationalists, or rather, those who consider the condition or actual use of rational ability relevant, admit the possibility that people 'in the strict sense' may freely decide to give embryos a high moral value in particular situations, for 'sympathy' (sharing the experience of others) or for 'charity' (will to do good for others). In these cases, however, the 'moral' duty of protection may be set aside when facing duties held to be more important, such as for instance the desire to experiment or to procreate.

94 See D'Agostino and Palazzani, *Bioetica. Nozioni Fondamentali*, cit., p. 79 s.; F. D. Busnelli, 'La tutela giuridica dell'inizio della vita umana', in R. Rossano and S. Sibilla (eds), *La tutela giuridica della vita prenatale*, Turin: Giappichelli, 2005, p. 64 et seq., who highlights that the adhesion to the aforementioned principle of autonomy leads to bringing together the fates of the unborn child and the infant, for whom the threshold for legal protection is not even birth.

penetration of the ovum,[95] it is a living being of the human species, a unique organism, integral and organized in every phase of its development, continuous and gradual, containing within itself all the genetic information of its fully developed body.[96] In this sense, it is clear that the human body is characterized by uninterrupted development with no particularly significant steps.[97] Thus, though they have not yet manifested completely and are not fully expressed, the individual characteristics of the embryo are already present and constitute the support for the dynamic process which will lead to their complete fulfillment.

According to this point of view, then, the human being is a person in virtue of his rational nature, and not in virtue of the effective exercise of certain functions on her/his part, or the existence of the conditions for their performance. This allows for the granting of the status of a person in the human being where it is in a 'potential' condition due to its incompleteness of development (zygote, embryo, foetus,[98] but also added to this are newborns and minors), or of 'deprivation', or of the momentary or permanent incompleteness of functions due to impediments from external factors (dying, elderly, handicapped, individual in a coma, and so on).[99]

Recognition of the embryo as a person leads to the attribution of strong intrinsic dignity to it: if the intrinsic aim of man is '*in nuce*' within the embryo, starting from this stage, human life must be fully respected, in

---

[95] In this regard, among others, see S. Buckle, K. Dawson and P. Singer, 'The syngamy debate: when precisely does a human life begin?', in P. Singer *et al.* (eds), *Embryo Experimentation. Ethical, Legal and Social Issues*, Cambridge: Cambridge University Press, 1990, p. 213 et seq.

[96] See, *ex multis*, R.P. George and C. Tollefsen, *Embryo: A Defense of Human Life*, New York: Doubleday, 2008, p. 3 et seq.

[97] The so-called 'Warnock Report' ('Report of the Committee of Inquiry into Human Fertilisation and Embryology', 1984, cit.), drafted by the committee established by the British government in view of the adoption of a regulation on the embryo, also stated that it is impossible to identify the true limits of the sphere of embryonic development, being that it is possible only to establish conventional limits for social reasons: 'Once the process has begun, there is no particular part of the developmental process that is more important than another; all are part of a continuous process. … However we agreed that this is an area in which some precise decision must be taken, in order to allay public anxiety' (M. Warnock, *A Question of Life. The Warnock Report on Human Fertilization and Embryology*, Oxford: Basil Blackwell, 1985, p. 65).

[98] It is the product of conception in development from the eighth week of inter-uterine life, until birth.

[99] See L. Palazzani, 'Persona e essere umano in bioetica e nel biodiritto', in *Idee*, 1997, no. 34/35, p. 133 et seq. (especially p. 143 et seq.).

order to guarantee its full expression. From this reconstruction is derived the absolute duty, and priority in balance with other interests, of respect of the embryo as an expression of human life.[100] In particular, the respect is to be understood in a dual sense, both negative, as a prohibition to destructive interventions, and positive, as an obligation to treat the embryo; the latter profile seems to allow, or rather encourage, therapeutic interventions on the embryo, on the condition that their aims are to heal, improve the condition of health or survival of the individual.[101]

In an intermediate position, then, are those who, considering it inappropriate or limiting to express an opinion on whether the embryo is to be considered a person, hold that in any case the embryo, destined to develop into a complete human being, should be strongly protected, on the basis of the Kantian categorical imperative and/or the golden rule.[102] On the basis of this viewpoint, the value of human life is so high that, even when in doubt regarding its being recognized as a person, it should in any case be thoroughly protected.

In this sense, in philosophical literature the thesis of Habermas is particularly interesting. He highlights the opportunity to recognize the unavailability of the pre-personal human life, except in order to distinguish it from the inviolability of the dignity of man, identifying then, in the well-known *continuum* of human life, unavailability which does not yet reach inviolability. As a consequence, according to the German philosopher, in order to establish the unavailability of prenatal human life it is not necessary to extend the argument of human dignity to the beginning of human life, given that 'our ethical self-understanding as beings-of-kind' is instead sufficient, and may in and of itself lead to regulatory limitation regarding the treatment of embryonic life.[103]

Always tending towards an approach of mediation, but of a significantly different nature given the different bioethical and constitutional principles respectively at the basis of European and North American experiences, the American philosopher Dworkin, starting from the premise of human liberty as the domain of life, states that the best way to face

---

[100] In this sense, among others, see A. Baldassarre, 'Le biotecnologie e il diritto costituzionale', in M. Volpi (ed.), *Le biotecnologie: certezze e interrogativi*, Bologna: Il Mulino, 2001, p. 15 et seq. (especially p. 25).

[101] On this point see also above, para. II.2.

[102] In this regard, refer to the above, para. II.1.

[103] See J. Habermas, *Die Zukunft der menschlichen Natur. Auf dem Weg zu einer liberalen Eugenik?*, Frankfurt: Suhrkamp, 2001, *passim*, who is in line with the reflections put forth by H. Jonas, *Das Prinzip Verantwortung. Versuch einer Ethik für die technologische Zivilisation*, *passim*.

the issue of sanctity of human life and the proper space of freedom in its domain is that of maintaining self-consciousness and self-respect[104] as central for our species. Based on this reconstruction, further, it is shown to be impossible to think that human beings have personal interest from the moment of conception, given that 'there is no such thought to be had'.[105]

An authoritative European scholar, similarly, has highlighted that the questions of whether the unborn child is or is not a person is not central, given that this does not condition the verification of relevance or protection of the unborn child itself.[106] In particular, it is said that it is necessary to start from the premise that within the embryo there is the beginning, all the beginning but only the beginning, of man: this is enough for its dignity and sanctity, but not yet for it to be considered, despite ontological diversity, man and therefore a person; the person to whom the law recognizes individuality.[107]

Therefore, given its uniqueness and its destiny, the unborn child cannot be the subject of laws in the hands of third parties. Private law highlights that there are values which are imposed 'immanently' with regard to subjective situations. Thus, private individuals cannot act contrary to the principle of the protection of life 'from its beginning' (except where it is permitted by law). On the contrary, any aim other than generation, even for scientific research or in particular in order to obtain stem cells for therapeutic reasons, which leads to the sacrifice of the embryo's natural destiny for life, is not legal in that it does not protect, but rather destroys the beginning of human life. In line with this reconstruction, neither the production of embryos for therapeutic purposes, nor the formation *in vitro* in the sphere of assisted fertilization of embryos in a superior

---

104 See R. Dworkin, *Life's Dominion. An Argument about Abortion, Euthanasia, and Individual Freedom*, New York: Vintage, 1994, p. 241.

105 Dworkin, *Life's Dominion. An Argument about Abortion, Euthanasia, and Individual Freedom*, cit., p. XI.

106 See G. Oppo, 'L'inizio della vita umana' (1982) *Rivista di diritto civile* 499 et seq. (p. 499).

107 G. Oppo, 'Scienza, diritto, vita umana' (2002) *Rivista di diritto civile* 11 et seq. (especially p. 13 et seq.), who highlights how, from the moment of fertilization the embryo is considered in international and western legal systems as the future recipient of personal rights of the man which is in formation of being (even if fertilization takes place outside the womb).

number not to be used for implantation are admissible, given that they lead to the destruction of the embryo and the life which it holds *in nuce*.[108]

At the level of principles established within regulatory acts, the Universal Declaration of Human Rights,[109] the International Covenant on Civil and Political Rights[110] and the International Covenant on Economic, Social and Cultural Rights[111] of the United Nations, in their respective preambles, indicated the principle of human dignity as fundamental to other human rights. Similarly, the Declaration of the Rights of the Child[112] ensures the child adequate legal protection 'before as well as after birth',[113] and the subsequent Convention on the Rights of the Child,[114] besides reiterating the provision cited above, provides specifications that 'every child has the inherent right to life' and obliges party nations to 'ensure to the maximum extent possible the survival and development of the child'.[115]

---

108 See again Oppo, 'Scienza, diritto, vita umana', cit., p. 24 et seq., who, to those who claim the prohibition of use of embryonic stem cells for therapeutic aims may lead *a fortiori* to the denial of abortion (and vice versa), replies that abortion is permitted by law only in defense of women, assuming that she cannot use the embryonic stem cells for any therapeutic use.

109 Approved on 10 December 1948 in Paris by the General Assembly of the United Nations.

110 Multilateral treaty adopted by the General Assembly of the United Nations on 16 December 1966 and entered into force on 23 March 1976.

111 Multilateral treaty adopted by the General Assembly of the United Nations on 16 December 1966 and entered into force on 3 January 1976.

112 Approved on 20 November 1959 by the General Assembly of the United Nations (following a similar document approved in 1924 by the General Assembly of the League of Nations).

113 Declaration cited, whereas 3.

114 The version, reviewed and extended the previous Declaration, approved in New York by the General Assembly of the United Nations on 20 November 1989 and entered into force as a binding tool of international law for the signing Nations (all the countries of the world except the United States and Somalia) on 2 September 1990. On the theme, among others, see J. Mason and T. Fattore (eds), *Children Taken Seriously: In Theory, Policy and Practice*, London/Philadelphia: Jessica Kingsley Publishers, 2005, *passim*.

115 'Convention on the Rights of the Child', art. 6.

The 'American Convention on Human Rights',[116] in line with the Latin American tradition,[117] provides that: 'Every person has the right to have his life respected. This right shall be protected by law and, in general, from the moment of conception.'[118]

In the United States, in the absence of legislation, provisions related to federal funding for stem cell research are relevant. In this regard, in 2009 President Barack Obama issued an Executive Order[119] removing limits to federal funding for research on stem cells which had been put in place by President George W. Bush in August 2001 and repeated in an Executive Order issued in June 2007.[120] In this way, the limitations on research to only those stem cell 'lines' existing before August 2001, which prohibited the allocation of public funds to create further lines, was overcome.

On the other hand, American legislation continues to prohibit the creation of embryos for research purposes, given that President Obama's order did not eliminate the so-called 'Dickey-Wicker Amendment' on the matter, which has been renewed annually by Congress since 1996, which prohibits the U.S. Department of Health and Human Services from using funds for the creation of human embryos for research purposes or for research where human embryos would be destroyed.

With specific reference to the concept of the embryo, in the 2009 version of the Amendment,[121] it is provided that 'the term "human embryo or

---

[116] Signed on the occasion of the Inter-American Specialized Conference on Human Rights, San José, Costa Rica, 22 November 1969, accessed 16 December 2014 at http://www.oas.org/juridico/english/treaties/b-32.html.

[117] See, in particular, the Argentinean civil code of 1871, where it is provided that the existence of a person begins from conception within the mother's womb (art. 70), and the Peruvian code of 1984, which gives the unborn child the quality of a legal individual (art. 1).

[118] American Convention on Human Rights, art. 4.

[119] 'Executive Order 13505 – Removing Barriers to Responsible Scientific Research Involving Human Stem Cells', 9 March 2009, accessed 16 December 2104 at http://www.whitehouse.gov/the_press_office/Removing-Barriers-to-Responsible-Scientific-Research-Involving-Human-Stem-Cells.

[120] Respectively, 'President Bush's Address to the Nation', 9 August 2001, and 'Executive Order 13435: Expanding Approved Stem Cell Lines in Ethically Responsible Ways', 22 June 2007, Federal Register, Vol. 72, No. 120, p. 34589, both accessed 16 December 2014 at http://stemcells.nih.gov/policy/guidelines.asp.

[121] Section 509 of the Omnibus Appropriations Act, 2009 (H.R. 1105, Pub.L. 111-8; 11 March 2009), which prohibited the U.S. Department of Health and Human Services use of federal funds for '(1) the creation of a human embryo or embryos for research purposes; or (2) research in which a human embryo or embryos are destroyed, discarded, or knowingly subjected to risk of injury or

embryos" includes any organism, not protected as a human subject under 45 CFR 46 (the Human Subject Protection regulations) … that is derived by fertilization, parthenogenesis, cloning, or any other means from one or more human gametes (sperm or egg) or human diploid cells (cells that have two sets of chromosomes, such as somatic cells)'.

In August 2010, in the context of the *Shirley v. Sebelius* case, Judge Royce C. Lamberth granted a preliminary injunction against research on embryonic stem cells funded at a federal level, on the basis of consideration according to which the Guidelines on stem cell research[122] 'clearly violate' the Dickey-Wicker amendment. The Department of Justice appealed to the U.S. Court of Appeals for the District of Columbia Circuit in order to obtain the withdrawal of the injunction, while awaiting the appeal of Judge Lamberth's decision, and the Court of Appeals allowed the appeal and issued the withdrawal on 29 April 2011 (essentially due to the lack of the 'likelihood of success on the merits'). As a consequence, Judge Lamberth was obliged to reverse his ruling, and subsequently closed the entire case on 27 July 2011.[123]

In particular, in the aforementioned decision, the Court of Appeals confirmed that the Dickey-Wicker amendment is ambiguous, and that it may reasonably be understood to prohibit the use of federal funding in order to 'directly' engage in the extraction of stem cells from an embryo which would lead to the destruction of the same; contrarily, according to the Court, the same amendment does not prohibit the funding of a research project which 'indirectly' uses previously extracted stem cells.

On this theme, moreover, one must also remember what was stated by the U.S. Supreme Court in the ruling on the case *Roe v. Wade*.[124] Even if the case concerned profiles of constitutionality the voluntary interruption of a pregnancy, in an *obiter dictum* the Court also noted: 'We need not resolve the difficult question of when life begins. When those trained in the respective disciplines of medicine, philosophy, and theology are unable to arrive at any consensus, the judiciary, at this point in the development of man's knowledge, is not in a position to speculate as to the answer'. In this perspective, the unborn child is not a person in the sense of the term contained in the XIV Amendment of the U.S.

---

death greater than that allowed for research on fetuses in utero under 45 CFR 46.208(a)(2) and Section 498(b) of the Public Health Service Act [1](42 U.S.C. 289g(b)) (Title 42, Section 289g(b), United States Code)'.

[122] In this regard, see below in para. (iii).

[123] The ruling is available at http://caselaw.findlaw.com/us-dc-circuit/1529053.html.

[124] U.S. Supreme Court, decision *Roe v. Wade*, January 22, 1973, cit.

Constitution,[125] and its protection is indirect, in as much as the right to privacy must be balanced with public interest, but not with the rights placed directly within its sphere.[126]

At the European level, on the subject, the first thing to stand out is the fundamental principle of human dignity, considered in article 1 of the Charter on the Fundamental Rights of the European Union[127] – which, following the Treaty of Lisbon, assumed the same legal value of a treaty[128] – on the basis of which: 'Human dignity is inviolable. It must be respected and protected'. Again in the Charter of Nice, then, article 2 reiterates the right of every person to life.[129] Article 3 then confirms the principle of integrity of the individual,[130] and in particular, in the sphere of medicine and biology, the following criteria must be respected:

- 'the free and informed consent of the person involved, according to the methods defined by law;
- the prohibition of eugenic practices, in particular those aimed at the selection of people;
- the prohibition of making the human body and its parts a source of profit;
- the prohibition of reproductive cloning of human beings'.[131]

---

[125] Which, in particular, at Section 2, provides that: 'All persons born or naturalized in the United States, and subject to the jurisdiction thereof, are citizens of the United States and of the State wherein they reside. No State shall make or enforce any law which shall abridge the privileges or immunities of citizens of the United States; nor shall any State deprive any person of life, liberty, or property, without due process of law; nor deny to any person within its jurisdiction the equal protection of the laws'.

[126] See the cited decision, 158 et seq.

[127] Proclaimed on 7 December 2000 in Nice, and subsequently, in an adapted version, on 12 December 2007 in Strasbourg.

[128] On the basis of art. 6 of the Treaty of the European Union, as modified by the 'Treaty of Lisbon which modified the Treaty of the European Union and the treaty which established the European Community', signed on 13 December 2007 and entered into force on 1 December 2009: 'The Union recognizes the rights, freedoms and principles sanctioned within the Charter of the Fundamental Rights of the European Union of December 7, 2000, adapted December 12, 2007 in Strasburg, which has the same value of the treaties'. With respect to the relevance to biotechnology of the principles of the Charter, see P. Šturma, 'The EU Charter of Fundamental Rights and the Governance of Biotechnology in the European Union', in F. Francioni and T. Scovazzi, *Biotechnology and International Law*, cit., p. 369 et seq.

[129] Expressly stating that: 'Every person has the right to life'.

[130] According to which: 'Every individual has the right to his own physical and psychological integrity'.

[131] 'Charter of Fundamental Rights of the European Union', art. 3.

The right to life is also covered in article 2 of the 'Convention for the safeguarding of the rights of man and fundamental freedoms',[132] on the basis of which: 'The right to life of every person is protected by law'.

The European Convention on Human Rights and Biomedicine, then, is particularly relevant,[133] which on the one hand provides that: 'Where the law allows research on embryos in vitro, it shall ensure adequate protection of the embryo', and on the other hand expressly prohibits '(t)he creation of human embryos for research purposes'.[134] What is more, such provisions are inserted in a context in which article 1 generally states that: 'Parties to this Convention shall protect the dignity and identity of all human beings and guarantee everyone, without discrimination, respect for their integrity and other rights and fundamental freedoms with regard to the application of biology and medicine', using then the ambiguous concept of 'human being' and mandating the party nations to 'take in its internal law the necessary measures to give effect to the provisions of this Convention'.[135]

In the 'Additional Protocol to the Convention on Human Rights and Biomedicine, concerning Biomedical Research',[136] again, in reference to

---

132 Or 'CEDU', signed in Rome on 4 November 1950 and entered into force on 3 September 1953. The Convention was ratified by all 47 Member states of the European Council, and established the European Court of Human Rights.

133 Oviedo, 4 April 1997, cited.

134 Convention, art. 18. The 'maximum extent possible', which in general is constituted by its balancing with the rights of the pregnant woman, the embryos *in vitro* seem then to be used without limit. On the other hand, in the debate on subjectivity, particularly heated for these embryos, there is a doubt on the possibility to attribute to them the right to life. This is, even when prenatal life protection is clearly granted as in the German law on the protection of embryos (*Gesetz zum Schutz von Embryonen – Embryonenschutzgesetz* of 1990), which provides strong protection criminally sanctioning the illegal use of human embryos; the *Bundesverfassungsgericht*, in fact, showed great caution regarding the issue of subjectivity, stating that it is not necessary to establish if the unborn child is the bearer of fundamental rights, or if instead, due to its lack of a pertinent legal capacity also regarding fundamental rights, it is protected by objective regulations of the Constitution in its right to life (see decision BverfGE39 of 25 February 1975). The same Italian law no. 40/2004 ('Norme in materia di procreazione medicalmente assistita'), which qualifies the unborn child as an individual and places a criminally sanctioned prohibition of suppression, provides exceptional cases in which use of cryopreservation is permitted, being that the hypothesis of excluding it completely is difficult to achieve.

135 For an analysis of provisions of the Convention, and of their relationship with those of the Charter of Nice and of the CEDU, refer to Part IV.

136 Strasbourg, 25 January 2005.

research during pregnancy, it has been stated that it is allowed as a matter of priority on condition that it poses a direct benefit to the health of the pregnant woman 'or to that of her embryo', but then adds – with a fairly generic provision which may be subject to different interpretations – that it may be carried out even when it does not meet such conditions, as long as the project creates 'only minimal risk', intended as 'a very slight and temporary negative impact on the health of the person concerned'.[137]

As far as stem cell research is concerned, the EU Seventh program framework for technological research and development (2007–2013), as well as the following Horizon 2020 Framework Programme for Research and Innovation (2014–2020), excludes: (a) funding of research activities whose aim is the production of inheritable modifications of the human genome, the creation of human embryos only for research goals or for the provision of stem cells, or the reproductive human clonation; and (b) that activity shall be funded in a Member State where such activity is forbidden.[138]

The various Member states of the European Union adopt, on the theme of the statute of the embryo, rather different solutions.

In the United Kingdom, the 'Donaldson Report'[139] on the legality of embryonic stem cell research gave a positive response, concluding that this research does not add any critical features with respect to the provisions of the Human Fertilisation and Embryology Act 1990,[140] which, in the modified version of 2008, allows for the use of human

---

[137] Protocol cited, articles 17 and 18.

[138] See Regulation (EU) 1291/2013 of the European Parliament and of the Council of 11 December 2013 establishing Horizon 2020 – the Framework Programme for Research and Innovation (2014–2020) and repealing Decision No 1982/2006/EC, art. 19, and Decision 1982/2006/EC of the European parliament and of the Council of 18 December 2006 concerning the Seventh Framework Programme of the European Community for research, technological development and demonstration activities (2007–2013) ([2006] OJ L412), art. 6. In this regard, see J.R. Herrmann and M. Rowlandson, 'The Role of Ethics and Morality in EU Law' (2008) 5 *Journal of International Biotechnology Law* 241 et seq. (247 et seq.).

[139] UK Department of Health, 'Stem Cell Research: Medical Progress with Responsibility', 16 August 2000, accessed 16 December at http://www.dh.gov.uk/en/Publicationsandstatistics/Publications/PublicationsPolicyAndGuidance/DH_4065084.

[140] Human Fertilisation and Embryology Act 1990, accessed 16 December 2014 at http://www.legislation.gov.uk/ukpga/1990/37/section/1; Human Fertilisation and Embryology Act 2008,' accessed 16 December 2014 at http://www.legislation.gov.uk/ukpga/2008/22/section/1.

embryos, regardless of how they are created (that is, also when they are not the result of a fertilization process),[141] on the condition that the appropriate license is obtained from an independent authority.[142]

In Spain, the 'Ley sobre técnicas de reproducciòn asistida' of 2006 distinguishes embryos from pre-embryos on the basis of the moment in which the primitive streak is formed (14th day after fertilization), allowing the use of embryos only prior to this moment. In particular, the Spanish law authorizes research and experimentation on pre-embryos as well as on gametes, with aims other than those for diagnosis or therapy, on the condition of a series of specific conditions.[143] The *Tribunal Constitucional*, with various decisions in the eighties and nineties of the previous century, stated that neither the unborn child nor the pre-embryo in vitro (nor *a fortiori* gametes), are qualified as people in the legal sense of the word. However, the unborn child is granted protection as a constitutionally protected legal asset, in compliance with article 15 of the Spanish Constitution, which provides the right to life, and physical and moral integrity.[144]

In Italy, law no. 40/2004 establishing 'Norme in materia di procreazione medicalmente assistita' provides broad prohibition of any experimentation on the human embryo, specifying that clinical and experimental research on any human embryo is permitted on the condition that it is in pursuit of exclusively therapeutic and diagnostic aims connected to it, in order to protect health and development of the embryo itself, and should alternative methods not be available for use. What is more, the law specifies that the following are in any case prohibited: (a) the production of human embryos for research or experimentation purposes, or in any case purposes which are different from those provided by the law itself; (b) every form of selection for eugenic aims of embryos and gametes, or operations which, through techniques of selection, manipulation or artificial procedures, are aimed at the alteration of the genetic makeup of the embryo or gamete or at the predetermination of its genetic characteristics, except in the aforementioned operations for diagnostic and therapeutic purposes; (c) operations of cloning through the transfer of nuclei or premature splitting of the embryo or of ectogenesis, for purposes of

---

141 Human Fertilisation and Embryology Act 2008, art. 1.

142 Human Fertilisation and Embryology Act 2008 (see Human Fertilisation and Embryology Act 2008, s. 3 et seq.).

143 '*Ley 14/2006, de 26 de mayo, sobre técnicas de reproducción humana asistida*', art. 14 et seq.

144 See, among others, decisions no. 116 of 17 June 1999 and no. 212 of 19 December 1996.

procreation or of research; and (d) the fertilization of a human gamete with a gamete of a different species and the production of hybrids or chimera.[145]

In Germany, in the 'Gesetz zum Schutz von Embryonen (Embryonenschutzgesetz – EschG)',[146] there is strict protection of the embryo, who is granted legal status nearing that of the individual. Cryopreservation is prohibited except for oocytes,[147] and the penalty for the alienation of an embryo produced outside the body or for the selling, purchase or use of it for aims other than its preservation is the term of three years imprisonment.[148] However, despite the ban on the production of stem cells from human embryos, the use of embryonic stem cells obtained in compliance with the legal situation of the country of origin prior to 1 May 2007 is permitted for research purposes.[149]

In conclusion, from the entirety of regulatory acts and legal and doctrinal interventions referenced, in view of the variety of the solutions adopted by single legal systems with respect to the complex question of the so-called statute of the embryo,[150] the approach to the theme seems to be founded on the observance of a parallelism between biology and law: the unborn child, who for the biological sciences is an individual who develops into a man, can be considered, also under the law, as an autonomous entity which develops into a person, or rather, according to a

---

145 Law no. 40/2004, cited, art. 13, paras 1–3.

146 Of 13 December 1990 (BGB1. I, p. 2746), then renewed in 2001, and in 2011 with the *Präimplantationsgesetz* (relative to the pre-implantation diagnosis). For a comment on the ESchG, see, among others, H.L. Günther, P. Kaiser and J. Taupitz, *Embryonenschutzgesetz. Kommentar zum Embryonenschutzgesetz*, Stuttgart: Kohlhammer Verlag, 2008.

147 Containing two derivative and different pronuclei from the gametes of the two parents, pronuclei that, influencing one another, themselves immediately trigger mechanisms for the duplication phase of DNA which will lead to the formation of the bi-cellular embryo.

148 ESchG, articles 1–2; 8.

149 'Gesetz zur Sicherstellung des Embryonenschutzes im Zusammenhang mit Einfuhr und Verwendung menschlicher embryonaler Stammzellen', of 28 May 2002 (BGB1. 2002I, p. 2277), as subsequently renewed in 2008, art. 4. For further critical observations with respect to the implantation of the ESchG, see E.W. Böckenförde, 'Die Würde des Menschen war unantastbar', in *Frankfurter Allgemeine Zeitung*, 3 September 2003, p. 33 et seq.

150 On the basis of the same principle of territoriality of law for which the States may allow the importation of human stem cells despite the prohibition of producing them within their own country (as noted, this occurs in Germany, where it is possible to use imported stem cells imported prior to 1 May 2007).

widely accepted definition, as an autonomous center of legal relations, in anticipation and while awaiting the person'.[151]

## II.3.3 The Conditions and Limits on Stem Cell Research

Human 'stem cells' are those cells which are not specialized and have two characteristics: (a) the ability of self-renewal, or rather of multiplication without differentiation; and (b) the ability to give rise, through division, to 'progenitor' cells, with limited proliferation ability, from which populations of 'daughter' cells or highly differentiated or specialized cells (nervous, muscular, blood, and so on) are descendant.[152]

These abilities change according to the origin of the cells and the stage of development of the organism from which they are extracted.

In general, based on their potential, four types of stem cells may be determined: totipotent, which may develop into an entire organism and even into extra-embryonic tissues; pluripotent, which may specialize into all types of cells of an adult individual, but not into cells which make up extra-embryonic tissues; multipotent, able to specialize only into some types of cells; or unipotent, which may generate only one type of specialized cell.

---

[151] See L. Bigliazzi Geri, U. Breccia, F.D. Busnelli and U. Natoli, *Diritto civile, Vol. 1: Norme soggetti e rapporto giuridico. Fatti e atti giuridici*, Turin: Utet, 1986, p. 94. The authoritative definition was given by: F. Santoro Passarelli, *Dottrine generali del diritto civile*, Naples: Jovene, 1983 (9th edn.), p. 26. In this regard, see as well, *Corte di Cassazione*, 22 November 1993, no. 11503.

[152] Considering stem cells, see, among others, A.R. Chapman, 'The Ethics of Patenting Human Embryonic Stem Cells' (2009) 19 *Kennedy Institute of Ethics Journal* 261 et seq.; E.A. Zerhouni (Director for the U.S. National Institute of Health), 'Stem Cell Science: The Foundation of Future Cures' – Testimony before Committee on Energy and Commerce Subcommittee on Health U.S. House of Representatives, 8 May 2008, accessed 16 December 2014 at http://www.hhs.gov/asl/testify/2008/05/t20080508c.html; B.R. Schaller, *Understanding Bioethics and the Law. The Promises and Perils of the Brave New World of Biotechnology*, Westport (USA)/London: Praeger, 2007, p. 133 et seq.; H.R. Schöler, 'The Potential of Stem Cells: An Inventory', in N. Knoepffler, D. Schipanski and S.L. Sorgner (eds), *Humanbiotechnology as Social Challenge. An Interdisciplinary Introduction to Bioethics*, Aldershot (UK)/Burlington (USA), Ashgate Publishing, 2007, p. 28 et seq.; European Group on Ethics in Science and New Technologies, 'Opinion n°16 – Ethical aspects of patenting inventions involving human stem cells', 7 May 2002 and 'Opinion n° 15 – Ethical aspects of human stem cell research and use', 14 November 2000.

Based on their origin, the distinction is made between embryonic stem cells, adult stem cells, fetal stem cells, umbilical cord stem cells and amniotic stem cells.

Embryonic stem cells are totipotent, which develop the embryo in the phase preceding its attachment within the uterus. Subsequently, they transform into pluripotent stem cells. For their totipotency, embryonic stem cells appear to be the most useful for the prevention and cure of various types of illnesses.[153] However, they raise significant ethical issues.

Foetal stem cells are multipotent stem cells which make up the foetus. They are usually gathered from abortions. The extraction and use of foetal stem cells, like those from embryos, raise serious ethical issues.

Adult stem cells are multipotent or unipotent cells present in the tissue of a fully-formed organism, with the task of developing it or providing substitute cells. They are more appropriately called somatic cells, given that they do not necessarily come from adults but also from children and umbilical cords. They are non-specialized cells found among specialized cells of a certain tissue. Unlike embryonic and foetal cells, their use in research does not raise any relevant ethical issue.

Umbilical cord stem cells have properties similar to the multipotent cells of an adult organism. There is, however, an important difference: in the fully formed organism stem cells are difficult to identify and extract, while in the umbilical cord they are quite numerous. These cells are withdrawn from the umbilical cord blood, through an extraction from the umbilical vein, both in the case of spontaneous and cesarean births.

Amniotic stem cells are found in the amniotic liquid which surrounds the foetus during pregnancy. They have characteristics which are similar to embryonic cells, without any sort of ethical contraindication which arises from the use of the embryonic cells. It is possible to obtain a residual sample extracted in order to carry out the optional prenatal exam called amniocentesis.

Recent studies, then, have greatly modified the concept of potentiality of adult stem cells, introducing the concept of plasticity. It was discovered, for instance, that bone marrow stem cells, which normally give rise to mature blood cells, may, under proper conditions and occasionally, produce skeletal muscle, cardiac or smooth cells, neurons, liver and epithelial cells. This phenomena, defined as plasticity and extended also

---

[153] See, for instance: President's Council on Bioethics, 'Monitoring Stem Cell Research', Washington DC, January 2004, accessed 16 December 2014 at https://www.bioethicsarchive.georgetown.edu/pcbe/reports/stemcell.

to other types of stem cells, has important applicative but also ethical and legal implications. In many genetic illnesses or in those which are acquired in old age, stem cells in the tissue affected may have been destroyed by the pathological process or exhausted their proliferative capacity and so may no longer be available for contribution to tissue regeneration. In these cases, it may be much more useful to be able to isolate stem cells from another tissue which is not affected by the illness and use them to repair the damaged tissue, after inducing them to differentiate into the type of cell needed.

On the other hand, the announcements of new possibilities for treatment connected to stem cell research, as relative to the efficacy and toxicity of medicines, the study of biological mechanisms of the development of certain pathologies, the production of organs or tissues for transplant, and so on, pose fundamental bioethical questions, in particular with regard to the origin and method of producing the cells themselves.[154]

The debate on the possibility to use embryonic stem cells is closely linked to that of the human embryo. In order to obtain embryonic stem cells, in this state, it is necessary to destroy the human embryo (whether it is produced specifically for this purpose[155] or cryopreserved in surplus).

In this case as well, as was said on the theme of the statute of the embryo, those who deny the personality of the embryo and do not hold it worthy of protection admit the possibility of using it in order to allow for scientific progress in the field of treatment. From this perspective, admitting that the embryo must be protected to some degree, it would in any case be subordinate to the duty to protect the ill.[156]

In an intermediate position are those who recognize the necessity for embryos to be treated with the maximum amount of respect, but at the same time find it worthwhile to balance the interest of the integrity of the

---

[154] See European Group on Ethics in Science and New Technologies, 'Opinion n°16 – 07/05/2002 – Ethical aspects of patenting inventions involving human stem cells' and 'Opinion n° 15 – 14/11/2000 – Ethical aspects of human stem cell research and use', cited.

[155] Through reproductive technology or through so-called therapeutic cloning, which consists in the production of embryos for the transfer of one nucleus of cells into an ovum from which the nucleus has been removed, with the possibility of obtaining stem cells with the same subjective genetic characteristics as those to be applied in treatment, so-called autologous embryonic stem cells (on the theme of cloning, refer below to para. (iv)).

[156] For instance, in Israel, China, South Korea and Singapore there is extremely lenient legislation.

embryo with the advantages that research on embryos would produce for those with illnesses. In this sense, they consider only the use of frozen embryos which are not suitable for implantation, or those which are in a state of abandonment and no longer destined for implantation legitimate, on the condition that they are knowingly donated by women or couples. This legitimization would in any case be subordinate to the verification of the absence of a reasonable possibility for implantation and of the free and informed consent of the women or couples, with the exclusion of any form of compensation or restriction on the destination of the cell lines which may eventually be produced.[157]

Those who believe that the embryo represents a person from the moment of fertilization, instead, consider experimentation on human embryonic stem cells seriously illegal in any case. On the basis of this view, the right to life of the embryo must have dignity equal to that of the person with the illness to be cured.[158] Even the use of frozen embryos, in this view, is considered to be illegal, given that it gives rise in any case to the direct and intentional suppression of live embryos; in this sense the difference between killing and leaving to die is emphasized. Moreover, the fact that the eventual permission to use frozen embryos may encourage the production and instrumental use of them is noted.[159]

---

157 In this regard, see T. Peters, 'Embryonic Stem Cells and the Theology of Dignity', in S. Holland, K. Lebacqz and L. Zoloth (eds), *The Human Embryonic Stem Cell Debate: Science, Ethics, and Public Policy*, Cambridge (USA): MIT Press, 2001, p. 127 et seq.; A.R. Chapman, M.S. Frankel and M.S. Garfield, *Stem Cell Research and Applications: Monitoring the Frontiers of Biomedical Research*, Washington DC: American Association for the Advancement of Science, 1999, *passim*.

158 See, among others, C.B. Cohen, *Renewing the Staff of Life: Stem Cells, Ethics, and Public Policy*, Oxford/New York: Oxford University Press, 2007, *passim*; M. Farley, 'Roman Catholic Views on Research Involving Human Embryonic Stem Cells', in S. Holland, K. Lebacqz and L. Zoloth, *The Human Embryonic Stem Cell Debate: Science, Ethics, and Public Policy*, cit., p. 113 *et seq.*; G. Meilaender, 'Some Protestant Reflections', therein, p. 141 et seq.

159 In this position there is, clearly, the Catholic Church; in this regard, see in particular, The Congregation for the Doctrine of Faith, 'Instruction Dignitatis Personae on certain bioethical issues', Rome, 8 September 2008, accessed 16 December 2014 at http://www.vatican.va/roman_curia/congregations/cfaith/documents/rc_con_cfaith_doc_20081208_dignitas-personae_en.html; John Paul II, 'Evangelium Vitae', Rome, 25 March 1995, accessed 16 December 2014 at http://www.vatican.va/holy_father/john_paul_ii/encyclicals/documents/hf_jp-ii_enc_25031995_evangelium-vitae_en.html, where in this sense it is stated, in particular (at no. 58), that 'procured abortion is the deliberate and direct killing,

From a regulatory standpoint, in the United States, as hinted at, in 2009 President Barack Obama adopted an Executive Order with which he removed limits to federal funding for research on embryonic stem cells put in place by President George W. Bush.[160] The Order of President Obama did not eliminate the so-called 'Dickey-Wicker Amendment', renewed annually by Congress since 1996, which prohibits the U.S. Department of Health and Human Services from using funds for the creation of human embryos for research purposes or for research in which human embryos are destroyed. However, in the ruling on the case of *Shirley v. Sebelius* of 29 April 2011, the Court of Appeals confirmed that the Dickey-Wicker amendment is ambiguous, and that it may reasonably be stated that it prohibits only the use of federal funds in order to 'directly' carry out stem cell extraction from an embryo which gives rise to its destruction.[161]

In the European context, the Seventh program framework for technological research and development (2007–2013), as well as the following Horizon 2020 Framework Programme for Research and Innovation (2014–2020), excludes: (a) funding of research activities whose aim is the production of inheritable modifications of the human genome, the creation of human embryos only for research goals or for the provision of stem cells, or the reproductive human clonation; and (b) that activity shall be funded in a Member state where such activity is forbidden. On the other hand, there is the possibility to fund research on human stem cells, both adult and embryonic, based on the contents of the scientific proposal and the legal context which exists in the Member state or states involved,

---

by whatever means it is carried out, of a human being in the initial phase of his or her existence, extending from conception to birth'.

[160] On the basis of the Executive order, the National Institute of Health adopted the 'Guidelines on Human Stem Cell Research', of 7 July 2009, accessed 16 December 2014 at http://stemcells.nih.gov/policy/Pages/2009 guidelines.aspx. As far, then, as research on stem cells outside federal funding is concerned, the National Academies published the 'Guidelines for Human Embryonic Stem Cell Research', Washinghton DC: The National Academies Press, 2005 (see also, subsequently, 'Final Report of the National Academies Human Embryonic Stem Cell Research Advisory Committee and 2010 Amendments to the National Academies Guidelines for Human Embryonic Stem Cell Research').

[161] It is not, as said above, the funding of a research project which 'indirectly' used embryonic stem cells previously extracted.

following particular indications concerning the measures adopted regarding the licenses and inspections by the competent authorities of the Member states, as well as the ethical authorizations which may be granted.[162]

In the single Member states the production of human embryos with the aim of obtaining stem cells for experimental or therapeutic uses, as noted, is accepted in the United Kingdom, where the Donaldson Report stated that these uses do not add critical aspects with regard to the existing regulation on assisted fertilization.[163] In Spain, the *Ley sobre técnicas de reproducciòn asistida* authorizes research and experimentation on pre-embryos, as well as on gametes, on the condition that a series of specific conditions are respected.[164] In Italy, the extraction of stem cells from a human embryo for research purposes is strictly forbidden.[165] In Germany, producing human embryonic stem cells is prohibited, but the use for research purposes of embryonic stem cells imported prior to 1 May 2007 is permitted.[166]

On this subject, clearly the debate on the possibility of producing embryonic stem cells without damaging embryos is particularly relevant. Starting from the *White Paper: Alternative sources of Pluripotent Stem Cells*, drafted by the President's Council on Bioethics in the United States, some methods have begun to be considered, such as: (a) the extraction of stem cells from dead embryos; (b) the extraction of stem cells through a biopsy of live embryos; (c) the extraction of stem cells from 'embryonic-like' entities called embryoids (with the transfer of altered nuclei); and (d) the reprogramming of specialized somatic cells.

---

[162] See Regulation (EU) 1291/2013 of the European Parliament and of the Council of 11 December 2013 establishing Horizon 2020 – the Framework Programme for Research and Innovation (2014–2020) and repealing Decision 1982/2006/EC', art. 19, no. 4, and Decision 1982/2006/EC of the European parliament and of the Council of 18 December 2006 Decision no. 1982/2006/EC of the European Parliament and of the Council of 18 December 2006 concerning the Seventh Framework Programme of the European Community for research, technological development and demonstration activities (2007–2013) cited, art. 6, no. 3. In this regard, see Herrmann and Rowlandson, 'The Role of Ethics and Morality in EU Law', cit., p. 247 et seq.

[163] Human Fertilisation and Embryology Act, cited, arts 3 et seq.

[164] 'Ley 14/2006 sobre técnicas de reproducción humana asistid', cited, arts 14 et seq.

[165] Law no. 40/2004, cited, art. 13, paras 1–3.

[166] 'Gesetz zur Sicherstellung des Embryonenschutzes im Zusammenhang mit Einfuhr und Verwendung menschlicher embryonaler Stammzellen' of 2002, renewed in 2008, cited, art. 4.

For the extraction of dead embryo stem cells, there is the problem of defining the death of an embryo. The proposed criteria is that of observation, in the arch of time conventionally defined, of the irreversible absence of cell division, with the ceasing of integral function of coordination, of growth and differentiation. The criteria has been criticized as it is lacking in animal experimentation, while experimentation on human embryos to verify its validity may be damaging.[167]

The extraction of stem cells for biopsy from live stem cells is criticized given that, beyond requiring a complex and costly procedure, it has risks for the embryo, and is not aimed at therapeutic uses, but for research with only hypothetical future benefits.

The extraction of stem cells from so-called embryoid bodies, through the transfer of altered nuclei, consists in a sort of cloning for the transfer of nuclei with genetic alterations.[168] The hypothesis has been criticized given that deactivation of the gene is an intentional act which deliberately produces a defective embryo, which does not have less dignity simply because it is destined to die; what is more, the result consists in anomalous cells which may be harmful upon the action of insertion in the ovum, producing a hybrid.

Finally, the attempt to reprogram specialized somatic cells, causing them to regress to a pluripotent state, is criticized given that it raises further ethical issues on the statute of reprogrammed totipotent cells, which are or could be embryos.

A similar solution has been explored in Italy by the *Relazione della Commissione di studio sull'utilizzo di cellule staminali per finalita' terapeutiche*,[169] chaired by Professor Renato Dulbecco, which hypothesized therapeutic cloning as a nuclear transfer for the production of autologous stem cells that would not automatically lead to the formation of an embryo, given the possibility to interrupt it prior to this event. Also in this case, the fact that the non-spontaneity of development does not

---

167 On this point see, for instance, in Italy, the disagreement between components of the National Committee for Bioethics expressed in a 'Parere sul destino degli embrioni derivanti da procreazione medicalmente assistita e non più impiantabili', of 26 October 2007, accessed 16 December 2014 at http://www.governo.it/bioetica/testi/parere061007.pdf.

168 In substance, the adult somatic cell is genetically altered (deactivating the gene which produces the protein for implantation), in order then to insert it into the ovum.

169 Presented to Minister Umberto Veronesi on 28 December 2000, accessed 16 December 2014 at http://www.europarl.europa.eu/comparl/tempcom/genetics/links/gene_links_it_commissione.pdf.

constitute a satisfactory element for its exclusion from the statute of the individual is noted.[170]

In reference to the use of stem cells originating from aborted foetuses, then, the directly relevant bioethical problem is not the interruption of pregnancy (which is a bioethical issue in and of itself), but rather the legality of the use of the stem cells from aborted foetuses for purposes of study, research or treatment. In this case, the legality of their use is generally admitted, as long as it is subordinate to the informed consent of the woman, the verification of the absence of complicity between he who carries out the abortion and he who extracts the stem cells of the aborted foetus, or of an incentive to abort deriving from the need to obtain stem cells, as well as the inability to sell or patent these cells.

The use of adult stem cells, on the other hand, does not raise bioethical issues.[171] In the last few years, the possibility of using these cells for therapeutic purposes has been discovered, with significant application potential, given their plasticity.[172] Moreover, there has been the discovery of the possibility to extract stem cells from the umbilical cord and to create autologous cell banks, which may be used even after decades, for each newborn.[173]

Also in these cases, evidently, the extraction must meet the requirements of a therapeutic aim, not be a danger to the health of the patient to be cured, and to follow informed consent.

---

[170] See F. D'Agostino and L. Palazzani, *Bioetica. Nozioni Fondamentali*, cit., p. 113.

[171] In this regard, the Pontifical Academy for Life, in a 'Declaration on the production and the scientific and therapeutic use of human embryonic stem cells' (accessed 16 December 2014 at http://www.vatican.va/roman_curia/pontifical_academies/acdlife/documents/rc_pa_acdlife_doc_20000824_cellule-staminali_en.html), stated that: 'The possibility, now confirmed, of using adult stem cells to attain the same goals as would be sought with embryonic stem cells – even if many further steps in both areas are necessary before clear and conclusive results are obtained – indicates that adult stem cells represent a more reasonable and human method for making correct and sound progress in this new field of research and in the therapeutic applications which it promises. These applications are undoubtedly a source of great hope for a significant number of suffering people'.

[172] Further, their use presents technical difficulties, in particular regarding the methods of extraction and in vitro growth.

[173] Regarding the legality of this practice see, at the European level: European Group on Ethics in Science and New Technologies, 'Opinion n° 19 – Ethical aspects of umbilical cord blood banking', 16 March 2004.

### II.3.4 The Prohibition or Admissibility of Reproductive or Therapeutic Cloning

In nature, cloning is a common form of reproduction in simple plants and animals. In plant, but overall in animal, life forms, the more advanced form of reproduction is sexual, which is more complex, but with the advantage that the synthesis of the genetic makeup of the two different individuals (male and female) creates a unique individual (except in the case of single zygote twin births).

In general, a clone is indicative of an artificially procreated individual to be the perfect genetic reproduction of the other, that is a child modelled on the genetic makeup of a single parent.[174]

The term cloning refers to the production of identical copies of molecules, cells, tissues or entire organisms.[175] The techniques of cloning are used in various sectors of research, constituting, in some cases, well established practices. In particular, there are three types:

(i) molecular cloning allows for the replication of single fragments of DNA, inserting them in a host cell (usually bacteria), the rapid multiplication of which increases the number of the fragment, the copies of which may then be extracted and purified. The cloned fragment may, then, be studied to understand the structure, function, any role it plays in a certain illness, or may be altered in order to study the function of single genes. In this way, it is possible to obtain plants or animals which are more resistant to certain illnesses, improve the effectiveness of drugs, and so on;

(ii) cellular cloning produces a cell line starting from a single cell cultivated in a laboratory with the same genetic composition as the original. The technique may also aid in the production of new medicines or antibodies;

(iii) the cloning of an entire organism leads to the creation of individuals with identical genetic makeup. This may occur in two ways. One is that of so-called embryo splitting, which consists in the division of embryos at the initial stage of development, when the cells which make it up are still totipotent. So, the same phenomenon which produces a twin birth in nature is artificially provoked.

---

[174] In biology, instead, a clone is intended as the grouping of individuals of a certain species which all have the same genetic makeup.

[175] See, among others, S. Bhattacharya, 'Cloning: The Ethical Concerns', in V. Kumari (ed.), *Issues In Applied Ethics,* New Delhi: Satyam Publishing House, 2009, p. 198.

> The other method is through nuclear cloning, in which the nucleus extracted from a somatic cell is transferred into an unfertilized ovum from which the nucleus has also been removed. Once it has been implanted in the uterus and allowed to develop, the ovum will produce an individual which is genetically identical to the donor.[176]

In 1997, in Edinburgh, a team of scientists created the first clone of a higher mammal, a sheep named Dolly. The technique used consisted in the extraction of a nucleus from a somatic (mammary) cell of an adult animal, which then was introduced into an ovum, and finally, after over 250 failed attempts, in the reactivation of the ovum with a new nucleus as a proper embryo. Then, the embryo was implanted in the uterus of a 'surrogate mother' to carry out the pregnancy, at the end of which an animal genetically identical to the one from which the adult cell was extracted was born.[177] After the cloning of the sheep Dolly, there were clones of cats, bulls, horses, and so on.

There are numerous advantages to animal cloning, according to its supporters: in particular, on one hand, once an animal in possession of extraordinary genetic characteristics is identified, it would then be possible to have it reproduce in order to obtain offspring with the same particular characteristics; on the other hand, the study of transgenic animals allows us to understand pathogenic mechanisms and to develop new treatment strategies.

In its current state, scientific knowledge cannot tell us which techniques may be used to safely clone human beings, but bioethics and bio-law have been dealing with this issue for some time.

There are two types of human cloning: (a) 'reproductive' cloning, which produces human embryos in a laboratory to then be transferred into a female uterus in order to produce cloned human babies; and (b) 'therapeutic' cloning, which consists in the creation of cloned human embryos for the purpose of developing stem cells which, being unspecialized and capable of dividing and renewing themselves over a long period,

---

[176] See National Bioethics Advisory Commission, 'Cloning Human Beings: Report and Recommendations of the National Bioethics Advisory Commission', 9 June 1997, p. 16 et seq., accessed 16 December 2014 at https://bioethics archive.georgetown.edu/nbac/pubs/cloning1/cloning.pdf.

[177] See I. Wilmut *et al.*, 'Viable offspring derived from fetal and adult mammalian cells' (1997) 385 *Nature* 810 et seq.

have the potential to develop into numerous and different types of cells, useful for the reparation of damaged tissues or organs.[178]

The possible practical applications of human cloning, then, may be related, on the one hand, to assisted reproduction: embryo splitting would increase the pregnancy rate in the case of failed in vitro embryo formation and nuclear cloning to satisfy the need for biological parentage, eliminating the need to resort to donations of gametes; on the other hand, to transplants and tissue regeneration: the availability of organs or tissues which are identical to those of the receiver could prevent the phenomena of rejection or incompatibility which is found with current allotransplants.[179]

The supporters of cloning maintain its utility in allowing scientific knowledge to progress, in guaranteeing the freedom of the individual to choose, and even to create a copy of himself, to the extent of limiting disadvantages by providing possible benefits.[180]

In an intermediate position, there is the stance of those who propose 'weak' prohibition of cloning, a legitimizing principle, to the extent that society should be open to such practices and that science should agree to applying it while avoiding or controlling negative consequences.[181]

A basic critique that is made regarding cloning concerns the fact that it carries out a process that not only is artificial, but that from an evolutionary point of view is against nature, in that reproducing species which reproduce sexually is equivalent to forcing them to unduly regress to a lower evolutionary level.

---

[178] In this regard, among others, see S. Bhattacharya, *Cloning: The Ethical Concerns*, cit. p. 198.

[179] See D. Solter and J.D. Gearhart, 'Putting Stem Cells to Work' (1999) 283 *Science* 1468 et seq.; J.D. Gearhart, 'New Potential for Human Embryonic Stem Cells' (1998) *Science* 282 1061 et seq.; J. Cohen and G. Tomkin, 'The Science, Fiction and Reality of Embryo Cloning' (1994) 4 *Kennedy Institute of Ethics Journal* 198 et seq.; H.W. Jones, 'Reflections on the Usefulness of Embryo Cloning', (1994) 4 *Kennedy Institute of Ethics Journal* 205 et seq.

[180] In this regard, see, in general, J. Harris, *Clones Genes and Immortality: Ethics and the Genetic Revolution*, Oxford; Oxford University Press, 1998, *passim*.

[181] In fact, scientific literature has shown data from experimentation on animals, which show the clone's high rate of mortality and the possibility of contracting illnesses (Dolly herself died prematurely of rheumatoid arthritis), the significant risks for life and the health of the pregnant individual, besides the predictable inherent social damage with regard to the possibility of choosing the individuals allowed to be cloned based on their genetic characteristics.

With respect to the protection of human life, in particular, cloning is considered one of the most abhorrent forms of manipulation and objectification, without considering the use of the results or the destiny of the human being created from it.

Cloning, further, would represent a new form of denial of freedom, through the genetic elimination of uniqueness determined by the genetic combination of paternal and maternal chromosomes, giving rise to a sort of 'technological totalitarianism'.[182]

The clone would be, basically, the copy of the original who lived (at least a part of his life) before him. In this way, the person would be limited in his freedom to discover his own identity and to allow it to be discovered by others, to be evaluated for how he her/himself is.[183]

Further, human cloning is seen as a sort of 'artificial selection' of human biological evolution, through the production of a series of individuals with pre-chosen characteristics. It would mean, basically, the risk of eugenics, which is the arbitrary selection of better characteristics according to social programming.[184]

---

[182] See F. D'Agostino and L. Palazzani, *Bioetica. Nozioni Fondamentali*, cit., p. 107.

[183] In this regard, Hans Jonas highlighted how with cloning the (hypothetical) cloned offspring knows (or believes she/he knows) too much about her/himself and others know (or believe they know) too much about her/him (H. Jonas, *Technik, Medizin und Ethik – Zur Praxis des Prinzips Verantwortung*, cit., p. 146). Jürgen Habermas added that the awareness of having been genetically programmed could disrupt the sense of naturalness of the clone's existence, transforming it into the perception of being a production of technique (J. Habermas, *Die Zukunft der menschlichen Natur. Auf dem Weg zu einer liberalen Eugenik?*, cit.).

[184] The concept of eugenics is in reference to the study of methods to perfect the human species through artificial selection carried out through the promotion of physical and mental characteristics held to be positive (positive eugenics) and the removal of those which are negative (negative eugenics), through the selection or modification of germinal lines, according to common techniques in the breeding of animals and in agriculture based on Mendelian genetics, and those which are currently or potentially available from biotechnology. Eugenism, instead, is the ideology that the solution to political, social, economic or health issues may be reached through the adoption of expected eugenic solutions, which found its most horrible expression in the so-called 'Aktion T4', the name given after the First World War to the Nazi program of euthanasia which, under medical observation, involved the suppression of people affected by incurable or genetic illness or by more or less serious physical malformations. On this theme, see Wikipedia.org, entries 'Eugenics' and 'Action T4', accessed 16 December 2014 at http://en.wikipedia.org/wiki/Eugenics and http://en.wikipedia.org/wiki/

What is more, replication in a series may also lead to the exploitation of man, in that one may decide to produce humans with certain physical characteristics for utilitarian purposes, or individuals destined for unpleasant work, giving rise to subhuman replicants, or even to produce individuals and freeze them in order to use them for organ or tissue reserves.

In the face of human cloning, therefore, bio-law is called upon to protect, in particular, the right to personal identity, the right to physical integrity, the non-manipulability of genetic makeup, genetic unpredictability, the right to a family (with two parental figures) and the ban on the exploitation and serialization of human beings.[185]

At the regulatory level, in 1997 the World Health Organization adopted a resolution[186] which, among other things, stated that the use of cloning for the replication of human individuals is ethically unacceptable and contrary to human integrity and the principles of morality.

In the same year, the Universal Declaration on the Human Genome and Human Rights of UNESCO, at article 2, stated that: '(a) Everyone has a right to respect for their dignity and for their rights regardless of their genetic characteristics. (b) That dignity makes it imperative not to reduce individuals to their genetic characteristics and to respect their uniqueness and diversity'. In article 11, then, the Declaration recommended the ban

---

Action_T4; Court of Nuremberg, '*Affidavit concerning the Nazi administrative system, the euthanasia program, and the sterilization experiments*', Document No. 426, Harvard Law School Library, Item No. 113, accessed 16 December 2014 at http://nuremberg.law.harvard.edu/php/search.php?DI=1&FieldFlag=11&NMTID =1&MTNo=426&MTNoSuff=; in doctrine: J. Glad, *Future Human Evolution: Eugenics in the Twenty-First Century*, New York: Hermitage Publishers, 2008; J. Joseph, *The Missing Gene: Psychiatry, Heredity, and the Fruitless Search for Genes*, New York: Algora, 2006; R.C. Engs, *The Eugenics Movement: An Encyclopedia*, Westport (USA): Greenwood Publishing Group, 2005; C.R. Browning. *The Origins of the Final Solution. The Evolution of Nazi Jewish Policy, September 1939–March 1942*, London: Arrow, 2005; R.J. Evans, *The Third Reich in Power, 1933–1939*: London, Allen Lane, 2005; D. Kuntz (ed.), *Deadly Medicine: Creating the Master Race*, Washington DC: United States Holocaust Memorial Museum, 2004.

185 See again, F. D'Agostino and L. Palazzani, *Bioetica. Nozioni Fondamentali*, cit., p. 109.

186 WHO, 'Cloning in human reproduction', WHA50.37, 14 May 1997, accessed 16 December 2014 at http://www.who.int/genomics/publications/governance/wha/wha050/en/index.html.

on all practices contrary to human dignity, among which was included, as an example but with no obvious significant value, the cloning of human beings.

In 2001, the General Assembly of the United Nations adopted resolution no. 56/93, entitled 'International Convention against the reproductive cloning of human beings',[187] which, in consideration of the research in progress regarding the reproductive cloning of human beings, in order to prevent such an attack on human dignity, established an *ad hoc* committee for the creation of a Convention against the reproductive cloning of human beings.[188] In 2005, the General Assembly adopted resolution no. 59/280, containing a 'United Nations Declaration on Human Cloning'[189], according to which Member states, in particular '(a) … are called upon to adopt all measures necessary to protect adequately human life in the application of life sciences; (b) … are called upon to prohibit all forms of human cloning inasmuch as they are incompatible with human dignity and the protection of human life; (c) … are further called upon to adopt the measures necessary to prohibit the application of genetic engineering techniques that may be contrary to human dignity; (d) … are called upon to take measures to prevent the exploitation of women in the application of life sciences'. The Declaration, though it may constitute an act of a more general and limited range than the initial hypothesis of the adoption of an international convention, contains in any case an important statement of common global principles, which highlight widespread agreement regarding the ban on cloning which is contrary to human dignity.

In the United States, though they were among the first to adopt a policy on cloning after the birth of the sheep Dolly, there are no federal laws on cloning. In 1997, President Clinton adopted the Executive Order[190] which prohibited federal funding for cloning, noting its profound ethical implications. Since then, despite various attempts, no federal legislation has been approved on the matter. Various states have

---

[187] Of 12 December 2001, accessed 16 December 2014 at http://www.un.org/law/cloning.

[188] On the debate under the committee's responsibility, also regarding the distinction between reproductive and therapeutic cloning, see, among others, E. Shuster, 'Human Cloning: Category, Dignity, and the Role of Bioethics' (2003) 17 *Bioethics* 517 et seq.

[189] Dated 8 March 2005, accessed 16 December 2014 at http://www.un.org/law/cloning, adopted with 84 votes in favor, 34 against and 37 abstentions.

[190] W.J. Clinton, 'Memorandum For The Heads Of Executive Departments And Agencies: Prohibition on Federal Funding for Cloning of Human Beings', 4 March 1997, accessed 16 December 2014 at http://clinton6.nara.gov/1997/03/1997-03-04-directive-on-cloning.html.

introduced regulations which prohibit reproductive cloning, and some also therapeutic cloning, which is conversely permitted by the other states.[191]

In Europe, community acts which inspired national legislation on the theme of assisted fertilization, as well as those subsequently adopted following the spread of news concerning cloning experiments, have repeatedly recommended the introduction of a ban on cloning,[192] given that the formation of embryos with the same genetic makeup as another living being was judged to be unacceptable from an ethical point of view and criminal from a legal one, putting the value of individuality at risk, which is a product of the casual combination of the genetic makeup of both parents, and therefore of factors which are independent from the will of other human beings.[193] European laws on assisted fertilization, then, provided bans, explicit or implicit, on cloning embryos.[194]

The European Convention on Human Rights and Biomedicine did not contain specific provisions regarding human cloning. What is more, considering that this practice could circumvent the bans provided by article 13 on the handling of the human genome given that in and of itself it does not lead to genetic manipulation in a strict sense,[195] a specific

---

[191] In this regard, see National Conference of State Legislatures, 'Human Cloning Laws', accessed 16 December 2014 at http://www.ncsl.org/issues-research/health/human-cloning-laws.aspx; C. Kuppuswamy, 'Cloning: National and International Law' (2006) 3 *Journal of International Biotechnology Law* 192 et seq. (at 205).

[192] See, in particular, European Parliament, Resolutions of ethical and legal problems of genetic manipulation and artificial fertilization *in vivo* and *in vitro*, both dated 16 March 1989; European Council, Recommendations no. 934 of 1982 on genetic engineering, no. 1046 of 1986 on the use of embryos and human fetuses for diagnostic, therapeutic, scientific, industrial and commercial purposes, and no. 1100 of 1989 on scientific research regarding embryos and human foetuses.

[193] See W. Rothley, 'Report on the ethical and legal problems of genetic engineering', in European Parliament – Committee on Legal Affairs and Citizens' Rights, 'Ethical and legal problems of genetic engineering and human artificial insemination', Luxembourg, Office for Official Publications of the European Communities, 1990, p. 25 et seq.

[194] Such bans, as noted, were provided in German, Spanish and Italian laws. In the United Kingdom, the Human Reproductive Cloning Act 2001 limited the ban on reproductive cloning.

[195] Art. 18 of the Convention, however, provided the ban on the formation of embryos for research purposes, and therefore may have already been sufficient to preclude either cloning experiments on human embryos or their use for therapeutic purposes.

Additional Protocol to the Convention for the Protection of Human Rights and Dignity of the Human Being with regard to the Application of Biology and Medicine, on the Prohibition of Cloning Human Beings[196] was arranged, which in article 1 bans: 'Any intervention seeking to create a human being genetically identical to another human being, whether living or dead.'[197]

The Charter of Fundamental Rights of the European Union, in article 3(2)(d), provides explicitly 'the prohibition of the reproductive cloning of human beings'. The criticism which is most often given to the declarations of principles referenced up to this point concerns the excessive use of the argument of the contrariety of cloning to human dignity, which is considered generic and lacks a binding regulatory basis.[198] On the other hand, these objections do not take into consideration the fact that, at least in the majority of European systems, human dignity has a strictly positive law significance.

What is more, with respect to the aforementioned acts, the tendency to unitarily evaluate a polifunctional technique like cloning has been criticized, and rightly so. To this end, it is worthwhile to distinguish in particular between (a) 'reproductive' cloning and (b) 'therapeutic' cloning.

Concerning reproductive cloning, in the first place it is important to note that in the analysis of this situation there is little consideration taken with regard to the possibility of cloning a series of individuals for the military needs of authoritarian nations, or to answer the desire of the wealthy to perpetuate themselves, or the desire to reproduce politicians, artists, and so on. These objectives, in fact, do not encounter any considerable interest from a legal point of view.

The analysis conducted on the theme has dealt rather with the use of cloning as a method of reproduction alongside techniques of assisted fertilization already in existence, in order to serve the interests of those

---

196 Paris, 12 January 1998, accessed 16 December 2014 at http://conventions.coe.int/Treaty/en/Treaties/Html/168.htm.

197 The initial proposal involved the protection of the 'human person', rather than the 'human being'. The first expression in the international legal system refers to the human being at the moment of its birth, while the second is more broadly considered. What is more, the 'Explanatory Report', at point 6, specifies that the identification of the extension of the concept of 'human being' for the purposes of application of the Protocol is left to national laws.

198 See WHO, 'Cloning in human reproduction', cit., p. 13; on this point, see also A. Kaufmann, 'Riflessioni giuridiche e filosofiche su biotecnologia e bioetica alla soglia del terzo millennio', cit., p. 214.

who are infertile or affected by genetic pathologies. In these cases, the hypothesis of cloning may be considered by legal systems, on the condition that other fundamental rights are respected, such as the health of the cloned individual, whose interest is to not be born affected by genetic illnesses.[199]

Thus, on one hand, it is shown that cloning, with respect to other reproductive techniques whose aim is making conception identical to the method in which it occurs in nature artificially possible, is characterized by its asexual nature, breaking significantly away from them. On the other hand, it is clear that, with respect to other techniques of assisted reproduction already being practiced, and in particular heterologous fertilization, cloning would lead to the 'advantage' of avoiding the involvement of a third unknown person, that is the semen or egg donor. These circumstances prevent the issue, then, from being included in the more general one of artificial reproduction.

From a legal point of view, first from the point of view of the qualification of the interest in resorting to cloning, it is necessary to emphasize that legal systems do not even incorporate techniques of assisted reproduction (especially of U.S. origin) within the frame of a broader right of self-determination with regard to procreation, although they enjoy more widespread social consensus;[200] rather they take them into consideration in the context of the fundamental right to health, allowing such techniques to be accessed as remedies for infertility.

With respect to the verification of compatibility of cloning with the interests of the potential cloned individuals, it must be noted that it is under this profile that the main objections to it are made. In particular, in regulatory acts on the matter it is stated that the practice gives rise to injury to the individual right to a genetic identity.[201] In this regard, it has been shown how the fact that this right is not provided for in today's constitutional charters does not mean that it does not exist, given that up

---

[199] In this regard, see e.g. *Comitato Nazionale per la Bioetica*, '*La clonazione*', 17 October 1997, accessed 16 December 2014 at http://www.governo.it/bioetica/pdf/30.pdf, p. 29.

[200] In this sense, among others, see J. Robertson, *Children of Choice: Freedom and New Reproductive Technologies*, Princeton: Princeton University Press, 1994, p. 22 et seq., who, in line with the decision of the U.S. Supreme Court in the case *Roe v. Wade* (cited), considers the right to procreation, similarly to the right to abortion, as a manifestation of the right to privacy.

[201] The right to a genetic identity is characterized by the exaltation of individual diversity with respect to the attempt at homologation. Thus, it is contrary to any arbitrary manipulation of the genome, including then also its complete definition by third parties, as is the case in cloning.

to now the genetic identity of an individual has been considered an indestructible fact, without there being a need to protect it as an inviolable right from possible manipulation.[202] Currently, instead, scientific progress causes the physical identity to be expressed, besides in somatic features, in an original genetic makeup for each human being.[203] As a consequence, reproductive cloning appears to violate the right to a personal and/or genetic identity, given that it works through the replication of the genome and therefore undermines the essential nature of genetic makeup as an identifying feature of the individual.

On the other hand, it has been claimed that cloning would not give rise to the violation of individuality of each person, because it is not at all determined by the possession of an unrepeatable genetic makeup; rather it depends on the complex interactions among genes and between these and the surrounding environment. The objection to this argument is that it is contradictory, since on one hand, it minimizes the risk of injury to personal identity inherent in reproductive cloning, on the other hand it implicitly notes the importance of chromosomal identity, justifying its use with the desire to transmit the 'family' genetic makeup to one's own children.[204]

Further arguments against reproductive cloning concern the previously mentioned injury to the rights of freedom of the cloned subject from a psychological point of view, given that it would undermine his capacity to consider himself a free being, to his right 'to not know', or rather to ignore his own future health and possible exposure to certain risks, and finally to the right to privacy of his genetic data with regard to third parties, especially employers, insurance companies, and so on.[205]

In any case, from a legal point of view the reproductive cloning technique represents perhaps the most important 'emerging problematic point of a deep rift between ethical concepts and constitutional principles dividing so-called 'Western Law'.

---

[202] See again *Comitato Nazionale per la Bioetica*, '*La clonazione*', cit., p. 73.

[203] In this regard, see C. Labrusse-Riou, 'L'enjeu des qualifications, la survie juridique de la personne' (1991) 13 *Droits* 19 et seq.; S. Rodotà, 'Privacy e costruzione della sfera privata. Ipotesi e prospettive' (1991) *Politica del diritto* 525 et seq.

[204] On this point, see K. Evers, 'The identity of clones' (1998) *Biblioteca della libertà* 25 et seq.

[205] See, among others, B.A. Lenox, 'Genetic Discrimination in Insurance and Employment: The Spoiled Fruits of the Human Genome Project' (1997) 23 *University of Dayton Law Review* 189 et seq.; N. Cooper Dreyfuss, 'The Jurisprudence of Genetics' (1992) 45 *Vanderbilt Law Review* 313 et seq.

The lack of a solid constitutional anchor for the illegality of the technique within the framework of the phenomena proposed by American doctrine is clear, thus one may even claim that human cloning as a reproductive technique is a protected freedom, 'guaranteed by the Due Process Clauses of the Fifth and Fourteenth Amendments'[206] and, on the contrary, claim that, except for the fundamental nature of the right to reproduce, the prohibition of cloning also holds up under the verification of its rational foundation.[207] What is more, in regard to reproductive cloning, in the American context the prohibition of such a form of cloning is supported more than anything by the basis of reasoning inherent to the principle of human dignity, rather than in virtue of the principle of precaution, given the risks associated with such techniques, whose effects on human health have not yet been adequately tested.

At an operative level, then, the ban of the practice is almost unanimous, but the relative underlying motivations are different, so that it appears on one hand definitive, being based on principles of respect for the dignity of the individual, of his identity and the free development of his personality;[208] on the other it appears provisional, dependent on the state of the art and subject to an evolution towards its opposite, where the

[206] In this regard, Note, 'Human Cloning and Substantive Due Process' (1998) 111 *Harvard Law Review* 2348; see also J.C. Kunich, *The Naked Clone: How Cloning Bans Threaten Our Personal Rights*, Westport (USA): Praeger, 2003; J.A. Robertson, 'Liberty, Identity and Human Cloning' (1998) 76 *Texas Law Review* 1372 et seq.

[207] Even though it has to be distinguished among the various rationales for banning the different kinds; see C.R. Sunstein, 'Is There a Constitutional Right to Clone?', University of Chicago, Public Law Research Paper No. 22, March 2002, accessed 16 December 2014 at SSRN: http://ssrn.com/abstract=304484 or http://dx.doi.org/10.2139/ssrn.304484.

[208] In this sense, see the Additional Protocol of the European Convention on Human rights and Biomedicine, cited, the resolutions of the European Parliament on human cloning of 7 September 2000, 15 January 1998 and 12 March 1997, as well as the national legislations cited previously. With regard to the reasons for the incompatibility of cloning with a constitutional model based on the protection of the dignity of the individual, see, for instance, the opinions of the French and Italian national committees on bioethics: *Comité Consultatif National d'Ethique pour les Sciences de la Vie et de la Santé*, 'Réponse au Président de la République au sujet du clonage reproductif', 22 April 1997, no. 54, accessed 16 December 2014 at http://www.ccne-ethique.fr/docs/fr/avis054.pdf; *Comitato Nazionale per la Bioetica*, '*La clonazione*', cit. In doctrine, see S. Bhattacharya, *Cloning: The Ethical Concerns*, cit., p. 197 et seq.; Kuppuswamy, *Cloning: National and International Law*, cit., p. 192 et seq.

indispensable conditions of safety and reliability of the technique are satisfied.[209]

As far as therapeutic cloning is concerned, at a legal level it exhibits regulatory inadequacy. Experimentation on embryos and tissue engineering based on the cloning of cells and embryonic or foetal tissues, with no reproductive purpose, but rather a therapeutic *lato sensu*, do not specifically fall under the prohibitions referenced above.

The bioethical and legal framework of reference is constituted in this case by the thought and legislation regarding experimentation on human embryos, which is significantly different even among European countries themselves, despite the attempt to unify them through the European Convention on Human Rights and Biomedicine.

It seems, then, that the legal response to such a technical possibility must be found in a balance of values: on one hand, the protection of the embryo, on the other, the value of such a form of research for the protection of collective health.[210]

In order to evaluate the admissibility of such practices, it seems worthwhile to refer in particular to article 18 of the European Convention on Human Rights and Biomedicine: the regulation, as noted, but without prejudice to the issue of the definition of the embryo, makes a distinction between the creation of embryos destined for research, which is explicitly prohibited, and the use of embryos which were initially created in the context of assisted reproduction, which in any case must be guaranteed protection, for purposes of experimentation.[211]

---

[209] See National Bioethics Advisory Commission, 'Cloning Human Beings: Report and Recommendations of the National Bioethics Advisory Commission', cit. In doctrine, see L.R. Kass, 'The Wisdom of Repugnance: Why We Should Ban the Cloning of the Humans' (1998) 32 *Valparaiso University Law Review* 687 et seq., who specifies, among other things, that 'repulsion is not an argument'.

[210] See F.D. Busnelli and E. Palmerini, entry 'Bioetica e diritto privato', in *Enciclopedia del diritto*, V, Milan: Giuffrè, 2001, p. 148, where it is further highlighted how once the mechanism for reprogramming adult cells is understood, such studies would allow scientists to overcome the need to use embryos as a source of stem cells (on this theme see also in Parts IV and V).

[211] For criticism of the distinction, see B. Mathieu, 'La recherche sur l'embryon au regard des droits fondamentaux constitutionnels', in *Recueil Dalloz-Sirey de doctrine, de jurisprudence et de législation*, 1999, *Chroniques*, p. 456. The illegality of cloning human embryos for research purposes was reiterated also in the Resolution of the European Parliament on human cloning dated 7 September 2000 (see below). On this theme, among others, see L.

Further, the Additional protocol to the Convention on cloning, which provides the broad prohibition of every action aimed at creating a human which is genetically identical to any other, alive or dead, is also relevant.[212]

What is more, in argument, the problem of the legitimacy of *in vitro* fertilization of an excessive number of embryos to that which is necessary for the single attempt at impregnation remains under examination and unresolved. Contrarily, the prohibition on the formation of embryos outside the sphere of a procreative project appears to be agreed upon. In fact, with regard to therapeutic cloning, this distinction is even accepted in the United States of America, where it is prohibited to use federal funds for research which leads to the formation of human embryos, while since 2009 funding has been reopened for research on stem cells obtained from embryo surplus.[213] Moreover, various states have issued legislation regarding the experimentation on embryos and fetuses, with the effect of limiting research conducted using private funding as well.[214]

At the European level, as noted, the EU Seventh program framework for technological research and development (2007–2013), as well as the

---

Cassiers, 'La dignité de l'embryon humaine' (2003) 54 *Trimestrielle des Droits de l'Homme* 403 et seq.; T. Lahalle, *Clonages et dignité humaine*, therein, p. 441 et seq.

[212] Additional Protocol to the Convention for the Protection of Human Rights and Dignity of the Human Being with regard to the Application of Biology and Medicine, on the Prohibition of Cloning Human Beings, cit., art. 1. What is more, some countries do not hold that the concept of 'human being' pursuant to art. 1 of the Protocol also refers to embryos. In this sense, for instance, in the act of signing the Protocol the Netherlands presented a declaration according to which: 'In relation to Article 1 of the Protocol, the Government of the Kingdom of the Netherlands declares that it interprets the term "human being" as referring exclusively to a human individual, i.e. a human being who has been born' (accessed 16 December 2014 at http://conventions.coe.int/Treaty/Commun/ListeDeclarations.asp?nt=168&cm=1&df=&cl=ita&vl=1).

[213] The possibility to finance research on stem cells, interrupted in 2001 by President Bush, had been introduced the previous year with the 'National Institutes of Health Guidelines for Research Using Human Pluripotent Stem Cells', of 25 August 2000, modified on 21 November 2000, Federal Register, Vol. 65, No. 166, p. 51975, No. 225, p. 69951, accessed 16 December 2014 at http://stemcells.nih.gov/policy/guidelines.asp. On this theme, see extensively point (iii).

[214] See National Bioethics Advisory Commission, 'Cloning Human Beings: Report and Recommendations of the National Bioethics Advisory Commission', cit., especially p. 89 et seq.

following Horizon 2020 Framework Programme for Research and Innovation (2014–2020), excludes: (a) funding of research activities whose aim is the production of inheritable modifications of the human genome, the creation of human embryos only for research goals or for the provision of stem cells, or the reproductive human clonation; and (b) that activity shall be funded in a Member state where such activity is forbidden.

In regard to the breadth of the prohibition on human cloning, the Resolution of the European Parliament of 7 September 2000 is relevant, in which it is stated that 'there is no difference between cloning for therapeutic purposes and cloning for the purposes of reproduction', and that therapeutic cloning 'poses a profound ethical dilemma, irreversibly crosses a boundary in research norms and is contrary to public policy as adopted by the European Union'.[215] In addition, the European Parliament called for research to refrain from using 'human artificial insemination techniques that do not produce an excess number of embryos in order to avoid generating superfluous embryos'; as a consequence, the Resolution appears to implicitly exclude the possibility of scientific research on surplus embryos.[216]

On the other hand, the European Group on Ethics in Science and New Technologies has condemned reproductive cloning,[217] while specifying that '(t)he respect for different philosophical, moral or legal approaches and for diverse national culture is essential to the building of Europe', and that '(i)t is not only legally difficult to seek harmonization of national laws at the Community level, but because of lack of consensus it

---

[215] Thus, the European Parliament has called for the British government to review its position on the cloning of human embryos.

[216] See F. Lenzerini, 'Biotechnology, Human Dignity and the Human Genome', cit., p. 313.

[217] European Group on Ethics in Science and New Technologies, 'Citizens Rights and New Technologies: A European Challenge. Report of the European Group on Ethics in Science and New Technologies on the Charter on Fundamental Rights related to technological innovation as requested by President Prodi on February 3, 2000', Brussels, 23 May 2000, accessed 5 January 2015 at http://www.pedz.uni-mannheim.de/daten/edz-du/gew/prodi_en_citizensrights.pdf, in which it is stated that in biomedicine the following key principles must be respected: (a) the prohibition of cloning human beings; (b) the ban on eugenic practices; (c) the prohibition of any discrimination based on genetic characteristics and health conditions; (d) the ban on the marketing of human body parts and their products; (e) the respect for the informed consent of the patient; (f) the right to the protection of personal information.

would be inappropriate to impose one exclusive moral code'.[218] What is more, the Group particularly stated that: 'A clear distinction must be drawn between reproductive cloning aimed at the birth of identical individuals … and non-reproductive cloning, limited to the in-vitro phase',[219] adding then that 'the philosophical and scientific debate on cloning is still open and it should be stressed that prohibition of a specific technique may prevent important discussions about human genetics'.[220]

In conclusion, then, while it may be said that human reproductive cloning is universally rejected, by virtually the entire international community, the debate on therapeutic cloning appears to remain completely open.[221] Beyond the states which do not support the ban on this form of cloning for ethical reasons, but rather in consideration of the risks associated with such a technique, whose effects on human health have not yet been adequately tested, or of those who highlight how once therapeutic cloning is allowed it would be practically impossible to prevent the exploitation of embryos for different, even illegal, purposes, it has been seen that a fundamental issue on the subject appears to involve the statute of the embryo. In particular, if the embryo is considered to be a human being, or, in any case, worthy of protection given that it is the

---

218 See 'Opinion n° 12 – Ethical aspects of research involving the use of human embryo in the context of the 5th framework programme', 23 November 1998.

219 'Opinion n° 9 – Ethical aspects of cloning techniques', 28 May 1997.

220 See again 'Citizens Rights and New Technologies: A European Challenge', cit. At the European level, as has been said, national legislations which prohibit, more or less extensively, research on human embryos are significant, reflecting the plurality of opinions on the statute of the embryo (see above, at point (ii)).

221 In general, in any case, on the subject of cloning, in order to avoid the spread of practices of 'law shopping', regulatory intervention at the international level is needed. In this sense, with reference to the American legal system, but on the basis of considerations which are also valid for other countries, see American Medical Association, 'The Ethics of Cloning: Report of the Council on Ethical and Judicial Affairs of the American Medical Association', June 1999, accessed 16 December 2014 at http://www.ama-assn.org/resources/doc/ethics/report98.pdf, p. 6, according to which: 'Even if the United States developed sound ethical guidelines and well-crafted regulations to address the practice of human cloning, some fear that human cloning would simply be forced into other locales. Individuals could travel to other countries where human cloning would be available and potentially unregulated. Because cloning technology is not limited to the United States, physicians should help establish international guidelines regarding human cloning'.

beginning of a human being, then non-reproductive cloning (carried out based on the current state of the technique) shall also be considered illegal.

# PART III

# The evolution of the discipline and problematic issues in the United States of America

## FOREWORD

In developed countries like the United States of America, the biotechnology industry has gradually grown to occupy a prominent position on the national economic scene. The patent system has played a fundamental role in this growth. Public and private institutions have recognized the significance of biotechnology in everyday life today and in the innovation of business models, and have invested considerable amounts in the research and development of biotechnological inventions, in the face of the legal system's clear preference for the encouragement of such innovations through the granting of exclusive patent rights.

In the United States, further, a continuous increase of research on human genetic material has been seen in recent decades, with increasing success and a consequent large number of patents on genes, DNA, etc. What is more, with the decoding of the human genome, medical science has been revolutionized, and individualized treatment is expected for many genetic and hereditary diseases. As a consequence, an extremely large number of patent requests for various segments of the human genome have been filed with the United States Patent and Trademark Office.[1]

---

[1] See, *ex multis*, N.S. Sreenivasulu and C.B. Raju, *Biotechnology and patent law. Patenting living beings*, Noida (India): Manupatra, 2008, p. 34 et

In general, the purpose of the American patent system is to promote scientific and technical progress, providing compensation to inventors as an incentive to reveal information which may benefit the public. Compensation is given as a concession by the federal government of the right to exclude third parties from the creation, use or marketing of the invention.

At first sight, then, the American patent system appears similar to the European one. However, there are differences between the two systems under which in the United States innovators receive greater protection than do those in Europe.

American public policy considers patents a means for the creation of new industries and jobs. The patent system is based on a *quid pro quo*: high protection is guaranteed, for a limited time, in exchange for complete disclosure of the invention. Through the granting of an exclusive right to the inventor to create and market the invention exclusively for a period of time, she/he receives an economic incentive to invest in the invention. This incentive is particularly important in sectors such as biotechnology and pharmaceuticals, where significant discoveries require extremely costly and lengthy research.[2]

In the American legal system, the roots of intellectual property law were introduced at the constitutional level, in order to 'promote the progress of science and the useful arts'.[3] Congress, then, has the power to legislate to this end. Further, Congress also has the power to regulate commerce. So, there is a distinction between patenting inventions and marketing the results.

In the United States, also, intellectual property has the important structural characteristic of being all inclusive, so the patent system does not contain a list of the exceptions to patentability. On the other hand, Title 35 of the United States Code,[4] at Section 101, deals with the

---

seq.; B.A. Brody, 'Intellectual Property and Biotechnology: The U.S. Internal Experience – Part I' (2006) 16 *Kennedy Institute of Ethics Journal* 1 et seq., and B.A. Brody, 'Intellectual Property and Biotechnology: The U.S. Internal Experience – Part II', therein, p. 105 et seq.

[2] See J.D. Wright, 'Implications of Recent Patent Law Changes on Biotechnology Research and Biotechnology Industry' (1997) 1 *Virginia Journal of Law and Technology* 1 et seq. In reference to the fundamental but complex relationship between exclusive patents and public health, see, among others, J.A. Burger and J. Brunner, 'A Court's Dilemma: When Patents Conflict with Public Health' (2007) 12 *Virginia Journal of Law & Technology* 71 et seq.

[3] United States Constitution, Article 1, Section 8(8). In this regard, see below.

[4] Hereafter also 'U.S.C.'.

identification of patentable subject matter, with clear favour for the granting of patent protection over life forms.[5]

In practice, different types of inventions, concerning transgenic microorganisms, plants, animals and human genetic material, are the subject of patent applications and, except in the case of transgenic human beings, are patentable. Currently, then, animals and plants visible to the naked eye, microorganisms, such as bacteria, cells, genes and gene sequences, DNA and DNA sequences, as well as fragments of DNA and genes which are visible only through the use of a microscope are all patentable. In addition, processes for the isolation and use of genetic material for certain purposes, processes for the genetic manipulation of living beings and those for the treatment of genetic and hereditary illnesses are also patentable.

In the United States, therefore, the evolution of patent law on biotechnological inventions, beginning with the patentability of microorganisms, has today reached the level at which only human beings are excluded from the context of patentability. In this evolution, legal rulings, rather than legislative initiatives, have played the main role.[6]

Given that the legislature has not provided for the development of biotechnology, patent legislation adopted decades ago does not contain provisions to provide protection for biotechnological inventions. Nor has the legislature been dynamic enough to react rapidly to the challenges posed by biotechnology.

In the absence of legal regulations on the discipline of biotechnological inventions, in light of their growing importance to the related industry, case law has assumed the task of recognizing and protecting such inventions, interpreting the existing legislation in an innovative and liberal way. Case law, thus, in view of the constitutional mandate for the promotion of 'progress of science and useful arts',[7] identified various principles and guidelines which have allowed for the protection of biotechnological inventions and which remain relevant today.

The U.S. innovative approach to case law, which has laid the foundation for the evolution of patent law for biotechnology, has found recognition and adhesion also at the international level in the context of

---

[5] On this point, beyond what will be said extensively hereafter, see A.J. Wells, 'Patenting New Life Forms: An Ecological Perspective' (1994) 3 *European Intellectual Property Review* 111 et seq.

[6] For the purposes of this book, 'case law' is also meant to include decisions of patent offices, given that they are – particularly with regard to biotechnology, as shall be seen – creators of law.

[7] In regard to which, see below.

the TRIPS Agreement, in which, in fact, the patentability of various biotechnological inventions such as microorganisms, plants, animals and human genetic material is provided.[8]

## III.1 BIOTECHNOLOGICAL INVENTIONS IN LEGISLATION

### III.1.1 The Evolution of Patent Regulation

The U.S. Constitution, Article 1, Section 8, provides that: 'The Congress shall have Power … To promote the Progress of Science and useful Arts, by securing for limited Times to Authors and Inventors the exclusive Right to their respective Writings and Discoveries'.[9]

The regulations provide, basically, legislation on the patent in order to provide protection to inventors and legislation on copyright to ensure protection of authors, in order to promote the progress of science and the useful arts.

In the case of *Kewanee Oil Company v. Bicron Corporation et al.*,[10] the Supreme Court took into consideration the purpose of patent law, stating that it consists in a modelled legal science to cover all emerging technology. In particular, in stating the opinion of the Court, Judge Berger specified that:

> The stated objective of the Constitution in granting power to Congress to legislate in the area of intellectual property is to 'promote the progress of science and useful arts'. The patent laws promote this progress by offering a right of exclusion for a limited period as an incentive for innovators to risk the often-enormous costs in terms of time, research and development. The productive effort thereby fostered will have a positive effect on society through the introduction of new products and processes of manufacture into the economy, and the emanations by way of increased employment and better lives for citizens.

The Supreme Court, then, highlighted that the considerations of constitutional policy are the foundation of patent law, and in particular that progress of science is a prerequisite for the granting of a temporary

[8] See TRIPS Agreement (Agreement on Trade Related Aspects of Intellectual Property Rights, cit.), arts 27 et seq., in regard to which refer above, to Part I.

[9] United States Constitution, Article 1, Section 8(8).

[10] Decision of 13 May 1974, 181 USPQ 673.

patent monopoly. This, though not definite, is held to be the tool to encourage the creative effort, which leads to the introduction of new products and procedures for manufacturing, new jobs and the improvement in the quality of life.

Congress has fulfilled its constitutional mandate working jointly with case law, and thus ensuring protection of new emerging technology over time.

With reference to patent legislation, it must be noted that the provision simultaneously contains an attribution of power and a limitation of its use in order to promote 'progress of science and the useful arts'.[11] Thus Congress, in the exercise of its power, may not surpass the limits imposed by this aim, nor may it widen the area of monopoly without considering the innovation reached and the progress it may provide with respect to the state of knowledge. In the context of the constitutional mandate, then, Congress must implement a policy which achieves its aim in the best manner possible.

Congress rapidly implemented these provisions through the United States Patent Act 1790,[12] through a first-to-invent patent system, rather than, as in general other legal systems are, through a first-to-file system.

The United States Patent Act, which entered into force in 1973, was designed by a board created in 1789 at the Department of State[13] under the leadership of Thomas Jefferson, Secretary of State at the time.[14] Jefferson had an instinctive aversion to monopolies; he did not believe it was appropriate to grant patents for small details or obvious improvements, but he had difficulty in drawing the line between innovation which was deserving of protection and that which was not. Thus, the board created various regulations, but, in accordance with Congress, the matter was delegated to case law, so that it could contribute to the formation of a system in which the patent applicant could be certain of the applicable

---

[11] See, in general, among others, D.S. Chisum, *Patents: A Treatise on the Law of Patentability, Validity and Infringement*, Newark (USA): LexisNexis, 1978–2012; D.S. Chisum *et al.*, *Principles of Patent Law: Cases and Materials*, University Casebook Series, New York: Foundation Press, 1998, as well as, with specific reference to the relationship between such aims and moral profiles, M.A. Bagley, 'A Global Controversy: The Role of Morality in Biotechnology Patent Law', in P. Yu (ed.), *Intellectual Property and Information Wealth: Issues and Practices in the Digital Age*, Westport (USA): Praeger Press, 2006, p. 317 et seq.

[12] Patent Act of 1790, Ch. 7, 1 Stat. 109-112, 10 April 1790.

[13] The board was known as Commissioners for the Production of the Useful Arts.

[14] In this regard, see R.P. Merges and J.F. Duffy, *Patent Law and Policy: Cases and Materials*, Newark (USA): LexisNexis, 2007 (4th edn).

discipline, in particular through the development of additional requirements for patentability.[15]

Though the Patent Act has been amended, reviewed or codified at various times – most recently through the Patent Reform Act of 2011,[16] which, in particular, replaced the 'first-to-invent' system with that of 'first-to-file'[17] – Congress has remained basically stable on the tests of novelty and utility formulated in the original version (except now for the necessary adaptations to the new 'first-to-file' system). The reasoning at the basis of the Act was that according to which 'the ingenuity should receive a liberal encouragement'.[18] To this end, the matter of patentability has been defined in broad terms as 'any new or useful art, machine, manufacture or composition of matter,[19] or any new or useful improvement thereof'.

The statute of 1793 was followed by others in 1836, 1870 and 1874, which repeated the conceptual approach of the first.

Subsequently, the regulation on patents was recodified in the Patent Act of 1952.[20] In it, in the definition of patentable subject matter the term 'art' was replaced with 'process' and novelty was no longer an alternative to but rather an addition to utility, resulting therefore in the new definition of patentable subject matter as 'any new and useful process, machine, manufacture or composition of matter, or any new and useful improvement thereof'.[21]

The aim of classifying patentable material in this way is to limit patent protection to the field of applied technology ('useful arts'). In this regard, the four classes are rather flexible, and have been interpreted in order to cover new technology that has developed in the past two centuries.[22]

In today's version of American legislation, the requirements for patentability – in line with that provided by Venetian law dated 1474 (the

---

15 See *Graham v. John Deere CO of Kansas City*, decision of 21 February 1966, 148 USPQ 464.

16 Or Public Law 112 – 29 – Leahy-Smith America Invents Act, 16 September 2011.

17 The changes are expected to enter into force in March 2013.

18 *Writings of Thomas Jefferson*, Washington, 1871 edition, p. 75 et seq.

19 Or rather, the 'great and distinct classes of invention' (*Ex parte Blythe* 1885 *Commission Decision* 1862, 86, *Commission Patent* 1885), not identified instead in European regulations (see Part IV).

20 The Act of July 19, 1952, c. 950, 66 Stat. 792.

21 The same definition is present in the version of Section 101 currently in force (as modified various times).

22 See M.A. Bagley, 'A Global Controversy: The Role of Morality in Biotechnology Patent Law', cit., p. 319.

so-called '*Parte Veneziana*') which historically remains the first systematic attempt to protect inventions[23] – are as follows: novelty, utility and non-obviousness. Novelty[24] and utility[25] may be traced back to the original Act, while non-obviousness is a requirement arising from case law which was first shown in the ruling on the case *Hotchkiss v. Greenwood* of 1850.[26] In the patent application, sufficient disclosure of the invention must also be provided.

---

[23] '*Parte*' of the Venetian Senate on 19 March 1474, on the basis of which: '*L'andarà parte che per auctorità de questo Conseio, chadaun che farà in questa Città algun nuovo et ingegnoso artificio, non facto per avanti nel dominio nostro, reducto chel sarà a perfection, siche el se possi usar, et exercitar, sia tegnudo darlo in nota al officio di nostri provveditori de Comun. Siando prohibito a chadaun altro in alguna terra e luogo nostro, far algun altro artificio, ad immagine et similitudine di quello, senza consentimento et licentia del auctor, fino ad anni X*' (Venetian State Archive, Senato terra, register 7, document 32).

[24] An invention does not respect the requirement of novelty when: (a) the inventor makes it known or allows the public to use it; (b) the invention is published (for instance in a patent application or in a news article); (c) the invention has been invented previously by another who did not abandon, suppress or hide it; (d) the invention has been described in a patent application filed by another, who afterwards obtained a patent. In the United States, there is anticipation, and therefore the invention is defective with regard to the requirement of novelty, when a reference to the state of the art, or an event, makes all characteristics of the invention known and allows the average expert in the field to create and use it.

[25] Three profiles of utility may be identified: (i) general utility is the requirement of usefulness; (ii) specific utility for which the invention must actually carry out its function; (iii) moral or beneficial utility requires that the invention not 'poison, promote debauchery, facilitate private assassination' (in this regard, see, in particular, *Lowell v. Lewis*, 15 F. Cas. 1018, 1019, C.C.D. Mass. 1817; *In re Fisher*, 421 F.3d 1365, 1371, 76 USPQ2d 1225, 1230, Fed. Cir. 2005, as well as that which will be said extensively below).

[26] 1850 11 How 248. In this, the Supreme Court stated, laying in fact the foundation for the requirement of 'non-obviousness', which: 'Unless more ingenuity and skill were required than were possessed by an ordinary mechanic acquainted with the business, there was an absence of that degree of skill and ingenuity which constitute essential elements of every invention'. Further, as stated by Thomas Jefferson in a letter dated 1813, American patent law had always required more than simple novelty ('Letter To Isaac McPherson Monticello', 13 August 1813, accessed 17 December 2014 at http://www.let.rug.nl/usa/P/tj3/writings/brf/jefl220.htm). See also, on this point, the approach adopted in *Great Atlantic & Pacific Tea Co. v. Supermarket Equipment Corp.*, 340 U.S. 147 (1950), following which the Patent Act of 1952 was issued, in order to also reduce the impact of the requirement of non-obviousness on patentability (eliminating the 'flash of genius' test). Subsequently, in the case of *Graham et al.*

At the policy level, in consideration of the importance of technology for the competitive market, the United States pursues the objective of ensuring that the patent system continues to maximize incentives for innovation and development. The success achieved in the past from the patent system in providing this stimulus is the result of the inherent capacity of the system itself to evolve and adapt to the changes of the times and to develop new technology.[27]

In particular, the United States has tried to harmonize the patent legislation of various states and to establish a minimum standard for the acquisition of rights and for the enforcement of them. In this sense, perhaps the most significant reform has been that concerning the creation, in 1982, of the Court of Appeals for the Federal Circuit.[28] The objective was to promote greater uniformity in the application of patent law and to reduce the possibility of 'forum shopping'.[29] The CAFC, which has exclusive jurisdiction over all appeals on the subject of patents, from its establishment has been decisively 'pro-patent'. In any case, it is debatable as to whether or not this is positive:[30] the result has

---

*v. John Deere Co. of Kansas City et al.*, 383 U.S. 1 (1966), the Supreme Court stated that the requirement should be identified on the basis of the following criteria: (a) the context and content of the state of the art; (b) the level of average experience in the sector; (c) the differences between the invention and the state of the art; (d) the objective proof of non-obviousness. This last point may be traced, for instance, to cases in which the invention had commercial success, providing the solution to needs which had remained for a long time without a solution, or third parties failed in attempting to reach the invention. In regard, further, to the issue of whether the combination of elements already previously known may be considered obvious, in the ruling of *Winner Int'l Royalty Corp. v. Wang*, 202 F.3d. 1340, 1348 (Fed. Cir., 2000), it was stated that there must be a suggestion or teaching in the state of the art, in order to combine already known elements and consider the invention as obvious (see however, for criticism regarding the relevance of this statement: *KSR Int'l Co. v. Teleflex, Inc.*, 550 U.S. 398, 2007).

[27] On the contrary, in the European Union patent policy is traditionally geared, mainly, towards the creation and reinforcement of the internal market (refer to Part IV).

[28] The 'CAFC', which was created by the merging of two pre-existing courts, the Court of Claims and the Court of Customs and Patent Appeals.

[29] In search of courts with more favourable positions. On this theme, see among others, Wright, 'Implications of Recent Patent Law Changes on Biotechnology Research and Biotechnology Industry', cit., p. 1 et seq.

[30] According to a part of the literature, the assumption that patent protection increases innovation has never been convincingly proven. Currently, a reason for the increase in the number of patents lies in the fact that the availability of

been an increase in patent protection and the number of disputes on the matter. The positive role carried out by the CAFC has been to provide clarification of the standards for the interpretation of patent rights and an increased predictability in the application of the legislation.

In all legislation referred to up to this point, neither biotechnology nor living organisms have been taken into consideration in the context of patentable subject matter. The provisions relevant in this regard are, as is known, contained in Title 35 of the United States Code, in Sections 100–105; in particular Section 101 indicates the context of patentable subject matter. At the time of their original formulation, the emergence of the science of biotechnology and its continuous developments were in no way predictable, therefore no provision on the matter was inserted into patent regulations. As far as living organisms are concerned, the lack of consideration was not due to ethical considerations or religious beliefs, but to the fact that living organisms were considered by multiple parties as the 'common heritage of mankind', and therefore not patentable.[31]

Following the revisions to federal patent policy on the matter of living organisms which was registered in the years following 1970,[32] there were some important legislative reforms. In 1980, the Bayh-Dole Act[33] encouraged the patenting and marketing of inventions funded by the government, allowing small businesses and non-profit organizations to maintain ownership of patents obtained through the research developed with such funding.[34] In 1983 the provision was extended, through an Executive Order, to large businesses.

---

complex patent portfolios represents an ever more widespread business strategy (see M. Blakeney, 'Biotechnological Patenting and Innovation', in W. Prinz *et al.* (eds), *Patents and Technological Progress in a Globalized World. Liber Amicorum Joseph Straus*, Berlin: Springer, 2009, p. 229 et seq.).

31 In this regard, among others, see K.R.G. Nair and A. Kumar, *Intellectual Property Rights*, New Delhi: Allied, 1994, p. 189.

32 In relation to which, see below.

33 Patent and Trademark Amendments Act of 1980, Public Law 96-620, amended in 1984.

34 With regard to the provision, and more generally on the theme of the risk that biotechnological patents granted 'upstream' may limit research 'downstream', where a single body is granted monopolistic control over 'basic' discoveries, see B. Williams-Jones and V. Ozdemir, 'Enclosing the "Knowledge Commons": Patenting Genes for Disease Risk and Drug Response at the University-Industry Interface', in C. Lenk, N. Hoppe and R. Andorno (eds), *Ethics and Law of Intellectual Property. Current Problems in Politics, Science and Technology*, Aldershot (UK): Ashgate, 2007, p. 137 et seq.; A.K. Rai and R.

The Federal Technology Transfer Act of 1986[35] granted federal sponsorship to the formation of unions with private interests.[36] In 1987, an Executive Order further encouraged the transfer of technology, permitting patent rights to be transferred to government assignees.[37]

Subsequently, considering the oscillating decisions of the Federal Circuit, the legislature intervened, amending the Patent Act in order to introduce a sort of 'compatibility clause' for biotechnology patents. In 1995, the Biotechnology Process Patents Act,[38] which amended the Patent Act regarding in fact the patentability of biotechnology processes, stated the compliance both of biotechnological products and processes with the requirement of non-obviousness where they produce a new and not obvious result, or are the result of the inventive step.[39] What is more, the amendment specified that, in particular, processes of genetic alteration of single or multi-cellular organisms, the expression, elimination or alteration of nucleotide sequences, the expression of physiological characteristics not usually associated with the aforementioned organisms, as well as the procedure of fusion of cell lines which express a protein as a monoclonal antibody, and the methods for using derivative products of biotechnological processes, are all patentable.[40] The Act showed, thus,

---

Eisenberg, 'The Public Domain: Bayh-Dole Reform and the Progress of Biomedicine' (2003) 66 *Law and Contemporary Problems* 289 et seq.

35 Public Law 99-502.

36 So bypassing the necessity for them to conform to antitrust law.

37 In regard to these provisions, see particularly Brody, 'Intellectual Property and Biotechnology: The U.S. Internal Experience – Part I', cit., p. 20 et seq.

38 Public Law 104-4. In this regard, see J. Zhe Zang, '*In Re Ochiai, In Re Brouwer* and the Biotechnology Process Patent Act Of 1995: The End of the Durden Legacy?' (1997) 37 *IDEA* 405 et seq.

39 See D. Chisum *et al.*, *Principles of Patent Law: Cases and Materials*, cit., p. 714.

40 In particular, as far as is significant here, following the amendments introduced in 1995, Section 103 of the U.S.C. provides the following:

> '(1) Notwithstanding subsection (a), and upon timely election by the applicant for patent to proceed under this subsection, a biotechnological process using or resulting in a composition of matter that is novel under section 102 and non-obvious under subsection (a) of this section shall be considered non-obvious if (A) claims to the process and the composition of matter are contained in either the same application for patent or in separate applications having the same effective filing date; and (B) the composition of matter, and the process at the time it was invented, were owned by the same person or subject to an obligation of assignment to the same person. (2) A patent issued on a process under paragraph (1)-(A) shall also contain the claims to the composition of matter used in or made by that process, or (B) shall, if such

the intention of Congress to provide legislative support to the patentability of biotechnology products and processes, endorsing the liberal interpretation of the existing law adopted by the Federal Circuit and the practices of the Patent Office to grant patents for them.[41]

Therefore, the context of American patent law suggests drawing on the most recent state of scientific knowledge, in order to interpret the criteria of patentability not specifically defined by the legislature and which by their very nature require determination by case law.

Given that the origin of patent law is constitutional, the language used by Congress for patent regulations is broad and general, and the Supreme Court has limited the exceptions to patentability created by the legal process. In this way, biotechnological inventions, even if they were not predictable at the time of the discipline's emergence, have been able to receive protection.

With reference to the non-patent legislation, then, as far as human cloning or the production of transgenic humans are concerned, the XIIIth Amendment to the U.S. Constitution,[42] stating the ban on slavery and therefore on the recognition of property rights over human beings, implies the ban of similar inventions. Contrarily, the invention of transgenic human beings through the cloning of another human being and the patenting of such inventions would give rise to the creation of property rights over human beings, which would be equivalent to reducing them to slavery, in violation of the ban of the XIIIth Amendment.

---

composition of matter is claimed in another patent, be set to expire on the same date as such other patent, notwithstanding section 154. (3) For purposes of paragraph (1), the term "biotechnological process" means- (A) a process of genetically altering or otherwise inducing a single- or multi-celled organism to (i) express an exogenous nucleotide sequence, (ii) inhibit, eliminate, augment, or alter expression of an endogenous nucleotide sequence, or (iii) express a specific physiological characteristic not naturally associated with said organism; (B) cell fusion procedures yielding a cell line that expresses a specific protein, such as a monoclonal antibody; and (C) a method of using a product produced by a process defined by subparagraph (A) or (B), or a combination of subparagraphs (A) and (B).'

41 In regard to which, see below.

42 On the basis of which: 'Neither slavery nor involuntary servitude, except as a punishment for crime whereof the party shall have been duly convicted, shall exist within the United States, or any place subject to their jurisdiction. Congress shall have the power to enforce this article by appropriate legislation'.

### III.1.2 Bioethical Profiles of Patentability

Research of human genetics, given that it is aimed at identifying the genetic structure of the human body in order to treat genetic and hereditary diseases, calls into question the boundaries of morality. Despite the fact that moral standards are different from place to place and from person to person, the patentability of life is generally held to be contrary to bioethics.[43]

The United States Constitution requires respect for the inherent dignity of life and its intrinsic values, providing equality to all individuals and guaranteeing human dignity; this is in particular, as noted, pursuant to Amendment XIII, which prohibits slavery and the commoditization of human life.[44] On the other hand, the actual importance of these principles in U.S. legislation and case law is generally rather limited.

American society is characterized by its flexibility and marked tendency to adapt to changing conditions. In this sense, in patent law, 'anything under the sun made by man'[45] is considered patentable. Within the U.S. patent discipline, contrarily, there is no specific provision with regard to morality.[46]

Under the profile of case law, on the other hand, the courts have sometimes rejected patents for inventions they considered to be – *lato sensu* – immoral. Despite the fact that patent legislation does not include

---

[43] In regard to these concepts, refer above, to Parts I and II.

[44] Concerning the reasons for this limited attention to the principle of human dignity in the American system, and to the relative differences with the European model, see G. Bognetti, 'The Concept of Human Dignity in European and U.S. Constitutionalism', in G. Nolte (ed.), *European and U.S. Constitutionalism*, Cambridge: Cambridge University Press, 2005, p. 85 et seq., who shows how this is the logical consequence of the neo-liberal approach with regard to state intervention; J.Q. Whitman, '"Human dignity" in Europe and the United States: The Social Foundations', therein, p. 108 et seq., who holds that the reason for such a 'weak' approach in regard to human dignity is the absence within the American model of a reaction to the fascist and Nazi experiences (at the basis, instead, of the 'strong' European approach; G.L. Neuman, 'Human Dignity in United States Constitutional Law', in D. Simon, M. Weiss (eds), *Zur Autonomie des Individuums: Liber Amicorum Spiros Simitis*, Baden-Baden: Nomos, 2000, p. 249 et seq.).

[45] See Supreme Court, *Diamond v. Chakrabarty*, 447 US 303 (1980), 309, where S. Rep. No. 82–1979, 5 (1952) is referenced; H.R. Rep. No. 82–1923, 6 (1952). With regard to the decision on the leading case *Chakrabarty*, see below.

[46] See, among others, M.A. Bagley, 'A Global Controversy: The Role of Morality in Biotechnology Patent Law', cit., p. 318.

a 'morality clause', moral regulations have often been made 'indirectly' in the courts through the utility requirement, in Sections 101 and 112 U.S.C.[47]

The American concept of utility is, at the same time, broader and stricter than the European notion of industrial applicability.[48] It is broader in that the same word includes elements of illegality and immorality.[49] It is stricter in that pure research is not considered to be equivalent to practical utility.[50]

The United States Patent and Trademark Office ('USPTO') has published guidelines for the examination of utility.[51] The guidelines are not legally enforceable and they do not mention moral or public policy issues. On the basis of the guidelines, utility must be 'specific, substantial and credible', that is, in the patent application the invention must be described as being useful for a specific practical aim, and the claim must be credible for the average expert in the field.[52]

---

[47] The latter in regard to the specification of the invention, which 'shall contain a written description of the invention, and of the manner and process of making and using it, in such full, clear, concise, and exact terms as to enable any person skilled in the art to which it pertains, or with which it is most nearly connected, to make and use the same, and shall set forth the best mode contemplated by the inventor of carrying out his invention'.

[48] In this regard, see, among others, N. Wee Loon, 'Patenting of Genes – A Closer Look at the Concepts of Utility and Industrial Applicability' (2002) 33 *IIC* 394 et seq.; M. Llewelyn, 'Industrial Applicability/Utility and Genetic Engineering: Current Practices in Europe and the United States' (1994) 11 *European Intellectual Property Review* 473 et seq.

[49] Concerning the 'intersection' between utility and morality, see S.R. Donnelly, 'The Patentability of Human Embryonic Stem Cells: Is the Inconsistent Application of the European Union Biotechnology Directive's Moral Exclusion Clause Undermining Investor Confidence in Europe, Providing a Competitive Advantage to the United States?' (2011) 20 *Dalhousie Journal of Legal Studies* 106 et seq. (123 et seq.); M.A. Bagley, 'A Global Controversy: The Role of Morality in Biotechnology Patent Law', cit., p. 320 et seq.; G.R. Hagen and S.A. Gittens, 'Patenting Part-Human Chimeras, Transgenics and Stem Cells for Transplantation in the United States, Canada and Europe' (2008) 14 *Richmond Journal of Law & Technology* 11 et seq. (46 et seq.).

[50] The rulings of *Brenner v. Manson*, 383 U.S. 519, 148 USPQ 689 (1966), *In re Joly*, F.2d 906, 153 USPQ 45 (CCPA 1967) and *In re Kirk*, 376 F.2d 936, 153 USPQ 48 (CCPA 1967), established that a composite which is only useful in research does not satisfy the requisite of utility, even if that research may lead to a useful product.

[51] In the most recent amended version: 'Examination Guidelines for the Utility Requirement', Federal Register, vol. 66, no. 4, 5 January 2001.

[52] 'Examination Guidelines' cit., Part B, point 3.

In order to be useful, the invention must be subject to a use which benefits society. The principle has been identified in two leading cases in 1817. In the ruling on the case *Lowell v. Lewis*,[53] Judge Story stated: 'All that the law requires is, that the invention should not be frivolous or injurious to the well-being, good policy or sound morals of society. The word "useful", therefore, is incorporated into the Act in contradistinction to mischievous or immoral'. The examples of immoral inventions mentioned below included poison, promotion of debauchery, and facilitating private assassination. In the subsequent case of *Bedford v. Hunt*,[54] then, Judge Story again stated that 'by useful invention, in the statute, is meant a one that may be applied to some beneficial use in society … (t)he law does not look to the degree of utility; it simply requires that it be capable of use'. Judge Story's vision of utility was later generally accepted by the courts, and applied to relative cases of fraud or gambling devices.[55]

The significance of the utility rule has been explained in the ruling on the case *Brenner v. Manson*,[56] in which the Supreme Court held that the creation of a product belonging to an important class of compounds, in the case of steroids, is not sufficient to satisfy the requirement. In particular, the Court stated that 'a process to produce a compound may be patented only if the compound has substantial utility or specific benefit … (i)n currently available form'. It must, then, be necessary to emphasize the function or utility of the effective compound produced. The decision is based on the reasoning according to which, until a procedure which is the subject of the application has not been directed at creating a product which has shown that it is useful, the limits of monopoly may not be determined with accuracy, and if a similar patent was granted entire areas of scientific development could be monopolized without the compensation of a benefit for the collective.

The position of case law on these matters, however, has changed over time. In the decision in *Juicy Whip Inc. v. Orange Bang Inc. and Unique Beverage Dispensers Inc.*,[57] the CAFC reformed the ruling of the District Court which had declared the patent invalid for lack of utility, on the basis of the fact that the patented invention was aimed at deceiving consumers through the imitation of another product, stating that the requirement of utility does not represent a directive for the Patent Office

53 15F Cas 1018 (no. 8568), Circuit Court, Massachusetts 1817.

54 3F Cas 37 (no. 1217), Circuit Court, Massachusetts 1817.

55 See, for instance, *In re Fisher*, 421 F.3d 1365, 1371, 76 USPQ2d 1225, 1230, Fed. Cir. (2005).

56 148 USPQ 689 (1966).

57 Decision of 6 August 1999, 185F 3d 1364.

or the courts to operate as referees of deceitful marketing practices. In particular, Judge Bryson stated that: 'Congress never intended that the patent laws should displace the police poker of the States, meaning by that term those poker by which the health, good order, peace and general welfare of the community are promoted'. Thus, the moral concerns examined previously – in the contextual limits noted – by patent law, must be faced elsewhere. In light of the *Juicy Whip* ruling, then, inventions used to defraud consumers, just as those relative to gambling devices, moral or not, may not be considered excluded from patentability on this basis alone.[58]

Though in time the American courts, as said, gradually rejected a utility based on moral profiles for traditional inventions of a mechanical nature, the link between utility and morality appears relevant for today's biotechnological inventions which concern gene sequences which are not identified in terms of structure or function. The patent applications for fragments of DNA, in fact, present special problems concerning the requirement of utility.

In the case of *Ex parte Tanksley*,[59] for instance, the applicants claimed to have identified the position of the cDNA clones on a chromosomal map of the tomato, but had not specified any specific use for such clones. Section 112(2) U.S.C. provides that: 'The specification shall conclude with one or more claims pointing out and distinctly claiming the subject-matter which the applicant regards as his invention'. The Board of Patent Appeals and Interferences specified that the examiner, for the sake of compliance with Section 112, may request that the invention be described in sufficiently detailed terms, as to structure or function, to allow for its comparison with the state of the art. This is justified by the *quid pro quo* principle, in that in order to inform the public of the development of the subject to which the exclusive right granted by the patent shall be applied, it is necessary to compare it with the state of the art. In this case, in particular, given that nothing of the structure and only a few functions of the clones of cDNA which were the subject of

58 See O. Mills, *Biotechnological Inventions. Moral Restraints and Patent Law*, Farnham (UK)/Burlington (USA): Ashgate Publishing, 2010 (Revised Edition), p. 48; M.A. Bagley, 'A Global Controversy: The Role of Morality in Biotechnology Patent Law', cit., p. 321 et seq., which defined the ruling as 'the death-knell for the moral utility requirement. … Thus, the judicially created moral utility requirement is now defunct, and no longer provides a basis for the rejection of morally controversial biotech patents, leaving the United States with a "patent first, ask (moral) questions later" system'.

59 Decision of 10 October 1991, 26 USPQ 1348.

the application were indicated, the Board did not recognize compliance with Section 112.

In the case of *Ex parte Thomas F. Deuel*,[60] concerning a purified growth factor isolated from prostate tissue and to a method of stimulating growth in fibroblast cells using the growth factor, the Board excluded utility pursuant to Sections 101 and 112, given that in the specification there was no declaration for the use, description or examples for the methods of use of the invention.

Given the decisions of the Board of Appeals in these cases, with respect to biotechnological inventions there is the issue of whether or not in the United States there is a veiled attempt to implement moral regulations through the utility requirement.

In this regard, the decision in which the USPTO rejected a patent request filed by Professor Stuart Newman and the sociologist Jeremy Rifkin is particularly interesting, concerning a method to create part human, part animal creatures.[61] The Patent Office declared in a press release that: 'It is the position of the PTO that inventions directed to human/non-human chimera could, under certain circumstances, not be patentable because, among other things, they would fail to meet the public policy and morality aspects of the utility requirement'.[62] In reality, the patent application was considered negatively given that it included a

---

[60] Decision of 28 January 1993, 27 USPQ 1360.

[61] U.S. Application Serial No. 08/993,564 (18 December, 1997), and divisional application U.S. Application Serial No. 10/308,135 (3 December 2002). The application, further, was not aimed at the creation of such a creature, but rather to spark debate and push Congress, the Patent Office or the courts to more clearly identify the limits of patentable subject matter. On the issues raised by the patent request, in addition to referring below, see S. Parthasarathy, 'Whose Knowledge? What values? The Comparative Politics of Patenting Life Forms in the United States and Europe' (2011) 44 *Policy Sciences* 267 et seq. (273 et seq.); R. Hagglund, 'Patentability of Human-Animal Chimeras' (2008) 25 *Santa Clara Computer & High Technology Law Journal* 51 et seq. S.M. Coughlin, 'The Newman's Application and the USPTO's Unnecessary Response' (2006) 5 *Chicago-Kent Journal of Intellectual Property* 90 et seq.; R. Weiss, 'Of Mice, Men and In-Between: Scientists Debate Blending of Human, Animal Forms', *Washington Post*, 20 November 2004, p. A1; S.A. Newman, 'The Human Chimera Patent Initiative (2002) 9 *Medical Ethics* 4 et seq.; A. Zitner, 'Patently Provoking a Debate: Two Friends Seek Rights to a Theoretical Human-Mouse, Thought up to Force Limits on Patenting Human Life', *Los Angeles Times*, 12 May 2002, p. A1.

[62] USPTO, 'Media advisory: facts on patenting life forms having a relationship to humans', 1 April 1998, available at http://www.uspto.gov/news/pr/1998/98-06.jsp.

human being, and the hybrid could not, therefore, fall under patentable subject matter.[63] On the other hand, with the aforementioned declaration the Patent Office expressed its willingness to continue to base its decisions on Judge Story's formula of utility.

The extension of this *revirement* of the application of the 'moral utility doctrine'[64] is uncertain, considering that, as noted, the Examination Guidelines for the Utility Requirement of 2001 did not mention moral or public policy issues.[65] Nevertheless, in February 2005 the patent application was rejected indicating the reasoning for the rejection as the lack of a written description and in fact the lack of utility. Among the principles referenced in regard to the rejection, were the risk of violation of the constitutional right to privacy and the possible conflict with the XIIIth Amendment.

In the subsequent ruling on the case *In re Fisher*,[66] then, the CAFC denied patentability of Expressed Sequence Tags,[67] given that they were 'only tools to be used along the way in the search for a practical utility', and thus were lacking in 'an immediate real world benefit' which would allow for the identification of substantial utility of the invention.[68]

From the decisions referenced, then, the requirement of utility and the methods in which through it moral profiles have gradually been implemented less often in American courts would suggest that moral concerns that appear within the patent system may be faced more appropriately

---

63 See R.P. Merges and J.F. Duffy, *Patent Law and Policy: Cases and Materials*, cit., p. 131; J. Wilson, 'Patenting Organisms: Intellectual Property Law meets Biology', in D. Magnus, A.L. Caplan and G. McGee (eds), *Who Owns Life?*, cit., p. 25 et seq. (54); PTO, 'Patent Application Is Disallowed as 'Embracing' Human Being' (1999) 58 *Patent, Trademark and Copyright Journal* 203.

64 Introduced with the ruling on the case *Lowell v. Lewis*, 15 Fed. Cas. 1018 (C.C.D. Mass. 1817). In this regard, see B.D. Enerson, 'Protecting Society from Patently Offensive Inventions: The Risk of Reviving the Moral Utility Doctrine' (2004) 89 *Cornell Law Review* 685 et seq.

65 See O. Mills, *Biotechnological Inventions. Moral Restraints and Patent Law*, cit., p. 50.

66 Decision of 7 September 2005, 421 F.3d 1365, 1371, 76 USPQ2d 1225, 1230, Fed. Cir.

67 Or 'ESTs', that is brief fragments of DNA transcribed and sequenced from a longer sequence of cDNA, which represent parts of expressed genes and can be used as an instrument for biotechnological research (on the theme, refer extensively above to Part II and below).

68 In this regard, see R. Fitt and E. Noddler, 'Specific, Substantial and Credible? A New Test for Gene Patents' (2008) 9 *Bio-Science Law Review* 16 et seq.

and easily in the context of *ad hoc* regulations. On the other hand, *de iure condito*, the recent case law relative specifically to biotechnological inventions indicates that there is limited space for the – 'indirect', as said – application of moral regulations.

In 1995, the TRIPS Agreement entered into force in the United States, which establishes globally that morality and public order constitute limits to the patentability of inventions.[69] The Agreement states, in fact, that inventions which are contrary to public order or morality can be excluded from the context of patentability. Being a nation party to the Agreement, the United States can exclude inventions from patentability based on reasons inherent to morality.

The Agreement also establishes, further, that inventions which may be damaging to the health of human beings, plants or animals, may also be excluded from patentability. Again, the Agreement considers the patenting of surgical, therapeutic and diagnostic methods for the treatment of humans or animals contrary to morality.

Those opposed to the patentability of life have stated that the TRIPS Agreement excludes plants, animals and human beings from patentability based on morality, for which patenting of said living organisms, including human genetic material, would violate ethical standards laid down by the Agreement. To the contrary, the supporters of the patentability of life noted that the ban on patentability concerns plants, animals and human beings but not non-natural plants or animals produced by man or those that are genetically modified, nor human genetic material which is isolated and purified into a non-natural form.

On the other hand, no objections have been made regarding the contrariety to morality of patenting surgical, therapeutic and diagnostic methods for the treatment of human beings and animals.[70]

Human genetic engineering, in particular, has led to research on the human genome, on the embryo and on stem cells, which raises ever more complex questions.[71] Research on the human genome, with the Human Genome Project, which was funded mainly by the United States government, creates significant possibilities for biomedical research, but leads to the manipulation of the genetic structure of the individual, which is highly controversial. Similarly, research on the embryo or foetus and the

---

69 TRIPS Agreement, art. 27, paras 2 and 3.

70 On this point, see N.S. Sreenivasulu and C.B. Raju, *Biotechnology and patent law. Patenting living beings*, cit., p. 205.

71 As seen above, in Part II.

extraction of stem cells from them in the initial phases of development of a human being have been strongly criticized.[72]

In consideration of these developments in human genetics, as noted, in 1997 the UNESCO Universal Declaration on the Human Genome and Human Rights was adopted,[73] which states that research on the genome must respect the ethical standards of society, promoting the development of ethical studies in light of scientific and technological progress in the fields of biology, genetics and medicine.[74] The Declaration recognizes the freedom of research as a part of freedom of thought, encouraging research in the aforementioned fields in order to guarantee the right of health and relief of suffering, but emphasizes that no research must prevail over respect for human rights and dignity.[75] In particular, the signing states are required to establish ethical committees which will deal with the moral, legal and social issues posed by research on the human genome and its applications, as well as with promotion of education in bioethics.[76] The Declaration, again, beyond identifying practices which may be contrary to human dignity and ethics, such as, for instance, action taken on germinal lines,[77] requires the states to adopt adequate legal measures in order to guarantee respect for ethics and human rights.[78]

In order to fulfil the obligations derived from the Declaration, in 1996 President Clinton implemented the National Bioethics Advisory Commission, whose main task is to present reports on bioethical profiles involved in research in the biological and medical fields, especially with regard to

---

72 On these themes, beyond referring to that which is above in Chapter 2, see among others, C. Kuppuswamy, *The International Legal Governance of the Human Genome*, London: Routledge, 2009, p. 1 et seq.; M. Rimmer, *Intellectual Property and Biotechnology: Biological Inventions*, Cheltenham, UK and Northampton, MA, USA, Edward Elgar, 2008, p. 138 et seq.; B.A. Brody, 'Intellectual Property and Biotechnology: The U.S. Internal Experience – Part II', cit.; F. Vandenabeele, *Patentability of Living Organisms: Legal and Ethical Aspects of the Question*, Montreal: McGill University, 2005, p. 79 et seq.

73 UNESCO, Universal Declaration on the Human Genome and Human Rights, cit.

74 See the preamble and arts 13 and 14 of the Declaration.

75 See N. Lenoir, 'UNESCO, Genetics, and Human Rights' (1997) 7 *Kennedy Institute of Ethics Journal* 31 et seq.

76 Declaration cit., arts 16 and 20.

77 Declaration cit., art. 26

78 In this regard, see G. Kamstra *et al.*, *Patents on Biotechnological Inventions: The EC Directive*, London: Sweet and Maxwell, 2002, p. 210.

the human genome and human cloning. Already in a report in 1997,[79] the Commission drew attention to the ethical and social questions arising from cloning, holding that of human beings as unethical.

In 1998, then, scientists were able to isolate human embryonic stem cells from a foetus within the uterus of a pregnant woman. Later, in 2001, American researchers announced that they had produced the first cloned human embryos.

By virtue of these developments, in 2001 President Bush believed it opportune to establish the President's Council on Bioethics, to provide opinions on bioethical issues connected to progress in science and biomedical technology; in 2009, President Obama replaced this body with the Presidential Commission for the Study of Bioethical Issues.[80] In its first report on ethics and human dignity in cloning, presented in 2002,[81] the Council highlighted how at the moment the majority of routine biological research and many important pharmaceutical applications depend on cloning, which, however, raised significant ethical issues. In conclusion, then, the report recommended the ban of cloning aimed at the production of human beings, specifying on the other hand how it must not interfere with other important areas of scientific research, including those relative to the cloning of DNA sequences and human cell lines or animal cloning.

Another significant ethical question faced by the President's Council on Bioethics has been that of utilizing biotechnology for the 'pursuit of

---

[79] *Cloning Human Beings. Report and Recommendations of the National Bioethics Advisory Commission*, June 1997, accessed 17 December 2014 at https://bioethicsarchive.georgetown.edu/nbac/pubs/cloning1/cloning.pdf.

[80] In particular, the President's Council, which succeeded the previous National Bioethics Advisory Commission established in 1995 by President Bill Clinton with Executive Order 12975, was established in 2001 by President George W. Bush, with Executive Order 13237. In 2009 President Barack Obama, stating that he believed it appropriate to replace the Council with a body which would provide practical policy options, rather than philosophical lines of thought (see N. Wade, 'Obama Plans to Replace Bush's Bioethics Panel', *New York Times*, 18 June 2009, p. A24), established the Presidential Commission for the Study of Bioethical Issues with Executive Order 13521. For further details on the history of further national commissions on bioethics, even prior to those cited, see A. McCoy, 'Biotechnology and Embryonic Stem Cells: A Comparative Analysis of the Laws and Policies of the United States and Other Nations (2009) 8 *Loyola Law and Technology Annual* 63 et seq. (68 et seq.) (as well as information at http://www.bioethics.gov).

[81] President's Council on Bioethics, 'Human Cloning and Human Dignity: An Ethical Inquiry', July 2002, accessed 17 December 2014 at https://bioethics archive.georgetown.edu/pcbe/reports/cloningreport/.

happiness', or for the insertion of desired characteristics or the elimination of unwanted characteristics in embryos or adult bodies (for instance, during pregnancy in order to have 'better children' or in old age to feel younger). In 2003, the Council presented a report focused on ethics in the manipulation of human embryos and the use of biotechnology for the pursuit of happiness.[82] The report did not completely oppose the manipulation of the human body or embryo in order to obtain the desired results, but investigated ethical profiles of such interventions and concluded by highlighting the opportunity for a close monitoring of them, in order to prevent their diffusion from provoking serious social inconveniences.

In 2008, Congress adopted the Genetic Information Nondiscrimination Act,[83] which prohibits the use of genetic information in the context of employment and insurance. In particular, the Act bans insurers from denying insurance coverage or imposing higher premiums on healthy individuals solely on the basis of a genetic predisposition for developing an illness in the future. What is more, employers are prohibited from using the genetic information of individuals in decisions regarding hiring, firing, promotions, and so on.

With reference to the theme of the so-called statute of the embryo, the Supreme Court, in the decision of *Roe v. Wade*,[84] in an *obiter dictum*, implicitly stated that the unborn child is not a person in the sense of the term contained in the XIVth Amendment to the U.S. Constitution,[85] and its protection is indirect, in as much as the right to privacy must be balanced with public interest, but not with rights superior to it.[86]

---

[82] President's Council on Bioethics, 'Beyond Therapy: Biotechnology and the Pursuit of Happiness', October 2003, accessed 17 December 2014 at https://bioethicsarchive.georgetown.edu/pcbe/reports/beyondtherapy/.

[83] 'Genetic Information Nondiscrimination Act of 2008' (Pub.L. 110-233, 122 Stat. 881, 21 May 2008).

[84] Supreme Court, *Roe v. Wade*, 22 January 1973, cit., regarding, as was said, profiles of the constitutionality of voluntary interruption of pregnancy.

[85] On the basis of which, in particular: 'All persons born or naturalized in the United States, and subject to the jurisdiction thereof, are citizens of the United States and of the State wherein they reside. No State shall make or enforce any law which shall abridge the privileges or immunities of citizens of the United States; nor shall any State deprive any person of life, liberty, or property, without due process of law; nor deny to any person within its jurisdiction the equal protection of the laws'.

[86] See decision cit., 158 et seq. With regard to the *obiter dictum*, refer above to Part II.

At the legislative level, the marketing of human organs is prohibited by the National Organ Transplant Act of 1984,[87] but the embryo is not considered to be part of the human body; therefore the prohibition does not apply to it,[88] leaving the possibility open then for the preservation and use of human embryos for research purposes (especially for the extraction of stem cells) and any marketing of related products, which gives rise to significant moral problems.

As far as stem cell research is concerned, first of all it is necessary to remember that they are non-specialized primitive cells with the ability to transform themselves into various other types of cells of the body through a process called cellular differentiation. In particular, those considered to be the most effective stem cells – as they are totipotent or pluripotent – may actually be those extracted from human embryos, alive or dead, at the stage of four or eight cells from the moment of fertilization and the formation of the zygote, or may be extracted from bone marrow; what is more, embryonic stem cells can be artificially produced through *in vitro* fertilization or embryo cloning.[89]

Stem cell research allows for the production of tissues which are useful in the transplant of organs and offer enormous potential for the treatment of illness. Stem cells can be used for the experimentation of potential pharmaceuticals and used as models in order to understand the causes of certain illnesses based on cells (for instance diabetes, Parkinson's disease, and so on). What is more, in order to avoid the crisis of rejection in organ transplants, tissues produced using stem cells can be used, which are compatible both with the transplanted organ and the body of the recipient.[90]

Research on embryonic stem cells, on the other hand, requires the isolation of human embryonic cells. This gives rise to a series of moral concerns, in that the stem cells are able to develop into a human being and the embryo is in fact in the initial stages of a human being.

---

87 Public Law 98-507, as amended in 1988 and in 1990.

88 On this point, see S. Krimsky and R. Hubbard, 'The Business of Research' (1995) 25 *Hastings Center Report* 41 et seq.; A. Alpers and B. Lo, 'Commodification and Commercialization in Human Embryo Research' (1995) 6 *Stanford Law and Policy Review* 39 et seq.

89 President's Council on Bioethics, 'Monitoring Stem Cell Research', January 2004, accessed 17 December 2014 at https://bioethicsarchive.georgetown.edu/pcbe/reports/stemcell/index.html. On this theme, refer extensively above, to Part II.

90 On this theme, among others, see S. Holland, K. Lebacqz and L. Zoloth (eds), *The Human Embryonic Stem Cell Debate: Science, Ethics and Public Policy*, Cambridge, MA: MIT Press, 2001, p. 1 et seq.

Bioethicists state that the destruction of embryos is equivalent, in effect, to the homicide of a potential human being. They claim that a life cannot be ended for the good of biomedical research. What is more, it is noted that research on stem cells which uses embryos is nothing more than a commoditization of human embryos. In this sense, it is necessary to determine the moral statute of the embryo, in order for it to be possible to evaluate if the use and destruction of embryos is ethically and legally proper or objectionable.[91]

In the United States, various institutions have expressed concern on the theme of the moral statute of the embryo and on ethical profiles involved in the use of embryos in stem cell research. The American Association for the Advancement of Science[92] and the Ethics Advisory Board of the Geron Corporation[93] highlighted that the issue of the moral statute of the embryo represents a basic ethical profile involved in the objections regarding stem cell research.

The National Bioethics Advisory Committee[94] described the ethical issues brought up by stem cell research as mainly connected to the sources and methods of their extraction. The National Academy of Sciences[95] showed that the basic ethical objection to stem cell research

---

91 With regard to these profiles, beyond referring *supra*, to Part II, see in particular, among others, A.R. Chapman, 'The Ethics of Patenting Human Embryonic Stem Cells' (2009) 19 *Kennedy Institute of Ethics Journal* 261 et seq.; E.A. Zerhouni (Director of the National Institute of Health), 'Stem Cell Science: The Foundation of Future Cures', Testimony before Committee on Energy and Commerce Subcommittee on Health U.S. House of Representatives, 8 May 2008, accessed 17 December 2014 at http://www.hhs.gov/asl/testify/2008/05/t20080508c.html; B.R. Schaller, *Understanding Bioethics and the Law. The Promises and Perils of the Brave New World of Biotechnology*, Westport (USA)/London: Praeger, 2007, p. 133 et seq.

92 AAAS, 'AAAS Policy Brief: Stem Cell Research', 1999, p. 11, accessed 28 December 2014 at http://www.aaas.org, in the most recently modified version in 2011.

93 Geron Ethics Advisory Board, 'Research with Human Embryonic Stem Cells: Ethical Considerations' (1999) 29 *Hastings Center Report* 31 et seq. (32). Geron is a state of the art company in the field of stem cell research, which retains some patent rights over their use, also because, at the end of the 1990s, the first cells from embryos were isolated in its laboratories.

94 NABC, 'Ethical Issues in Human Stem Cell Research', 1999, accessed 17 December 2014 at http://bioethics.georgetown.edu/nbac/pubs.html, p. 45.

95 'Report of the National Academy of Sciences: Stem Cells and the Future of Regenerative Medicine', 2002, accessed 17 December 2014 at http://www.nap.edu/openbook.php?isbn=0309076307, p. 44.

using embryos lies in the fact that it deprives the embryo of the possibility of developing into a human being.

On the other hand, the National Institute of Health,[96] an agency funded at the federal level, through its Human Embryonic Research Panel, stated that the isolation of human embryonic stem cells is a research area subject to the production of significant scientific progress, which may be considered worthy of public funding.[97]

The Bush administration, with regard to the controversy on stem cell research and unresolved issue of the statute of the embryo, released the guidelines of the National Institute of Health on the funding of embryonic stem cell research.[98] These guidelines provided for the respect of high ethical and moral standards in human genetic research. The guidelines declared that federal funding could be granted for research on existing stem cells[99] which were extracted from an embryo, on the

---

[96] NIH, 'Human Embryonic Research Panel Report', 1994, accessed 17 December 2014 at http://bioethics.georgetown.edu/pcbe/reports/past_commissions, ch. 2.

[97] See also below, extensively, NIH, 'Stem Cells: Scientific Progress and Future Research Directions', 2001, accessed 17 December 2014 at http://stem cells.nih.gov/staticresources/info/scireport/PDFs/fullrptstem.pdf.

[98] 'Notice of Criteria for Federal Funding of Research on Existing Human Embryonic Stem Cells and Establishment of NIH Human Embryonic Stem Cell Registry', 7 November 2001, accessed 17 December 2014 at http://grants.nih.gov/grants/guide/notice-files/NOT-OD-02-005.html. The previous Clinton administration issued guidelines on the same theme ('National Institutes of Health Guidelines for Research Using Human Pluripotent Stem Cells') dated 17 August 2000.

[99] In this regard, the case of embryonic stem cells which were isolated in 1998 by Doctor James Thomson of the University of Wisconsin is particularly interesting. The non-profit organization of the University to which Doctor Thomson transferred the patent, the Wisconsin Alumni Research Foundation (WARF), filed three applications for patent coverage with the Patent Office, subsequently agreed upon, on three distinct features of a technique which allowed for the obtaining of embryonic stem cells. WARF, itself, granted Geron Corporation the right to develop stem cells of some cell lines, limiting the possibility, in the relevant agreement reached in 2002 after a complex negotiation, to those of the brain, heart and pancreas. The issue was raised as to researcher access to stem cells. The desire of WARF had always been to encourage such access, while Geron was evidently opposed to it. WARF undertook – through a 'Memorandum of understanding' signed in 2001 by its affiliate WiCell Research Institute with NIH (accessed 17 December 2014 at http://www.nih.gov/news/pr/sep2001/od-05.htm) – to make its own cells available for didactic or non-commercial research for a royalty fee under 5,000 dollars, while in regard to commercial products obtained through research a

condition that they no longer had the ability to develop into a defined human being. What is more, the guidelines stated that for every embryo donated, the informed consent of the donor had to be obtained, without any form of financial incentive. In view of the guidelines, then, stem cells could be extracted from dead embryos, but no specific indication for the determination of when an embryo may be considered dead was provided; that is when losing the ability to replicate becomes equivalent to death. Contrarily, it has been held that the extraction of stem cells from a living embryo may damage its development into a defined human being. Thus, the extraction of stem cells from living embryos has been considered unethical.[100]

In 2009, the Obama administration adopted an Executive Order[101] with which the limits on federal funding for embryonic stem cell research decided by President George W. Bush in August 2001 and subsequently repeated in an Executive Order of June 2007[102] were removed. In this way the limitation of research solely to those stem cell 'lines' in existence prior to August 2001, which had prohibited the allocation of public funds to create others, was overcome. On the basis of the new Executive Order, then, in 2009 again, the National Institute of Health adopted specific 'Guidelines on Human Stem Cell Research',[103] which establish the rules and procedures on the basis of which the NIH grants

---

subsequent agreement was provided on conditions which were not less favourable than those of other commercial licenses. The Geron, for its part, had freedom of action (and therefore to limit research) on its own cell lines. In the case of the agreement, see Brody, 'Intellectual Property and Biotechnology: The U.S. Internal Experience – Part II', cit., p. 122 et seq.

[100] President's Council on Bioethics, 'Monitoring Stem Cell Research', cit., p. 23 et seq.

[101] 'Executive Order 13505 – Removing Barriers to Responsible Scientific Research Involving Human Stem Cells', 9 March 2009, accessed 17 December 2014 at http://www.whitehouse.gov/the_press_office/Removing-Barriers-to-Responsible-Scientific-Research-Involving-Human-Stem-Cells.

[102] Respectively 'President Bush's Address to the Nation', 9 August 2001, and 'Executive Order 13435: Expanding Approved Stem Cell Lines in Ethically Responsible Ways', 22 June 2007, Federal Register, Vol. 72, No. 120, p. 34589, accessed 17 December 2014 at http://stemcells.nih.gov/policy/guidelines.asp.

[103] NIH, 'National Institutes of Health Guidelines on Human Stem Cell Research', July 2009, available at http://stemcells.nih.gov/policy/Pages/2009 guidelines.aspx. In particular, it must be shown that in the guidelines the NIH clarifies inspiring principles of its policy on the matter from the beginning, stating that: 'These guidelines are based on the following principles: 1. Responsible research with hESCs has the potential to improve our understanding of human health and illness and discover new ways to prevent and/or treat illness. 2.

funding for such research, and which, in particular, are meant to 'ensure that NIH-funded research in this area is ethically responsible, scientifically worthy, and conducted in accordance with applicable law'.

President Obama's Order, on the other hand, did not eliminate the so-called 'Dickey-Wicker Amendment', which bans the U.S. Department of Health and Human Services from using funds for the creation of human embryos for research purposes or for research in which human embryos are destroyed.[104]

In 2010, in the decision on the case *Shirley v. Sebelius*, the Court of Appeals for the District of Columbia Circuit revoked the injunction previously granted by Judge Lamberth on the basis of the consideration according to which the Guidelines on Human Stem Cell Research[105] 'clearly violate' the Dickey-Wicker amendment, stating that it may be held that the Dickey-Wicker amendment prohibits the use of federal funds for the 'direct' extraction of stem cells from an embryo which gives rise to its destruction; contrarily, according to the Court, the same amendment does not prohibit the funding of research projects which 'indirectly' use embryonic stem cells which have previously been extracted.[106]

With regard to the relevance of the regulations on federal funding for the discipline of the matter, what is more, it must be noted that research on stem cells, with or without public funding, may in any case continue. Where the government limits its funding on cells extracted from certain embryos, research on cells extracted from other embryos may continue with private funds. In this sense, it is said that the most appropriate solution with regard to moral issues brought up by research on human stem cells, rather than that of establishing criteria for the delivery of governmental funding, would be to develop *ad hoc* regulations aimed at

---

Individuals donating embryos for research purposes should do so freely, with voluntary and informed consent'.

104 The amendment, renewed annually since 1996, in the version in which Section 509 of the 'Omnibus Appropriations Act, 2009' (H.R. 1105, Pub.L. 111-8; March 11, 2009), prohibited the U.S. Department of Health and Human Services from using public federal funds for '(1) the creation of a human embryo or embryos for research purposes; or (2) research in which a human embryo or embryos are destroyed, discarded, or knowingly subjected to risk of injury or death greater than that allowed for research on fetuses in utero under 45 CFR 46.208(a)(2) and Section 498(b) of the Public Health Service Act [1](42 U.S.C. 289g(b)) (Title 42, Section 289g(b), United States Code)'.

105 In regard to which refer below, to para. (iii).

106 Concerning these profiles refer extensively above to Part II.

regulating such research, which would provide specific measures regarding ethical profiles.[107]

Again, on the theme of stem cell research, it must be added that embryos may also be 'produced' with congenital defects and insufficient genetic structures. In these cases, the embryo lacks the ability to develop into a defined human being. A similar method of producing defective embryos also raises significant moral issues. The deliberate production of defective, ill or disabled embryos, through the manipulation of the human genome and inserting destructive components, is ethically unacceptable and has been severely criticized.[108] The creation of defective embryos involves the donation of eggs by a woman and their artificial fertilization. There is the possibility of women being forced or obliged to donate eggs, involuntarily violating ethical barriers; this has occurred, for instance, in the case of a professor of the University of Pittsburgh who gathered eggs to be used in research without obtaining required consent, in violation of the principles of the NIH guidelines.[109]

In consideration of the growing controversy regarding stem cell research, various federal nations have adopted different legislation to regulate it. This legislation provides, in general, that stem cells may be extracted from embryos fertilized *in vitro* only prior to the 14th day from conception and prior to the embryo's implantation in the uterus of the mother. After the 14th day, regardless of the nature of fertilization, stem cells cannot be extracted. In contrast, some nations have completely banned *in vitro* fertilization for the creation of embryos from which stem cells will be extracted. Therefore, there is no uniformity in the views among nations with regard to stem cell research. Thus, it would seem appropriate to issue federal legislation on the matter, which allows the balancing of biotechnological developments and the moral issues connected to them.

---

[107] See N.S. Sreenivasulu and C.B. Raju, *Biotechnology and Patent Law. Patenting Living Beings*, cit., p. 213.

[108] See again President's Council on Bioethics, 'Monitoring Stem Cell Research', cit., p. 36 et seq.

[109] In this regard, see L. Kim, 'Explaining the Hwang Scandal: National Scientific Culture and its Global Relevance' (2008) 17 *Science as Culture* 397 et seq.

## III.2 BIOTECHNOLOGICAL INVENTIONS IN CASE LAW

### III.2.1 The Evolution of Patent Protection

In the American legal model, case law took the initiative in interpreting existing patent law in a liberal and innovative way, in order to offer protection to biotechnological inventions.

In the United States, legislative provisions on patent law are designed to be implemented through developments in case law, with a consequent 'proactive' dynamic between the legislature, case law and the patent office. The result is that the identification of patentable material comes about due to case law, and in this regard the Patent Office and the courts have adopted a broad approach. More than legislative provisions, the decisions of the PTO and the judges were the ones to lay the foundation for the evolution of biotechnological patents. The case law has not changed or altered existing provisions; rather it has adopted a liberal attitude in its interpretation: as a consequence, currently biological inventions are considered patentable subject matter.

The flexibility of this interpretation of the traditional criteria of patentability has made it possible for biotechnological inventions in the United States to receive greater protection than that which is provided in Europe.[110]

#### III.2.1.1 The origins of the path: Patentability of microorganisms

Microorganisms represent a fundamental tool for scientific research, and they play a key role, in particular, in the preparation of new medicines.

In 1873, for the first time the United States Patent Office granted a patent to Louis Pasteur on yeast which was resistant to organic germs of diseases.[111] On this occasion, the first in which a patent was granted for living matter, the invention was considered an article of manufacture pursuant to Section 101 U.S.C, and thus fell within the sphere of patentability, regardless of whether it consisted in a living process.

---

110 With regard to these characteristics in the patent system in general and of the biological patent discipline in particular in the American model, see, *ex multis*, R. Chisum, *Patents: A Treatise on the Law of Patentability, Validity and Infringement*, cit., *passim*; S.N. Sreenivasulu and C.B. Raju, *Biotechnology and Patent Law. Patenting Living Beings*, cit., p. 19 et seq.; B.A. Brody, 'Intellectual Property and Biotechnology: The U.S. Internal Experience – Part I', cit., p. 1 et seq.

111 US Patent No. 141,072.

Traditionally, living organisms were considered as 'products of nature', and as such were not patentable, given that they belonged to the collective and therefore were not subject to individual monopoly.[112]

Later, in the case of *Shell Development Co. v. Watson* the expression 'composition of matter' in Section 101 was interpreted by the District Court of the District Of Columbia as 'all compositions of two or more substances and all composite articles, whether they be results of the chemical union or of mechanical mixture or whether they be gases, fluids, powder or solids'.[113] On the basis of this interpretation, biotechnological inventions fall under the meaning of 'composition of matter'.

Biotechnological inventions involve the isolation of genetic material of a species and its incorporation into a designed species, or the isolation and purification of genetic material of different parts of the human or animal body. In this sense, biotechnological inventions consist in the composition of the genetic material of various species or parts of the body of a living organism.

Following the decision of the Court on the case *Shell v. Watson*, evidently aimed at a 'liberalization' of patent law in order to encourage new technology, Congress adopted the Patent Act of 1952, codifying pre-existing laws. The Act extended the reasoning according to which 'the ingenuity should receive a liberal encouragement', broadening the sphere of patentable material through a substitution of the term 'art' with 'process', and laying a critical foundation for the inclusion of biotechnological products and processes in the area of patentable subject matter.[114]

The discovery of DNA by Crick and Watson in 1953 initiated modern biotechnology, which since then has developed unceasingly, with continuously new possibilities for manipulation, identification and even production of living organisms.

In any case, until the subsequent decade, no particular developments with regard to the patentability of biotechnological patents were registered in the United States. Case law was not required to face issues relative to the patenting of living organisms.

---

[112] See, in particular, 1889 Comm'n 13 December 1889, *Ex parte Latimer*; 33 US 127 (1948), *Funk Brothers Seed Co. v. Kalo Inoculant Co.* A 'product of nature' cannot be patented, given that it does not fall under one of the four classes of statutory subject matter. On the other hand, patent protection may be granted for the process in which it is used (Title 35, Section 103 U.S.C.).

[113] 149 F. Supp. 279, 280 (D.C. 1957).

[114] See S.N. Sreenivasulu and C.B. Raju, *Biotechnology and Patent Law. Patenting Living Beings*, cit., p. 23.

In the years following 1970, instead, important decisions were taken regarding the patentability of living organisms, which stimulated the effective evolution of the biotechnological patent.

In the case of *In Re Bergy, Coats and Malik*,[115] the issue concerned a biologically pure and manmade culture, not existing in nature. The Board of Appeals of the Patent Office rejected the application on the basis of the fact that Section 101, in defining the categories of patentable subject matter, did not include living organisms. The United States Court of Customs and Patent Appeals[116] overturned the decision, stating the subject of the invention was not a 'product of nature', given that the culture in its pure form did not exist in nature, and had been produced solely in the context of carefully controlled laboratory conditions. What is more, the CCPA showed that the reference to a previous decision allowed for applications concerning chemical procedures involving living organisms, and the product under discussion in the application in this case, though living,[117] was '(a)n industrial product used in an industrial process, a useful or technological art if ever there was one'.

The decision was based on analogical logical-legal reasoning. Given that patents were granted for procedures including living organisms, by analogy they could not be rejected for products which included such organisms. The CCPA concluded that it was in the public's interest to include microorganisms in the terms 'manufacture' and 'composition of matter'.[118]

Subsequently, in the leading case of *Diamond, Commissioner of Patents and Trademarks v. Chakrabarty*,[119] the patent application concerned a genetically modified microorganism, obtained through genetic engineering techniques which unified within a single bacteria the operative abilities of four different strains of pre-existing bacteria,[120] giving it

---

[115] Decision of 6 October 1977, 195 USPQ 344.

[116] Hereafter also 'CCPA'.

[117] This was a fact which represented a 'decision without legal significance'.

[118] The pragmatic approach adopted by the courts allowed biotechnological inventions to receive patent protection. On the other hand, the effect of the decision was limited, given that the CCPA itself showed that it did not concern the patentability of living beings in general, unlike microorganisms.

[119] Decision of 16 June 1980, 206 USPQ 193.

[120] It did not involve, however, recombinant DNA techniques. For further analysis of the *Bergy* and *Chakrabarty* cases and their implications, see, *ex multis*, R.S. Eisenberg, 'The Story of Diamond v. Chakrabarty: Technological Change and the Subject Matter Boundaries of the Patent System', in R.C. Dreyfuss and J.C. Ginsburg (eds), *Intellectual Property Stories*, New York: Foundation Press, 2006, p. 327 et seq.; B.A. Brody, 'Intellectual Property and

the ability to dispose of oil leaks through the scission of hydrocarbon chains, making crude oil biodegradable in sea water. In 1972 Doctor Ananda Mohan Chakrabarty, an Indian microbiologist residing in the United States, working at General Electric, requested the patent both for the process for producing the bacteria and the bacteria itself, but the USPTO rejected the application for the latter, stating that it could not consider it a patentable invention given that it was (a) a 'product of nature'; and (b) a living being, despite it being new.[121] The Board of Patent Appeals and Interferences afterwards confirmed this decision. The CCPA, instead, nullified the decision of the USPTO.[122] Finally, in 1980 the Supreme Court, by a majority of five to four, confirmed the decision of the CCPA.[123]

The Court recognized that the Constitution had granted Congress extensive powers for the promotion of scientific progress, and the contents of the legislation on patents was such as to include living organisms under such protection. In particular, the Supreme Court put at the center of its evaluation the two concepts of 'manufacture' and 'composition of matter' as per Section 101 U.S.C., basing its interpretation of them respectively on the decisions of *American Fruit Growers*[124] and *Shell v. Watson*.[125] On the first point, it was believed that Doctor Chakrabarty had genetically modified the natural bacteria in order to give

---

Biotechnology: The U.S. Internal Experience – Part I', cit., p. 1 et seq. A. Lumelsky, '*Diamond v. Chakrabarty*: Gauging Congress's Response to Dynamic Statutory Interpretation by the Supreme Court' (2005) 39 *University of San Francisco Law Review* 641 et seq.; M.B. Tropper, 'Patentability of Genetically Engineered Life-Forms: Legal Issues and Solutions' (1991) 25 *John Marshall Law Review* 119 et seq.; R.A. Kulseth, 'Biotechnology and Animal Patents: When Someone Builds a Better Mouse' (1990) 32 *Arizona Law Review* 691 et seq.; P.B. Maggs, 'New Life for Patents: Chakrabarty and Rohm & Haas Co.' (1982) 14 *Intellectual Property Law Review* 285 et seq.; J. Carroll, 'Bergy, Flook, and Microorganisms as patentable Products' (1980) 12 *Intellectual Property Law Review* 261 et seq.; E.W. Guttag, 'The Patentability of Microorganisms: Statutory Subject Matter and Other Living Things' (1979) 13 *University of Richmond Law Review* 247 et seq.; W.F. Eberle, 'Bergy, Chakrabarty and Flook: is a "living" Article of Manufacture patentable Subject Matter under 35 USC § 101?' (1979) 11 *Intellectual Property Law Review* 381 et seq.

[121] See, *ex multis*, B.A. Brody, 'Intellectual Property and Biotechnology: The U.S. Internal Experience – Part I', cit., p. 1 et seq. (10 et seq.).

[122] See N. Wade, 'Supreme Court to Say if Life is Patentable' (1979) 206 *Science* 664.

[123] *Diamond v. Chakrabarty*, 447 US 303.

[124] 283 US 1.11.51 S.Ct, 328, 330, 75 L.Ed. 801 (1931).

[125] 149 F. Supp. 279, 280 (D.C. 1957), cit.

it the ability to dispose of crude oil, producing a result of human ingenuity and intellectual work deserving of incentive. On the second point, the Court observed that the bacteria under discussion presented a group of characteristics of various known species, reunited in a single species.

The Court rejected, then, the exceptions according to which the living organism was a product of nature and could not be patented, holding instead that the microorganism had been produced by man, in that the inventor had added his own ingenuity to the natural bacteria in order to render it capable of disposing of oil leaks with accuracy and speed. According to the Court's opinion, given that patents are necessary to guarantee inventive activity, essential for economic development, the exception of patentability applied to the discovery of natural products must be interpreted restrictively.[126]

Finally, the Court rejected the argument according to which microorganisms cannot be considered patentable up until the moment in which Congress expressly authorizes such protection. To the contrary, the Court showed that Congress could not predict the development of biotechnology when it approved Section 101 on patentable subject matter, and in any case in the outlining of the Section Congress had used broad and general language. Thus, in light of the constitutional objective to promote scientific progress and the reasoning of incentive for human ingenuity cited above, the Court held it opportune to confirm the inclusion of living organisms produced by man through biotechnology within the sphere of patentable subject matter.[127]

On the basis of the decision of the Supreme Court, then, a patent may be granted for any invention; whether or not the invention is a living being is without importance, as long as it has been created – that is isolated and purified – by man.[128] The criteria of distinction is 'not between living and inanimate things, but between products of nature, whether living or not, and human-made inventions'. In particular, according to the formula used by the Court and which since then has become

---

[126] See F. Vandenabeele, *Patentability of Living Organisms: Legal and Ethical Aspects of the Question*, cit., p. 33.

[127] See, among others, S.N. Sreenivasulu and C.B. Raju, *Biotechnology and Patent Law. Patenting Living Beings*, cit., p. 25.

[128] In this regard, among others, see L. Demaine and A. Fellmeth, 'Reinventing the Double Helix' (2002) 55 *Stanford Law Review* 303 et seq.; M. Davis, 'The Patenting of Products of Nature' in (1995) 21 *Rutgers Computer and Technology Law Journal* 293 et seq.

well known, 'anything under the sun that is made by man' constitutes a patentable invention.[129]

What is more, the Supreme Court, recognizing the possible repercussions of its decision, added that:

> We are told that genetic research and related technological developments may spread pollution and disease, that it may result in a loss of genetic diversity, and that its practice may tend to depreciate the value of human life. These arguments are forcefully, even passionately, presented; they remind us that, at times, human ingenuity seems unable to control fully the forces it creates – that, with Hamlet, it is sometimes better 'to bear those ills we have than fly to others that we know not of'.[130]

On the other hand, the Court declared no competence with regard to taking into consideration similar 'high policy' arguments regarding morality, claiming rather that its role was limited to 'the narrow one of determining what Congress meant by the words it used in the statute; once that is done, our powers are exhausted … (U)ntil Congress takes … action, this Court must construe the language of § 101 as it is'.[131]

The ruling in *Chakrabarty*, which for the first time directly considered the patent application for a new living being, created from an inventive activity of man through the use of biotechnological science, was later a fundamental reference point, not only in the United States but around the world.[132]

---

[129] See *Diamond v. Chakrabarty*, cit., 309.

[130] *Diamond v. Chakrabarty*, cit., 316.

[131] *Diamond v. Chakrabarty*, cit., 318. On these aspects of the ruling, see in particular S.R. Donnelly, 'The Patentability of Human Embryonic Stem Cells', cit., p. 124; M.A. Bagley, 'A Global Controversy: The Role of Morality in Biotechnology Patent Law', cit., p. 319 et seq.

[132] See, *ex multis*, C.A. Fowler, 'Ending Genetic Monopolies: How the TRIPs Agreement's Failure to Exclude Gene Patents Thwarts Innovation and Hurts Consumers Worldwide' (2010) 25 *American University International Law Review* 1073 et seq.; E.R. Gold and B.M. Knoppers (eds), *Biotechnology IP & Ethics*, Markham: Canada, LexisNexis, 2009, p. 22 et seq.; M. Rimmer, *Intellectual Property and Biotechnology: Biological Inventions*, cit., p. 24 et seq.; I.J. Churchill, 'Patenting Humanity: The Development of Property Rights in the Human Body and the Subsequent Evolution of Patentability of Living Things' (1994) 8 *Intellectual Property Journal* 249 et seq. Despite this, the concerns expressed, in particular, by various internal groups of the religious community of the United States itself; in this regard, see A.R. Chapman, 'The Ethics of Patenting Human Embryonic Stem Cells' cit., p. 261 et seq.(268), and A.R.

### III.2.1.2 Patentability of animals

The ability of animals to think and express themselves makes them different from inferior life forms (microorganisms and plants) and causes them to be categorized, together with humans, as higher life forms. The evolution of biotechnology patent law has, over time, come to include animals and related inventions within the sphere of patentability. Following the granting of patents regarding genetically modified plants,[133] the testing of the production of transgenic animals, new species of animals with genes coming from external sources, was begun.[134]

In 1987, on this subject, the fundamental decision on the case of *Ex parte Allen* was adopted. The case concerned an application for a patent on the animal, a lobster with a triple chromosomal structure which made it sterile and therefore edible all year, given that it did not lose body weight for reproduction during the breeding season. Those opposed to the application based their opposition on the criteria of the 'product of nature', claiming that the animal, being a divine creation, was in fact a product of nature and therefore could not be patented, in that it was the property of God and could not be owned by men. The USPTO rejected the patent application, on the basis of the fact that the subject of the

---

Chapman, *Unprecedented Choices: Religious Ethics at the Frontiers of Genetic Science*, Minneapolis (USA): Fortress Press, 1999, p. 125 et seq.

[133] The patentability of which was established five years after the decision of *Chakrabarty* by the Board of Patent Appeals and Interferences, which extended the possibility of obtaining a patent to the inventions related to multi-cellular plants not existing previously in nature and produced by man: ruling of *Ex Parte Hibberd*, 227 USPQ 443 (1985). The theme goes beyond the subject of this book, but is also the subject of extensive debate (see for instance, *ex multis*, C. Schneider, 'The Battle for Patent Rights in Plant Biotechnology: Evidence from Opposition Fillings' (2011) 36 *Journal of Technology Transfer* 565 et seq.; C.N. Stewart (ed.), *Plant Biotechnology and Genetics: Principles, Techniques, and Applications*, New York: Wiley & Sons, 2008; C.B. Onwuekwe, 'Plant Genetic Resources and the Associated Traditional Knowledge: Does the Distinction between Higher and Lower Life Forms Matter?', in J. Gibson (ed.), *Patenting Lives: Life Patents, Culture and Development*, Farnham (UK)/Burlington (USA): Ashgate Publishing, 2008, p. 139 et seq.; P. Christou and H. Klee (eds), *Handbook of Plant Biotechnology*, New York: Wiley, 2004; G. Van Overwalle, 'Patent Protection for Plants: A Comparison of American and European Approaches' (1999) 39 *Journal of Law & Technology* 143 et seq.).

[134] For a reconstructive framework on this theme, see D. Kevles, *A history of patenting life in the United States with comparative attention to Europe and Canada: a report to the European Group on Ethics in Science and New Technologies*, Luxembourg: Office for Official Publications of the European Communities, 2002.

invention was a living organism existing in nature and therefore not in line with the requisite of non-obviousness. On appeal, the Board of Patent Appeals and Interferences confirmed the decision, rejecting the application on the grounds that the information on fertility was so well known by experts in the sector as to make the invention void of the inventive step. In any case, the Board avoided stating that the lobsters were products of nature, as a consequence leaving the issue of the patentability of living organisms open. Contrarily, the Board accepted that the lobster referred to in the application was not a living organism produced by nature, further animating the debate on the patentability of animals.[135]

Meanwhile, patent applications with reference to animals, genes or animal tissues, and other inventions concerning animal genetic material, began to multiply. The biotechnology industry exerted considerable pressure to achieve the granting of patents for animals, in light of the enormous potential and utility of such inventions.

In this scenario, shortly after the ruling in *Ex parte Allen*, the Patent Office adopted a notice which detailed its position on the patentability of animals, stating that: 'The Patent and Trademark Office now considers non-naturally occurring non-human multi-cellular living organisms, including animals, to be patentable subject matter within the scope of 35 U.S.C. Section 101', excluding, on the contrary, human beings given that 'the grant of a limited, but exclusive, property right in a human being is prohibited by the Constitution'.[136] In the notice the constitutional provision of the ban was not specified, but subsequently was held to be based on the ban on slavery provided by the XIIIth Amendment to the U.S. Constitution.[137]

---

[135] Decision of 3 April 1987, 2 USPQ 2d 1425. In this regard, among others, see F. Vandenabeele, *Patentability of Living Organisms: Legal and Ethical Aspects of the Question*, cit., p. 34.; B.A. Brody, Intellectual Property and Biotechnology: The U.S. Internal Experience – Part I', cit., p. 13.

[136] '*Animals – Patentability*', 1077 Official Gazette USPTO 24, 21 April 1987. In doctrine, see H.T. Makley, 'Patentability of Animals in the United States' (1987) 20 *IIC* 372 et seq.; P.A. Rac, 'Patentability of Living Subject Matter' (1993) 10 *CIPR* 41 et seq.; J. Chambers, 'Patent Eligibility of Biotechnological Inventions in the US, Europe and Japan: How Much Patent Policy is Public Policy?' (2002) 34 *George Washington International Law Review* 223 et seq.

[137] See, among others, M.A. Bagley, 'A Global Controversy: The Role of Morality in Biotechnology Patent Law', cit., p. 324 et seq. (where the irrelevance of it is shown in cases of other morally controversial biotechnological inventions regarding humans); D.S. Bennett, 'Comment: Chimera and the Continuum of

By virtue of this position, then, in 1988 the USPTO granted the first patent for an animal, a genetically modified mouse, an 'Onco-Mouse', made to contract cancer through the incorporation of an extraneous gene within the mouse (through genetic engineering), an onco-gene, which made the mouse incredibly useful for experimentation on cancer. The Patent Office granted the patent to the University of Harvard, holding that the invention was a non-natural living organism produced by man.[138]

The decision, then, introduced animals and findings related to them to the sphere of patentable subject matter. From then on,[139] the granting of patents for genetically modified animals, from mice to livestock with milk rich in particular proteins, to pigs with low cholesterol meat, and so on was begun.

---

Humanity: Erasing the Line of Constitutional Personhood (2006) 55 *Emory Law Journal* 347 et seq. (esp. p. 386); T.A. Magnani, 'The Patentability of Human-Animal Chimera' (1999) 14 *Berkeley Technology Law Journal* 443 et seq. (448); R.H. Walker, 'Note: Patent Law – Should Genetically Modified Human Beings Be Patentable?' (1991) 22 *Memphis State University Law Review* 101 et seq. (esp. p. 110).

[138] In regard to the decision, see, in particular, R. Robins, 'Inventing Oncomice: Making Natural Animal, Research Tool and Invention Cohere' (2008) 4 *Genomics, Society and Policy* 21 et seq.; E.R. Gold and B.M. Knoppers, *Biotechnology IP & Ethics*, cit., p. 22 et seq.; C. Ezzell, 'First Ever Animal Patent Issued in United States' (1988) 332 *Nature* 668; K. Schneider, 'Harvard Gets Mouse Patent, A World First', *New York Times*, 13 April 1988, p. A1.

[139] Despite strong criticism generated by the new policy, in particular regarding the risk that such patents would encourage the development of transgenic animals, called into question the value and dignity of life, damaged traditional family farms and the environment and increased the suffering of animals. In regard to such criticism, among others, see B.A. Brody, 'Intellectual Property and Biotechnology: The U.S. Internal Experience – Part I', cit., p. 14 et seq.; U.S. Office of Technology Assessment, 'New Developments in Biotechnology – Patenting Life', Washington DC, U.S. Government Printing Office, 1989, accessed 17 December 2014 at ota.fas.org/reports/8924.pdf, p. 134 et seq., where, further, there is the declaration of the National Council of Churches of 1988 according to which: 'The ethical, environmental, socioeconomic and theological ramifications of genetic engineering and patenting of life are profound. They point to the probability that the integrity and future of creation will be placed in even greater jeopardy if our power over the genes of life is not exercised prudently and with reverence to help to restore the covenant: to heal the Earth and ourselves'; M.A. Bagley, 'A Global Controversy: The Role of Morality in Biotechnology Patent Law', cit., p. 322 et seq.; R. Dresser, 'Ethical and Legal Issues in Patenting New Animal Life' (1988) 28 *Jurimetrics Journal* 399 et seq.

### III.2.1.3 Patentability of genetic material and human cells

The new advanced biotechnology techniques, such as that of recombinant DNA, of linking or manipulating genes, have made the isolation and purification of cells, genes, gene sequences, human DNA and DNA sequences possible.

The evolution of patent law has reached a point, recently, of recognizing the patentability of human genetic material, such as cells, genes or DNA (with the exception of the aforementioned patentability of human beings).[140]

In this regard, research carried out by various research groups between the end of the 1960s and the beginning of the 1970s, aimed essentially at the creation of cells with hybrid genes, are particularly important. Among these, a team led by Doctor Stanley Cohen of the University of Stanford and Doctor Herbert Boyer of the University of California, San Francisco, managed to insert genes for kanamycin resistance into a plasmid, and introduced hybrid genes into Escherichia coli bacteria, which replicates and expresses the genes. The University of Stanford presented a patent application for the results of the research. Due to the doubts which existed regarding the patentability of living material, the application was divided into two parts: one regarding the process used in the research, the other regarding the transformed biological material. The patent for the process was granted in December 1980, that for the product in 1984.[141] In this regard, the University of Stanford adopted an open access approach to its findings, by requiring only that other non-profit organizations obtain a license for the use of the technology and by applying, in any case, inexpensive licensing fees.[142]

In the case of *John Moore*, the first patent application for a human cell line was registered. The cell line was from Mr. Moore's spleen, who

---

[140] In this regard, among others, see F. Vandenabeele, *Patentability of Living Organisms: Legal and Ethical Aspects of the Question*, cit., p. 35 et seq.; E.S. van de Graaf, *Patent Law and Modern Biotechnology*, Amsterdam: Gouda Quint, 1997, p. 119 et seq.

[141] In this regard, see B.A. Brody, 'Intellectual Property and Biotechnology: The U.S. Internal Experience – Part I', cit., p. 1 et seq. (3 et seq.); T. Beardsley, 'Cohen-Boyer Patent Finally Confirmed' (1984) 311 *Nature* 3; D. Dickson, 'Stanford Ready to Fight for Patent' (1981) 292 *Nature* 573 et seq.

[142] Concerning this, see A.R. Chapman, 'The Ethics of Patenting Human Embryonic Stem Cells', cit., p. 275; M.P. Feldman, A. Colaianni and C.K. Liu, 'Lessons from the Commercialization of the Cohen-Boyer Patents: The Stanford University Licensing Program', in A. Krattiger *et al.* (eds), *Intellectual Property Management in Health and Agricultural Innovation: A Handbook of Best Practices*, Oxford: MIHR/Davis (USA): PIPRA, 2007, p. 1797 et seq.

subjected himself to a therapeutic treatment for hairy cell leukemia, a terminal illness affecting the spleen. His doctors discovered that the patient's cell lines, taken in fact from his spleen, were useful for the production of anti-bacterial and anti-cancer proteins. In 1984, the USPTO granted a patent on the cell line, thus making human genetic material patentable subject matter. The decision was confirmed in 1990 by the Supreme Court of California, which stated that Mr. Moore did not have any right to a share of the profits earned from the marketing of inventions developed from 'discarded' parts of his body following therapeutic treatment.[143]

Given that ever more advanced research, which is producing the identification of genetic material with significant uses, is continuously being carried out, a 'race' has begun for the application of patents for specific genetic material such as DNA and genes.

In the case of *Amgen v. Chugai*,[144] the application concerned a sequence of isolated and purified DNA. The DNA codified a protein called erythropoietin, which accelerates the production of red blood cells. The invention was useful for people affected by anaemia with red blood cell deficiency. Opponents based their exceptions on the effective isolation, possession and conception of the invention. The Court concentrated its analysis on the verification of whether or not the inventors had really thought up the invention, defining the conception as the actual knowledge of the DNA sequence with its physical and chemical properties and the

---

[143] Supreme Court of California, *John Moore v. Regents of the University of California*, 9 July 1990 (51 Cal. 3d 120; 271 *Cal. Rptr.* 146; 793 P.2d 479). With regard to the decision, among others, see B.A. Brody, 'Intellectual Property and Biotechnology: The U.S. Internal Experience – Part I', cit., p. 15 et seq., who highlights in particular how the Court, on one hand, has stated that doctors may satisfy the requirement of informed consent, obtaining the consent of the patient in broad terms 'to any conceivable subsequent research use of excised cells'; on the other, that the patient is not the owner of any property right over the cells extracted from his body, in that this recognition could hinder the free circulation of biological material among researchers; see also S. Zullo, 'From the Patentability of Living Matter to the Ethics of Biotechnological Innovation: The Person-Body Relationship', in R. Bin, S. Lorenzon and N. Lucchi (eds), *Biotech Innovations and Fundamental Rights*, Milan: Springer Verlag Italia, 2012, p. 87 et seq. (92 et seq.).

[144] 927 F.2d 1200, 18 USPQ 2d 1016 (Fed. Cir. 1991).

consequent effective experimentation of the DNA itself.[145] Upon completion of the investigation, it was ascertained that the inventors did in fact create the DNA which was the subject of the application and had effectively experimented on it: therefore, the patent was granted.

Following the *Amgen* ruling, in the case of *In re Bell*[146] the applications concerned DNA and RNA which codify human insulin as a growth factor, which carry out a key role in the regulation of the growth of somatic cells for the management of growth hormones. Considering that in *Amgen* it had been decided that human genetic material was patentable, the opposition did not claim to be against this patenting, but rather concentrated on the fulfilment of the requirement of non-obviousness, and therefore on the inventive step of the invention. It concentrated, in particular, on the state of the art which suggested the general method for the isolation of genes and DNA. The Federal Circuit held that a general method for the isolation of genes or DNA would not have made a specific method of isolation of particular DNA which codified certain growth factors of human insulin obvious, and thus granted the patent.

The decision, then, stated that a general suggested method for the isolation of genes, DNA or other genetic material does not deny the inventor of a specific method the right to a related patent;[147] it in fact encourages biotechnologists to continue to develop new methods.

In the case of *In re Deuel*,[148] the application dealt with molecules of DNA and cDNA which codified growth factors linked to heparin, a protein which stimulates cell division and facilitates the reparation and replacement of damaged or diseased tissue. In this case as well those opposed to the patent based their argument on the state of the art, which suggested a general method for the isolation of genes and DNA. The Federal Circuit reiterated the same principles dictated in the case of *In re Bell*, granting the patent.

Subsequently, there was the case of the patent for the gene responsible for Canavan disease, a congenital neurological disease which causes the

---

[145] Given that biotechnological inventions require the conception of different genetic material or different methods for the isolation and use of it, the definition of conception is incredibly important.

[146] 991 F .2d 781 (Fed. Cir. 1993).

[147] A general method may not always be effective, or even if it is effective the result can be inaccurate. Contrarily, specific methods provide accurate results, in particular for the isolation of a specific gene or DNA, in consideration of the enormous number of genes and DNA in our bodies.

[148] 51 F.3d 1552 (Fed. Cir. 1995).

degeneration of the myelin sheath – isolating protection of the nerve cells of the brain – and death of patients in infancy. Research for the isolation of the gene had been conducted at the Miami Children's Hospital, whose research foundation had thus obtained a patent in 1997. The foundation, however, had then begun to request payment for the license, irritating the parents who had donated material which had made the research possible.[149]

In the case of patients affected by pseudoxanthoma elasticum, a genetic disease which causes the calcification of connective tissue, they formed a group, which established its own blood and tissue bank. Researchers wishing to use material in the bank had to consent to the sharing of ownership of any patent which may result from the research. The objective of the group was that the tests and treatments based on patents derived from research be accessible at a reasonable price. In this regard, it is necessary to highlight how similar agreements, on one hand, do not modify the principle of full ownership of the first inventor, which is modified only where the parties preventively agree to it; on the other, they demonstrate the possibility for an alternative which reconciles the interests of the various parties.[150]

Currently, then, the patentability of isolated and purified human genetic material represents a consolidated principle. The interpretation of patentable subject matter has accepted human genetic material within the sphere of patentability in an innovative and liberal way, granting relative protection for microscopic human genetic material such as genes, DNA and the like.[151]

*(i) Patentability of genetic sequences*

DNA, as seen previously, is composed by sequences. A gene is a portion of DNA, which expresses a certain protein. In turn, the gene is also composed by sequences. The isolation and division of DNA requires isolation and division of its various sequences, and the same concept applies to genes.

While, in the decade subsequent to 1980 and in the years immediately following 1990 there was the development of patents on DNA and genes,

---

[149] See J. Gillis, 'Gene Research Success Spurs Profit Debate', in *Washington Post*, 30 December 2000, p. A-01.

[150] See B.A. Brody, 'Intellectual Property and Biotechnology: The U.S. Internal Experience – Part I', cit., p. 19; P. Smaglik, 'Tissue Donors Use Their Influence in Deal over Gene Patent Terms' (2000) 407 *Nature*, 821.

[151] See S.N. Sreenivasulu and C.B. Raju, *Biotechnology and Patent Law. Patenting Living Beings*, cit., p. 30.

at the end of the 1990s[152] applications for DNA and gene sequences and fragments multiplied.

Expressed Sequence Tags (ESTs), as noted, are part of a sequence extracted from a molecule of cDNA, which can be used, for instance, for the discovery of new genes, the identification of their sequence and their position in the genome, and so on. The patenting of ESTs has been controversial in that it has been challenged by a variety of entities, such as the Human Genome Organization, according to which in the hypothesis that there be a lack in the inventive step, there is not the individuation of substantial and credible utility. Given the strong opposition, from the beginning, the USPTO had not decided to grant patents on ESTs. On the other hand, the main argument in support of patentability of ESTs is that they may be used to identify genetic material, in that they are useful for the individuation of an entire DNA or gene, given that in fact the isolation of every gene or DNA begins with the isolation of its sequences.[153]

In 1997 the Clinton administration announced that the Patent Office would begin to grant patents on ESTs based on the evaluation of their utility to the aforementioned purpose. Having received such input from the government, the USPTO began to take the patent applications for ESTs into consideration,[154] and in 1998 granted the first patent for an

---

152 After an initial attempt of the NIH in 1992, with negative results (see F. Vandenabeele, *Patentability of Living Organisms: Legal and Ethical Aspects of the Question*, cit., p. 36 et seq.; B.A. Brody, 'Intellectual Property and Biotechnology: The U.S. Internal Experience – Part II', cit., p. 106 et seq.; L. Roberts, 'Genome Patents Fight Erupts' (1991) 254 *Science* 184 et seq.).

153 In this regard, in addition to referencing that which is noted above, in Part II, see, among others, A. Oser, 'Patenting (Partial) Gene Sequences Taking Particular Account of the EST Issue' (1999) 30 *IIC* 1 et seq., as well as again, F. Vandenabeele, *Patentability of Living Organisms: Legal and Ethical Aspects of the Question*, cit., p. 36 et seq.

154 See B.A. Brody, 'Intellectual Property and Biotechnology: The U.S. Internal Experience – Part II', cit., p. 109. What is more, in doctrine, the risks of such an approach have been shown, in particular with respect to the possibility that such patents may lead to multiple applications for a single gene, with consequent significant hindering of further research. This criticism has been summarized in the concept of '*tragedy of anticommons*', which consists in the under use of a resource due to the fact that there are too many exclusive rights and the transaction costs for the purchase of the user rights are too high. In this regard, see in particular: M. Heller, 'The Tragedy of Anticommons' (1998) 111 *Harvard Law Review* 621 et seq.; M. Heller and R.S. Eisenberg, 'Can Patent Deter Innovation? The Anticommons in Biomedical Research' (1998) 280 *Science* 698 et seq., as well as, on the '*gridlock economy*' arising from it: M.

EST to the company Incyte Pharmaceuticals.[155] In the years to follow, the controversy on patents for sequences of DNA influenced many decisions of the CAFC regarding the exceptions of experimental use,[156] and convinced the Patent Office to modify its Utility Examination Guidelines substantially:[157] in particular, in 2001 the Patent Office explained its position stating that genetic inventions are patentable, in that and in the

---

Heller, *The Gridlock Economy. How Too Much Ownership Wrecks Markets, Stops Innovation, and Costs Lives*, New York, Basic Books, 2008, *passim* (on this theme refer also to Part V). In general, with regard to the 'enclosure' phenomena at work in the intellectual property sector – a concept which indicates the continued extension of the field of exclusive rights and the relative reduction of the sphere of access to knowledge and intangible resources – refer, among others, to G. Resta, 'The Case against the Privatization of Knowledge: Some Thoughts on the Myriad Genetics Controversy', in R. Bin, S. Lorenzon and N. Lucchi, *Biotech Innovations and Fundamental Rights*, cit., p. 11 et seq.; G. Ghidini, 'Introduzione' in G. Ghidini and A. Stazi (eds), *Accesso a informazione e conoscenza nell'era multimediale. Libertà di espressione, libertà di concorrenza, proprietà intellettuale*, Rome: Luiss University Press, 2011, p. 7 et seq., and G. Ghidini, *Innovation, Competition and Consumer Welfare in Intellectual Property Law*, Cheltenham, UK and Northampton, MA, USA, Edward Elgar, 2010, p. 19 et seq.; C. May, *The Global Political Economy of Intellectual Property Rights. The New Enclosures*, London/New York: Routledge, 2009 (2nd edn), p. 12 et seq.; P. Drahos, *The Global Governance of Knowledge. Patent Offices and their Clients*, Cambridge: Cambridge University Press, 2010, p. 3 et seq.; G. Krikorian and A. Kapczynsky (eds), *Access to Knowledge in the Age of Intellectual Pròperty*, New York: Zone Books, 2010, *passim*; J. Boyle, *Public Domain. Enclosing the Commons of the Mind*, New Haven (USA): Yale University Press, 2008, p. 42 et seq., and J. Boyle, 'The Second Enclosure Movement and the Construction of the Public Domain' (2003) 66 *Law and Contemporary Problems* 33 et seq.; A. Kapczynsky, 'The Access to Knowledge Mobilization and the New Politics of Intellectual Property' (2008) 117 *Yale Law Journal* 804 et seq.; M. Boldrin and D.K. Levine, *Against Intellectual Monopoly*, Cambridge: Cambridge University Press, 2008, p. 68 et seq.; N. Kranich, 'Countering the Enclosure: Reclaiming the Knowledge Commons', in C. Hess and E. Ostrom (eds), *Understanding Knowledge as a Commons. From Theory to Practice*, Cambridge (USA): The MIT Press, 2006, p. 85 et seq.; J.H. Reichman and P.F. Uhlir, 'A Contractually Reconstructed Research Commons for Scientific Data in a Highly Protectionist Intellectual Property Environment' (2003) 66 *Law and Contemporary Problems* 315 et seq.

155 US Patent No. 5,817,419, *Human Kinase Homologs*.

156 See, in particular: *Embrex Inc. v. Service Engineering Corp.*, 216 F.3d 1343 (2000); *Madey v. Duke University*, 307 F.3d 1351 (2002); *Integra Lifesciences Ltd. v. Merck KGaA*, 331 F.3d 860 (2003).

157 First in 1999 and then in 2001: *Utility Examination Guidelines*, 66 Federal Register 1092 (5 January 2001).

limits within which they are the result of human intervention such as to satisfy the 'made by man' standard established by the *Chakrabarty* decision, specifying that this can come about also (as demonstrated by a series of previous decisions in case law referred to within the guidelines) in the case of the isolation of a DNA sequence existing in nature.[158]

An international perspective on the approach of the theme arrived in 2000 through a comparative study of the so-called 'Trilateral Co-operation', a stable cooperative initiative initiated in 1983 between the EPO, the USPTO and the JPO (Japanese Patent Office). The study[159] highlighted that the three patent offices did not consider a sequence of genes patentable without an indication of their specific function or alleged utility. What is more, it was shown that any inventions relative to molecules of nucleic acid,[160] without an indication of the substantial and credible, specific function or utility, did not satisfy the requirements of industrial application (utility) or sufficient description.

In 2005, the Supreme Court intervened in the case of *Merck v. Integra*,[161] concerning the use of patented inventions in pre-clinical

---

[158] See C.M. Holman, 'Gene Patents Under Fire: Weighing the Costs and Benefits', in E. Arezzo and G. Ghidini (eds), *Biotechnology and Software Patent Law: A Comparative Review of New Developments*, Cheltenham, UK and Northampton, MA, USA, Edward Elgar, 2011, p. 260 et seq. (262 et seq.); Chapman, 'The Ethics of Patenting Human Embryonic Stem Cells', cit., p. 268 et seq. In this sense, the decisions of the Patent Office in general consider the sequences of DNA as chemical compositions (rather than as forms of life), patentable in that they are isolated and purified with respect to the naturally existing gene (see R.S. Eisenberg, 'Structure and Function in Gene Patenting' (1997) 15 *Nature Genetics* 125). A different possible framework on the theme suggests, instead, that in this case the subject of the patent is the knowledge of the DNA, a type of information, rather than the gene itself (see T. Peters, 'DNA and Dignity: A Response to Baruch Brody', in A.R. Chapman (ed.), *Perspectives on Genetic Patenting: Religion, Science, and Industry in Dialogue*, Washington DC: American Association for the Advancement of Science, 1999, p. 129).

[159] 'Trilateral Project B3b, Comparative study on biotechnology patent practices – Theme: Patentability of DNA fragments', accessed 17 December 2014 at http://www.trilateral.net/projects/biotechnology/patentability.pdf. In doctrine, see E.R. Gold and B.M. Knoppers, *Biotechnology IP & Ethics*, cit., p. 46; M.J. Howlett and A.F. Christie, 'An Analysis of the Approach to the European, Japanese and United States Patent Offices to Patenting Partial DNA Sequences (ESTs)' (2003) 34 *IIC* 581 et seq.

[160] Including entire fragments of DNA and SNP.

[161] *Merck KGaA v. Integra Lifesciences I, Ltd.*, 545 U.S. 193 (2005).

research,[162] adopting an important change of course on the issue of the patentability of research tools. In particular, the Court issued an unfavourable judgment with regard to the owner of a patent of a biological compound, and provided a broad interpretation of the legal exception regarding the purposes of research. Though the decision did not specifically face the theme of peculiarity of the inventions regarding research tools, it led to the reduction of the level of legal protection available to the owners of such patents.[163]

The restrictive tendency was later confirmed by the decision in *Fisher v. Lalgudi*,[164] in which the CAFC clarified the state of the discipline on the patentability of gene sequences endorsing the changes introduced to the utility guidelines of the Patent Office and adopting the 'substantial utility' test designed by the Supreme Court in the decision in *Brenner v. Manson*.[165] According to the principle stated by the Court in this ruling, in fact:

> The basic quid pro quo contemplated by the Constitution and the Congress for granting a patent monopoly is the benefit derived by the public from an invention with substantial utility. Unless and until a process is refined and developed to this point – where specific benefit exists in currently available form – there is insufficient justification for permitting an applicant to engross what may prove to be a broad field.
>
> On the basis of such developments, however, millions of patent applications were registered; as a consequence, the majority of human genes may (at least partially) already be the subjects of patents.[166]

---

[162] That is that it refers to the presence of a pathological condition prior to its being seen by the naked eye.

[163] E.R. Gold and B.M. Knoppers, *Biotechnology IP & Ethics*, cit., p. 44 et seq.; S. Rubin, 'Merck KGaA v. Integra Lifesciences I, Ltd.: Greater Research Protection for Drug Manufacturers' (2006) 1 *Duke Journal of Constitutional Law & Public Policy* 1 et seq.

[164] *In re Dane K. Fisher and Raghunath v. Lalgudi*, 421 F.3d 1365 (2005).

[165] 383 U.S. 519, 148 USPQ 689 (1966).

[166] See C.A. Fowler, 'Ending Genetic Monopolies: How the TRIPs Agreement's Failure to Exclude Gene Patents Thwarts Innovation and Hurts Consumers Worldwide', cit., p. 1084; L.G. Restaino, S.E. Halpern and E.L. Tang, 'Patenting DNA-related inventions in the European Union, United States and Japan: A trilateral approach or a study in contrast?' (2003) 2 *UCLA Journal of Law and Technology* 1 et seq. According to a 2005 study, already at least 20 per cent of genes of the human body known at that time were the subjects of patent applications; see K. Jensen and F. Murray, 'Intellectual Property Landscape of the Human Genome' (2005) 310 *Science* 239 et seq.; on the other hand, for further considerations on the actual consequences of this data, more or less

*(ii) … of isolation and incorporation processes and methods*

Not only human genetic material, but also the methods or procedures for its isolation are the subjects of patent applications. One may apply, then, for a patent for methods to create specific proteins *in vitro*, that is outside the body of the animal or man, through the incorporation of genes which codify them.

In order to produce certain codified proteins of a particular gene, the latter is isolated and inserted into the target organism, a bacterium for instance. As a consequence, the bacteria will produce, as a regular biological process, the protein for which it has been genetically modified. Similarly, the methods for producing antibodies[167] from the genetic material are also the subject of patent applications.

In the case of *Hybertech Inc. v. Monoclonal Antibodies Inc.*,[168] the application concerned an immunoassay method,[169] which used monoclonal antibodies to measure the concentration of (virus) antigens.[170] The objective was to develop a process to measure the quantity of certain antigens of a given solution. The invention consisted in a method for the identification of antigens using antibodies produced within the body, which would combat the antigens in a given solution. Upon conclusion of the investigation, the inventors obtained a patent, the first granted by the Patent Office for a method of use for antibodies produced by the immune system of the body to combat antigens.

This patent gave rise to a large number of applications regarding various methods of obtaining antibodies and using them against antigens.

---

exclusionary according to the specific case, see C.M. Holman, 'Gene Patents Under Fire: Weighing the Costs and Benefits', cit., p. 267 et seq., and C.M. Holman, 'The Impact of Human Gene Patents on Innovation and Access: A Survey of Human Gene Patent Litigation' (2007) 76 *UMKC Law Review* 295 et seq.

167 Antibodies are a class of protein produced by the immune system in order to combat against antigens, that is external bodies which cause illness, defending then the body against invaders such as viruses and bacteria. See, *ex multis*, D. Chisum *et al.*, *Principles of Patent Law: Cases and Materials*, cit., p. 292.

168 802 F.2d 1367 (Fed. Cir. 1986).

169 An immunoassay method identifies and measures antigens using antibodies as reactors.

170 See D. Chisum *et al.*, *Principles of Patent Law: Cases and Materials*, cit., p. 287 et seq.; B.C. Cannon, 'Toward a Clear Standard of Obviousness for Biotechnology Patents' (1994) 79 Cornell Law Review 735, accessed 5 January 2015 at http://scholarship.law.cornell.edu/cgi/viewcontent.cgi?article=2515&context=clr.

In *In re Wands*,[171] the application concerned the immunoassay which used monoclonal antibodies, high affinity immunoglobulins. The invention consisted in immunoassay methods for the identification of the superficial antigen of hepatitis B, in fact through the use of high affinity antibodies; that is, basically, it was a method to produce protein against a specific virus. The Federal Circuit, reforming the decision of the USPTO, based its decision on the reasoning of the decision in *Hybertech* and granted the patent, though further practical experimentation of the invention was still needed. This decision represents, then, another example of a patent for the methods of using antibodies against antigens.

In the case of *In re o Farrell*,[172] the invention dealt with a method to control the expression of a protein by cloned heterologous genes inserted into a transformed species of bacteria. The Federal Circuit agreed to grant the patent, in particular, on the basis of the consideration for which, though there was uncertainty regarding the results of the practical application of the invention, it met the requirement of non-obviousness, in that the state of the art provided detailed methodology which allowed for its use, and thus a stimulus for the experimentation of the invention and evidence which suggested a reasonable expectation of success.

In *In re Vaeck*,[173] the application concerned the manipulation of a bacterium in order to insert a gene. The inventors had isolated the gene and had combined it with a strain of DNA of the bacteria, creating in this way a chimerical gene. The patent was also granted in this case.[174]

Thus, the isolation of genetic material, which as noted is not easy, requires first the isolation of various fragments of DNA or of the gene. After isolation, the material must be purified and incorporated into a specific organism, in order to carry out certain tasks such as, for instance, the production of a particular protein. On the basis of recent decisions of the USPTO, biotechnological processes or methods for the isolation and incorporation of genetic material of one species to another constitutes patentable subject matter.

*(iii) ... of genetic treatments and tests*

Biotechnology has developed new methods for the diagnosis and treatment of genetic or hereditary diseases, transmitted from parent to offspring.

---

171 858 F.2d 731 (Fed. Cir. 1988).

172 853 F.2d 894 (Fed. Cir. 1988).

173 947 F.2d 488, 20 USPQ2d 1438 (Fed. Cir. 1991).

174 With regard to the case, see B.C. Cannon, 'Toward a Clear Standard of Obviousness for Biotechnology Patents', cit., p. 735 et seq.

Gene therapy consists in a biotechnological process which acts on such diseases suppressing or removing the gene which causes the disease from the body. Once the gene has been removed, the subsequent generation will not receive the gene from the parents, and thus the disease cannot be transmitted.

Similar processes are considered, today, patentable subject matter pursuant to Section 101 of the U.S. Code.

In particular, the first experiment of gene therapy in the United States approved by the Food and Drug Administration[175] was recorded in 1990,[176] and the Patent Office granted the first patent on gene therapy in 1995.[177] Since then the USPTO has granted dozens of patents for gene therapies.[178]

With reference to genetic tests, in 2009 opposition to patents on genes was manifested in an action brought by the American Civil Liberties Union and Public Patent Foundation,[179] in representation of a variety of interested individuals, which put into discussion the validity of a group of genetic patents regarding diagnostic tests for the predisposition to breast and ovarian cancer.[180] The respondents were the Patent Office, Myriad Genetics and the director of the University of Utah Research Foundation, co-owners of the patents. The subjects of these patents were methods and reactors used in genetic diagnostic examinations for the mutations in genes displaying the predisposition for breast and ovarian cancer BRCA1 and BRCA2, including the claims on isolated polynucleotides in the sequence corresponding to genes existing in nature. The mutations of

---

175 The federal agency which deals, also, with medicines.

176 When a little girl named Ashanti DeSilva was treated for serious immunodeficiency together with severe combined immunodeficiency from adenosine deaminase deficiency (ADA-SCID).

177 U.S. Patent No. 5,399,346 (1995), so-called 'Anderson Patent', from the name of the inventor. On this theme, see OECD, 'Competition Policy and Intellectual Property Rights', 209 (1998), accessed 17 December 2014 at http://www.oecd.org/dataoecd/34/57/1920398.pdf, 860; E.S. Grace, *Biotechnology Unzipped: Promises & Realities*, Washington DC: Joseph Henry Press, 1997, p. 213 et seq.

178 In this regard, see J. Chambers, 'Patent Eligibility of Biotechnological Inventions in the US, Europe and Japan: How Much Patent Policy is Public Policy?' cit., p. 223 et seq.

179 Hereafter 'ACLU'.

180 *Association of Molecular Biology et al. v. United States Patent and Trademark Office et al.*, Doc. No. 09-4515 (S.D.N.Y. 2009).

such genes are associated with a higher risk of developing breast or ovarian cancer.[181]

Myriad was at the center of the debate on the matter for many years, due to the diffused perception that the company had an excessively aggressive approach in protecting its patent rights.[182] The Patent Office was called into question given that it had granted the patents, and the ACLU claimed that the PTO was wrong not only in granting the patents in question, but more generally in its consolidated practice of allowing the patentability of genes.

The reasons for the action taken by the ACLU were based, first of all, on the consideration according to which Myriad protected its patents in a way which illegally hindered competition for the test based on the BRCA genes, a fact which had slowed the development of better versions of the tests, and restricted patients' access to them, especially uninsured

---

[181] At the beginning of the 1990s, on the basis of studies conducted on families in which young women had been affected by breast and/or ovarian tumors, the BRCA1 and BRCA2 (Breast Cancer 1 and 2) genes were identified. In ill women belonging to these families there was a hereditary mutation of BRCA1 or BRCA2, which was not visible in the healthy women. This led to believe that women who were carriers of the mutation of the BRCA genes and belonged to that type of family are at high risk for developing a breast or ovarian tumor. Such studies allowed for the creation of genetic tests, capable of identifying hereditary alterations of the aforementioned genes, for clinical purposes.

[182] In this regard, see T. Minssen and D. Nilsson, 'Standing on shaky ground: US patent-eligibility of isolated DNA and genetic diagnostics after AMP v USPTO – Part I' (2011) 1 *Queen Mary Journal of Intellectual Property* 223 et seq., *Part II* (2012) 2 *Queen Mary Journal of Intellectual Property* 136 et seq., *Part III* (2012) 2 *Queen Mary Journal of Intellectual Property* 225 et seq., and *Part IV* (2013) 3 *Queen Mary Journal of Intellectual Property* 118 et seq.; A. Lauer, 'The Disparate Effects of Gene Patents on Different Categories of Scientific Research' (2011) 25 *Harvard Journal of Law & Technology* 179 et seq.; A.S. Kesselheim and M.M. Mello, 'Gene Patenting – Is the Pendulum Swinging Back?' (2010) *New England Journal of Medicine* 1855 et seq.; R. Gold and J. Carbone, 'Myriad Genetics. In the Eye of the Policy Storm' (2010) 12 *Genetics in Medicine* 39 et seq.; C. Heaney *et al.*, 'The Perils of Taking Property Too Far' (2009) 1 *Stanford Journal of Law, Science and Policy* 46 et seq.; T. Caulfield *et al.*, 'Evidence and Anecdotes: An Analysis of Human Gene Patenting Controversies' (2006) 24 *Nature Biotechnology* 1091 et seq.; B. Williams-Jones, 'History of a Gene Patent: Tracing the Development and Application of Commercial BRCA Testing' (2002) 10 *Health Law Journal* 123 et seq.

women.[183] It was requested, then, to invalidate a group of specific patent claims regarding isolated DNA molecules which corresponded in structure to the natural types and mutated forms of the BRCA genes, as well as to methods of analyzing the BRCA genes of a patient to evaluate the mutations regarding the predisposition for cancer, and to the methods of using recombinant BRCA genes in analysis based on cells to identify potential anti-cancer medicines. In particular, it was stated that such claims were not subject to patenting pursuant to Section 101 U.S.C., and invalid for the violation of the constitutional principles of 'free speech' and the 'progress of science'.[184]

In March 2010, the District Court for the Southern District of New York adopted a decision with which it granted the ACLU a summary judgment and invalidated all the patents in question, for not falling within Section 101 U.S.C. and in particular with the rule of non-patentability of 'product(s) of nature' or 'abstract mental process(es)'. In the BRCAnalysis test, the action carried out is a reading of real data concerning the mutations of BRCA1 and BRCA2 and afterwards a comparison with other real reference data. Given that the initial material is not modified and new elements are not introduced, Judge Sweet concluded that the patents meant to protect isolated codifying DNA for sequences present in nature are not valid.[185]

It was pointed out by several parties that the reasoning at the basis of the decision, where confirmed definitively, would have undermined the

---

183 On this theme, see Secretary's Advisory Committee on Genetics, Health, and Society (U.S. Secretary of Health and Human Services), 'Revised Draft Report on Gene Patents and Licensing Practices and Their Impact on Patient Access to Genetic Tests', February 2010, accessed 17 December 2014 at http://oba.od.nih.gov/oba/SACGHS/SACGHS%20Patents%20Report%20Approved%202-5-20010.pdf.

184 In this regard, see C.M. Holman, 'Gene Patents Under Fire: Weighing the Costs and Benefits', cit., p. 265 et seq.

185 See in particular: A.S. Kesselheim and M.M. Mello, 'Gene Patenting – Is the Pendulum Swinging Back?', cit., where it is specified that: 'Judge Sweet also rejected Myriad's patent claims relating to the use of these DNA sequences in evaluating a patient's susceptibility to cancer. Building on the principles outlined in a 2008 federal appeals court decision, *In re Bilski*, he found these claims to be inappropriate subject matter for a process patent because they related to "abstract mental processes" of "analyzing" or "comparing" and did not involve a transformative step in which matter was turned into a different state or thing. That is, the patented method consisted only of comparing the nucleotide sequence in a test sample to the reference sequence'; J.D. Jackson, 'Something Like the Sun: Why Even «Isolated and Purified» Genes Are Still Products of Nature' (2011) 89 *Texas Law Review* 1453 et seq.

validity of a significant portion of patent portfolios of businesses working on biotechnological research.[186]

An appeal was brought against the decision before the CAFC. In October 2010, the Department of Justice expressed its opinion with an important amicus brief[187] on the matter.[188] There were two issues which were faced: if the molecules of manipulated DNA, like complementary DNA, can be the subject of a patent based on 35 U.S.C. § 101; and if isolated, but not modified, DNA can be patented in accordance with Section 101 itself.

Based on the note from the Department of Justice, the mere identification of DNA sequences within the genome 'that has merely been isolated from the human body, without further alteration or manipulation' should not be patentable. This is due to the fact that: 'The chemical structure of native human genes is a product of nature'. Contrarily, allowing for their patentability would be equivalent to granting a patent on 'cotton fibers that have been separated from cotton seeds or coal that has been extracted from the earth'. On the other hand, the patentability of DNA sequences which are manipulated in a laboratory, such as complementary DNA, was confirmed.

The opinion showed that the decision of the District Court represented a true change of direction with respect to the policy followed up to that moment, putting it in contrast with the practices of other public structures – from the USPTO to the National Institute of Health – which in recent years has obtained patents for the isolation of DNA.

On the other hand, the Department of Justice added that the impact on the biotechnology industry would not be so serious, given that the manipulation of DNA – such as that used to create transgenic products or particular gene therapies – may continue to be patented, in that it represents 'the synthetic results of scientists' manipulation of the natural

---

[186] See, among others, C.A. Holman, 'Gene Patents Under Fire: Weighing the Costs and Benefits', cit., p. 266, who showed also that, beyond the result of the appeal, the debate on the subject of genetic patents was destined to continue, and there was a strong push towards the imposition of at least some limits to such patents, by the Courts or Congress; the author himself, on the other hand, specified that some concerns at the basis of the action brought against Myriad do not have proof in certain empirical data (therein, p. 274 et seq.).

[187] Or rather a non-formally binding opinion.

[188] 'Brief for the United States as amicus curiae in support of neither party', No. 2010-1406, 29 October 2010.

laws of genetics … man-made compositions of matter that are eligible for patent protection'.[189]

In July 2011, the CAFC partially overturned the decision of the District Court, confirming it in part. The Federal Circuit reformed the previous decision stating that the claims covering molecules of isolated DNA are valid, in that they do not exist in nature. What is more, it denied that the claim on the method for monitoring potential anti-cancer treatments on the basis of the rates of cellular mutation may be traced to an unpatentable scientific principle. To the contrary, the Federal Circuit confirmed that the claims on diagnostic methods which are limited to merely comparing or analyzing DNA sequences, with no transformative step, are invalid.[190]

In December 2011, the ACLU presented an appeal to the Supreme Court, whose decision is clearly significant not only for the purposes of the case, but in general as a clarifying contribution of last resort on the

---

[189] For comments with regard to the evaluations expressed in the decision of the District Court and in the amicus brief of the DoJ, see C.S. Chuang and D.T. Lau, 'Patenting Human Genes; The Myriad Controversy' (2010) 32 *Clinical Therapeutics* 2054 et seq., where, in particular, the opportunity is shown to 'critically consider whether controlling an isolated gene sequence translates into de facto control of the genetic information encoded by that sequence'; J. Pila, 'Patents for Genes and Methods of Analysis and Comparison' (2010) 126 *Law Quarterly Review* 534 et seq.; M. Yoon, 'Gene Patenting Debate: The Meaning of Myriad' (2010) 9 *John Marshall Review of Intellectual Property Law* 953 et seq.; M. Cho, 'Patently Unpatentable: Implications of the *Myriad* Court Decision on Genetic Diagnostics' (2010) 28 *Trends in Biotechnology* 548 et seq.; R.A. Bohrer, 'Much Ado About Something' (2010) 29 *Biotechnology Law Report* 607 et seq.; A.J. Morrison, 'Rethinking the Gene Patent' (2010) 29 *Biotechnology Law Report* 609 et seq.; C.A. Fowler, 'Ending Genetic Monopolies: How the TRIPs Agreement's Failure to Exclude Gene Patents Thwarts Innovation and Hurts Consumers Worldwide', cit., p. 1087 et seq.; J. Paradise, 'European Opposition to Exclusive Control Over Predictive Breast Cancer Testing and the Inherent Implications for U.S. Patent Law and Public Policy: A Case Study of the Myriad Genetics' BRCA Patent Controversy' (2004) 59 *Food and Drug Law Journal* 133 et seq. (145 et seq.).

[190] In particular, the court explained that 'we reverse the district court's decision that Myriad's composition claims to "isolated" DNA molecules cover patent-ineligible products of nature under § 101 since the molecules as claimed do not exist in nature. We also reverse the district court's decision that Myriad's method claim to screening potential cancer therapeutics via changes in cell growth rates is directed to a patent-ineligible scientific principle. We, however, affirm the court's decision that Myriad's method claims directed to "comparing" or "analyzing" DNA sequences are patent ineligible; such claims include no transformative steps and cover only patent-ineligible abstract, mental steps'.

subject.[191] Any decision of the Supreme Court in line with the positions expressed by the District Court and by the Department of Justice, in fact, immediately appeared likely to lead to the assertion of the principle according to which genes in and of themselves would be excluded from patenting; as a consequence it is difficult to imagine how any test in which gene sequences appear could still be patentable, given that it does not involve any inventive activity. The businesses dealing with the discovery of genes and the development of genetic tests which compare DNA sequences could in any case put the latter on the market, having, however, to face competition much sooner than they might have in the past; in this way, there could be the positive effect of promoting the development of less expensive and therefore more accessible tests for patients. At the same time, having a lower number of patent claims at the level of basic science, unprotected sequences may freely be the subject of studies, in order to accelerate the evolution of knowledge and improve the prospects of interested businesses in creating treatments based on this information.[192]

In March 2012, the Supreme Court intervened through the granting of a certiorari and the return of the case to the Federal Circuit, in order that it should reconsider its decision in light of the aforementioned decision of the Supreme Court on the case of *Mayo v. Prometheus*,[193] This case, though it did not directly deal with the issues raised in Myriad, is

---

[191] In the meantime, Professor Steven Salzberg of the University of Maryland released online open source software which would be directly in competition with the BRACAnalysis method. The program is said to be able to identify 78 damaging mutations in the BRCA1 and BRCA2 genes in a few hours of calculations. The only data needed for it to work is the rough sequences generated by the sequencing of a human genome (for which the cost is always lower, tending to be around a thousand dollars; see G. Sinha, 'Academia's $1,000 genome' (2011) 29 *Nature Biotechnology* 960), without the need for assembly (see S.L. Salzberg and M. Pertea, 'Do-it-yourself Genetic Testing' (2010) *Genome Biology* 7 October, where in particular it is added that: 'If free software can be used to diagnose human genetic mutations, then individuals will be able to run their own tests in the privacy of their own homes. Fundamentally, this seems no different from measuring one's temperature or blood pressure, but because of gene patents, the act of reading one's own genome may require the permission of a private company').

[192] See T.L. Russell, 'Unlocking the Genome: The Legal Case Against Genetic Diagnostic Patents' (2012) 16 *Marquette Intellectual Property Law Review* 81 et seq.; A.S. Kesselheim and M.M. Mello, 'Gene Patenting – Is the Pendulum Swinging Back?', cit., p. 1858.

[193] Supreme Court, *Mayo Collaborative Services, DBA Mayo Medical Laboratories, et al. v. Prometheus Laboratories, Inc.*, 566 U.S. (2012).

interesting for the relative evaluations, in that a patent for a method aimed at the optimization of healing efficacy of a medicine for the treatment of a gastrointestinal illness was declared invalid, maintaining that the claim actually concerned the 'law of nature' underlying the method itself,[194] and appears then likely to have important implications for the patentability not only of isolated genes, but more generally of natural purified products and other innovations based on biological material existing in nature.[195]

In particular, as shown also in the subsequent amicus brief filed by the Department of Justice in June 2012,[196] the *Mayo* decision seemed to provide important indications regarding the central theme relative to the doubt as to whether the differences between isolated DNA and natural DNA are significant enough to make the former perceptibly different from the latter for the purposes of Section 101. In this sense, the Department of Justice showed that the central issue in the *Mayo* case was whether the claims limited the use of the underlying 'law of nature'; thus, also in *Myriad* there is the question of whether the claims concerning isolated molecules of DNA effectively limit third parties from research on and in the use of a 'product of nature'. Based on the opinion of the Department of Justice, the response to this question must be based on the consideration according to which 'patents on isolated but otherwise unmodified DNA would significantly impair the public's ability to study and make use of genomic DNA'. As a consequence, the Department of Justice specified that 'unlike cDNA, isolated but otherwise unmodified DNA molecules are not patent-eligible because they are "products of nature", not "human-made inventions". … Patent protection is not available to those who simply discover existing aspects of nature, even if

---

[194] In the Court's decision (adopted unanimously), in particular, it is stated that the method – related essentially to the measure of metabolite (produced by the process of metabolism) following the assumption of medicines – concerned in effect a 'well-understood, routine, conventional activity previously engaged in by scientists in the field'.

[195] Vice versa, the diagnostic methods of Myriad appear characterized by an even smaller incentive step than that of Prometheus (which, as previously said, the Supreme Court excluded from patentability), given that the former does not even have the characteristic (typical of the latter) to physically analyze a DNA molecule.

[196] 'Brief for the United States as amicus curiae in support of neither party', No. 2010-1406, 15 June 2012.

the discovery requires arduous work, represents keen scientific insight, and is of great value to society'.[197]

Following the path laid down by the *Mayo* decision, the Federal Circuit could have held that the codified information in DNA represents a 'law of nature', that DNA is a 'product of nature', and that the isolation of DNA consists in a process which is already well known and predictable at the moment of the application, for which isolated DNA would not be patentable given that the claims would concern in effect a 'law of nature' and a 'product of nature'.

In fact, beyond economic interests, *lato sensu* moral profiles relevant to the matter are also to be considered, that is of collective well-being, as well as the freedom of research and competition. In this regard, for instance, recent studies at Duke University showed that many hospitals cannot offer often fundamental, diagnostic services – such as that which identifies a predisposition or the presence of a serious pathology – given that otherwise they would have to answer to the businesses which hold the patents on the single genes of each specific case.[198] The refusal of Myriad to grant licenses for its tests, in particular, has produced prejudicial effects for patients and researchers. As a consequence of this, for the women for whom the test gave a positive result there was no possibility of turning to other facilities to receive a second opinion regarding the results. What is more, the exclusive expertise held by Myriad has excluded those with insufficient insurance coverage or those without any coverage from the test, or forced them to pay with their own

---

[197] Further, according to the reading of the DoJ, the *Mayo* decision worked a balance between the relevant interests on the matter, holding that the need for incentive for the development of genetic innovations through the granting of patents on isolated DNA is subordinate to the limits of patentability established within Section 101: 'The potential incentive effects … do not alter the boundaries the Supreme Court has set between unpatentable products of nature and patentable creations of man'.

[198] See Institute for Genome Sciences & Policy – Duke University, *Patently Complicated: Case Studies on the Impact of Patenting and Licensing on Clinical Access to Genetic Testing in the United States*, in *Genetics in Medicine*, 2010 (Suppl.), p. 1 et seq.; on this theme, see also R. Cook-Deegan, 'Gene Patents: What do they do? What policy options have been proposed? What are the implications of the Myriad ruling?' (Institute for Genome Sciences & Policy – Duke University, Issue Brief for Congressional Staff), September 2011, accessed 17 December 2014 at http://www.genome.duke.edu/centers/cpg/Myriad/documents/Gene%20Patents%20Issue%20Brief%2015.Sept.2011.pdf; J.M. Mueller, 'Facilitating Patient Access to Patent-Protected Genetic Testing' (2011) 6 *Journal of Business & Technology Law* 81 et seq.

savings. The exclusive right, finally, has limited the ability of university centers and research laboratories to carry out research in order to develop new, more inclusive and possibly more accurate tests.[199]

However, in August 2012, the Federal Circuit Court of Appeals confirmed the validity of Myriad's patents on 'isolated' genes related to breast and ovarian cancer.[200]

Thus, in June 2013 the Supreme Court confirmed that Myriad's DNA claim falls within the law of nature exception, and so is not patentable.[201] According to the Court's decision, in fact, Myriad did not create or alter either the genetic information encoded in the BCRA1 and BCRA2 genes or the genetic structure of the DNA. It found important and useful genes,[202] but innovative or brilliant discovery alone does not satisfy the U.S.C. § 101 requirement.

The Court specified, then, that the decision does not include, on one hand, cDNA, which is not a 'product of nature'. So, it is eligible for a patent under § 101, as its creation results in a not naturally occurring

---

[199] In regard to these profiles, see, in particular, C.A. Fowler, 'Ending Genetic Monopolies: How the TRIPs Agreement's Failure to Exclude Gene Patents Thwarts Innovation and Hurts Consumers Worldwide', cit., p. 1073 et seq. (1091 and 1097 et seq.).

[200] More specifically, the Federal Circuit upheld Myriad's composition claims to isolated DNA (reversing the District court), invalidated Myriad's method claims for comparing or analyzing gene sequences (affirming the District court), and upheld Myriad's method claims for screening potential cancer therapeutics (reversing the District court); see M. Textor, 'Gene Patents at Home and Abroad: Should the WTO Take Action in Light of Myriad?', accessed 17 December 2014 at http://ssrn.com/abstract=2282676, May 2013, p. 9.

[201] In this sense, the Court recalled the wording used lastly in *Mayo* decision, according to which: 'laws of nature, natural phenomena, and abstract ideas ... "are basic tools of scientific and technological work"' that lie beyond the domain of patent protection. As the Court had held there, in fact, without this exception, there would be considerable danger that the grant of patents would 'tie up' the use of such tools and thereby 'inhibit future innovation premised upon them'; this consequence, evidently, would be at odds with the very point of patents, which, as known, exist to promote creation. In this respect, it is worth noting how a recent study affirmed that sequence patents would already cover the entire human genome; see J. Rosenfeld and C.E. Mason, 'Pervasive Sequence Patents Cover the Entire Human Genome' (2013) 5 *Genome Medicine* 27 et seq.

[202] In fact, Myriad's principal contribution was uncovering the precise location and genetic sequence of the BRCA1 and BRCA2 genes within chromosomes 17 and 13.

exon-only molecule.[203] On the other hand, the decision itself does not cover method claims, patents on new applications of knowledge about the BRCA1 and BRCA2 genes, or the patentability of DNA in which the order of the naturally occurring nucleotides has been altered.[204]

Therefore, every patent drafted similarly to Myriad's broadest claim – an isolated DNA that codes for a specific protein – is now invalid. Vice versa, claims related to cDNA versions of genes continue to pass the threshold test, though they are still subject to scrutiny under all the other patentability requirements.[205]

---

[203] This distinction has its roots mainly in the decision of the case *Amgen v. Chugai* (927 F.2d 1200, 18 USPQ 2d 1016, 1991), where the Federal Circuit focused the subject matter of the claim on a purified and isolated DNA sequence encoding human erythropoietin, interpreting the term 'purified' as meaning essentially only the coding regions, that is, only the novel purified and isolated sequence which coded for EPO (see also for the other relevant precedents, M. Textor, 'Gene Patents at Home and Abroad: Should the WTO Take Action in Light of Myriad?', cit., pp. 23–6; D.M. Gitter, 'International Conflicts over Patenting Human DNA Sequences in the United States and the European Union: An Argument for Compulsory Licensing and a Fair-Use Exemption' (2001) 76 *New York University Law Review* 1623 et seq.

[204] Supreme Court, *Association for Molecular Pathology et al. v. Myriad Genetics Inc. et al.*, 569 U.S. (2013).

[205] As far as concerns Myriad, in particular, about three-quarters of its BRCA-related patents have not been invalidated, including claims to cDNA and some of its methods. Moreover, the invalidated patents would have begun to expire in the next year. Last but not least, Myriad's most valuable asset may be the proprietary database it has built up through its testing monopoly: this database consists of test results, i.e. DNA sequences, and associated health outcomes, and gives Myriad a relevant advantage in interpreting BRCA gene mutations, especially the lesser-known one (see J. Conley, '*Myriad*, Finally: Supreme Court Surprises by not Surprising' (2013) *Genomics Law Report*, accessed 17 December 2014 at http://www.genomicslawreport.com/index.php/2013/06/18/myriad-finally-supreme-court-surprises-by-not-surprising, 18 June 2013, p. 1; and, also for an interesting analysis of the previous European decisions and related developments concerning Myriad's patents, G. Matthijs, I. Huys, G. Van Overwalle and D. Stoppa-Lyonnet, 'The European BRCA Patent Oppositions and Appeals: Coloring Inside the Lines' (2013) 31 *Nature Biotechnology* 704 et seq., especially 709). In this respect, then, a new episode of the 'Myriad saga' before antitrust authorities could be projected (regarding the profiles involved in the relationship between access to databases and antitrust, and the relative case law, see among others, D. Lim Tze Wei, 'Regulating Access to Databases Through Antitrust Law: A Missing Perspective in the Database Debate' (2006) *Stanford Technology Law Review* 7 et seq.; E. Derclaye, 'The IMS Health Decision: A Triple Victory' (2004) 27 *World Competition* 397 et

## III.2.2 Bioethical Profiles of Protection

### III.2.2.1 Bioethics in the patentability of microorganisms

The Supreme Court's decision on *Diamond v. Chakrabarty*[206] raised moral issues regarding the patentability of microorganisms for the first time.

The application concerned, as was seen, a bacterium capable of disposing of oil leaks. Following the initial rejection of the application, the Supreme Court considered the invention not natural and produced by man, holding that the patent could be granted through a liberal interpretation of the language of the Patent Act in force at the time.[207]

The entire world was surprised by the decision and severe criticism was given on the appropriateness of the decision with respect to moral and bioethical standards. A lively worldwide debate was sparked, and some commentators stated that the patentability of living organisms went against those standards.

Following the decision in the *Chakrabarty* case, however, a significant number of applications for patents regarding various unnatural and manmade life forms were filed, highlighting the benefits of such inventions and minimizing the moral implications.

The USPTO, following the path beaten by the Supreme Court, granted a large number of patents over various microorganisms, feeding the fire of debate on morality regarding the patentability of life.[208]

### III.2.2.2 Bioethics in the patentability of animals

Animals are considered superior life forms due to the fact that they can suffer and express feelings, unlike inferior life forms such as plants or other living beings such as microorganisms. Thus, the moral concerns which arise for animals are more important than those relative to other such life forms.

---

seq.; D.M. Gitter, 'Strong Medicine for Competition Ills: The Judgment of the European Court of Justice in the IMS Health Action and Its Implications for Microsoft Corporation' (2004) 15 *Duke Journal of Comparative & International Law* 153 et seq.).

[206] 447 US 303 (1980).

[207] See D. Chisum *et al.*, *Principles of Patent Law: Cases and Materials*, cit., p. 287 et seq.; J. Chambers, 'Patent Eligibility of Biotechnological Inventions in the US, Europe and Japan: How Much Patent Policy is Public Policy?', cit., p. 223 et seq.

[208] See S.N. Sreenivasulu and C.B. Raju, *Biotechnology and Patent Law. Patenting Living Beings*, cit., p. 200.

In *Ex parte Allen*,[209] as was said, the patent application concerned a genetically modified lobster which held a triple chromosomal structure, which made it sterile and therefore edible all year long. The application was rejected based on the fact of the requirement of non-obviousness, but the decision, which did not deny patentability based on the principle of the animal, opened the way for such a development.

A short time later the Patent Office decision, which held that ethics could not hinder the patentability of animals, published a notice in which it expressed its opinion on this theme, stating that from that moment on multi-cellular, non-human, living organisms, including animals were considered patentable.[210]

The position taken implied that the USPTO had decided that in regard to the patentability of animals, moral considerations could be put aside.[211] On the other hand, the notice gave rise to vehement opposition, in particular on the part of associations and groups in defense of animal and farmer's rights, which brought forth legal actions protesting the power of the Patent Office to allow the patentability of animals. They complained that the issue was of public interest and that this could not be denied by minimizing moral concerns. The genetic manipulation of animals, making them predisposed to illness and causing their suffering, was held by those groups and associations as contrary to morality. On the contrary, the scientific community stated that the potential benefits from genetic manipulation for society had to be considered as more important than moral concerns.

The following year, the USPTO granted the first patent concerning animals, on a transgenic mouse useful for cancer research. The so-called 'Harvard mouse' contained a gene which caused cancer, inserted through genetic engineering, which made it predisposed to cancer. The patent received significant criticism, in that it was considered immoral. Those opposed to it stated that the patenting of an animal capable of expressing itself and showing feeling, was in and of itself contrary to morality, and could lead, in the future, to the cloning and patenting of human beings.[212]

---

[209] 1987 USPQ 2d 1425.

[210] '*Animals – Patentability*', 21 April 1987, cit.

[211] See Rac, 'Patentability of Living Subject Matter', cit., p. 41 et seq.

[212] In this regard, besides referring above, see in particular, E.R. Gold and B.M. Knoppers, *Biotechnology IP & Ethics*, cit., p. 22 et seq.; L.G. Restaino, S.E. Halpern and E.L. Tang, 'Patenting DNA-related inventions in the European Union, United States and Japan: A trilateral approach or a study in contrast?, cit., p. 1 et seq.

In fact, the moral issues raised in opposition to the patentability of animals are significant, in comparison with those for plants and other inferior life forms.[213] The Patent Office, in any case, held that the benefits of the invention had to be considered more important than moral issues. In particular, in the aforementioned case, it was considered that cancer is one of the most diffused illnesses in the world, from which millions of people suffer. Given that the invention was an aid in the experimentation on cancer, it represented a chance, and therefore its potential benefits were worth the granting of the exclusive patent.

### III.2.2.3 Bioethics in the patentability of genetic material and human cells

Biotechnology is also capable of manipulating human beings, isolating, removing, suppressing and incorporating genes from and towards a human body. Further, it has made the identification, isolation and commercial production of proteins and hormones which the human body produces naturally possible.

It has been said that the alteration of the natural structure of the human body is against its integrity. Bioethicists believe that the patentability of human genetic material such as cells, DNA and genes, gives rise to the exercise of a property right over the human body or life, similar to slavery, which violates human dignity.

Debates on the morality of manipulation of the human body and its patentability began the moment patents were granted for living beings, given that the developments for man were already predictable at that time.

As soon as it was possible to manipulate the human body through the isolation or the removal of genetic material (such as genes and DNA), or the incorporation of foreign genetic material into the body, there were patent applications for human genetic material, giving rise to extremely lively debates on the morality of manipulating the human body and the ownership of human genetic material.

In the *John Moore* case,[214] for the first time a patent was requested for a human cell line, which had been subject to medical treatment for hairy cell leukemia, a terminal illness which affects the spleen. The patient's

---

[213] As such, *ex multis*, see extensively: R. Robins, 'Inventing Oncomice: Making Natural Animal, Research Tool and Invention Cohere', cit., p. 21 et seq.; M.A. Bagley, 'A Global Controversy: The Role of Morality in Biotechnology Patent Law', cit., p. 323 et seq.; J. Watal, *Intellectual Property Rights in the WTO and Developing Countries*, Oxford: Oxford University Press, 2002, p. 164.

[214] Supreme Court of California, 9 July 1990, cit.

doctors discovered that his cell lines were useful in the preparation of an antibacterial and anti-cancer protein, and obtained a patent for them in 1984. Mr. Moore opposed the patenting of his cell line (most recently) before the Supreme Court of California. He stated that the cells came from his body, that they were his property and therefore he alone could claim rights. On the other hand, independently from the opinion of Mr. Moore, issues were raised regarding the morality of granting patents on a cell line which was part of a human body. In particular, it was observed that the possession of a part of a human body through a patent gave rise to a clear violation of human dignity and affected the moral roots of humanity. The court rejected the appeal and held that no property right could be recognized for one's own body, in that it would be contrary to human dignity. In any case, facing the moral situations involved in patenting a human cell line, the court considered that the cell lines were not immediately available for isolation, but rather isolation took considerable effort. In this perspective, it was taken into consideration that the patent was not requested for the natural cell lines of the body, but on those which had been isolated outside of it. Thus, such isolated cell lines, resulting from laborious research, were worthy of a patent. Moreover, it was held that the use of the invention could go beyond ethical concerns relative to it. In conclusion, the court confirmed the patentability of human cells, genes and DNA in their isolated form outside the body, so putting ethical issues aside.[215]

However, after the decision of the Supreme Court of California, the debate on the matter continued and further intensified. In 1993, a new patent application was filed with the Patent Office for human cell lines.[216] The cell line in this case was developed from the blood of a female patient of the Guarani Indian tribe of Panama, because it had been discovered that it was useful in AIDS and cancer research. This time the patent application was not from a single individual, but from the government of the United States. In this way, it was clear that the American government itself was in favour and interested in the patenting of human cell lines and genetic material, putting aside the moral issues it involved.

There were fervent protests at an international level and before the USPTO against the moral implications of granting such a patent.

---

[215] See S.N. Sreenivasulu and C.B. Raju, *Biotechnology and Patent Law. Patenting Living Beings*, cit., p. 203; B.A. Brody, 'Intellectual Property and Biotechnology: The U.S. Internal Experience – Part I', cit., p. 15 et seq.

[216] Patent application No. 9208784 A 1, 1993.

Non-governmental organizations, such as the Rural Advancement Foundation International and tribal communities forcefully objected, bringing under discussion the ethics of the patentability of a cell of a woman who belonged to a tribe.[217] The American government stated that the patent had to be granted, by virtue of the fact that the cell line was useful in research against the two most dangerous illnesses in the world. Given that the government itself was involved in the application, on the other hand, those in opposition criticized its position in requesting the patent on a cell line which would have led to the commoditization of a human life. Finally, in consideration of the vehement opposition and international criticism, the American government in the end decided to withdrawal the patent application.

In the years since 1990, however, significant progress was made in research on human genetic material in the United States, which gave rise to a large number of patents over such materials, in particular for DNA and genes. Once the moral issues were overcome and patents on human genetic material and cells had been granted, the USPTO received a large number of applications regarding similar patents; to recall a few, think for instance of the cases of *Amgen v. Chugai*, *In re Bell*, *In re Deuel* and *In re o Farrell*,[218] the first three regarding[219] DNA molecules and the last a method for the production of a protein in a transformed species of bacteria.

Starting from that point, then, patent protection was obtained for a significant number of inventions related to genes, DNA and other human genetic material.

*(i) The issue of embryonic stem cells*

With regard to research on human embryos,[220] in January 2011 a decision of particular significance was issued:[221] the Court of Appeals for the Fourth Circuit decided that the frozen embryo does not have the legal standing to assert constitutional rights. The request for a writ of certiorari was presented by the National Association for the Advancement of

---

217 See B.K. Dhadda, 'Patenting Human Genetic Information. Is Nothing Sacred?' (1998) L. Bently and S.M. Maniatis (eds), *Intellectual Property and Ethics*, London: Sweet and Maxwell, 1998, p. 89 et seq.

218 All cited previously.

219 As seen above.

220 With regard to the so-called statute of the embryo, refer above and in particular to Part II.

221 *Mary Scott Doe, et al., v. Barack H. Obama, President of the United States, et al.*, 21 January 2011, 180 L. Ed. 2d 226, 131 S.Ct. 2938.

Preborn Children and others, in order to block federal funding of research on stem cells obtained from human embryos. The plaintiff was a frozen embryo named 'Mary Scott Doe', as a representative of the entire category of such embryos.

The issue presented by the appellants was whether the District Court had made a mistake in holding that Mary Scott Doe and all other frozen embryos were people, entitled to rights and protection under the XIVth Amendment to the U.S. Constitution. The implications of the decision concerned the constitutionality of the ruling on the case *Roe v. Wade*[222] and every other decision which denied the personality of the embryo, together with rights and protection guaranteed by the constitution, including the right to appear in court. The Court of Appeals, in particular, held that: 'We cannot identify a particularized harm because the complaint does not identify any of the named plaintiff's particularized characteristics. Instead, it leaves us only with questions such as whether the embryo will ever be used for research and whether that research will be funded (by National Institute of Health)'. What is more, the court added that the potential adoptive parents could not prove that all the frozen embryos would be threatened by research on stem cells.[223] Finally, in order to clarify that the decision does not provide any limiting principle for future cases, the court explained that 'we do not suggest that no party would ever have standing to assert similar claims. … A complaint that provided more concrete information about the identity of the named plaintiff embryo or the plaintiff parents' plans for adoption would at least address more directly what the Supreme Court has identified as serious constitutional concerns'.[224]

On the subject of the patentability of embryonic stem cells, the result is particularly interesting in regard to the case of patents filed since 1998 and then obtained by WARF for a technique which allowed the obtainment of embryonic stem cells.[225] In 2006, two foundations – the Foundation for Taxpayer and Consumer Right and the Public Patent Foundation – presented an appeal to the PTO for the patents to be

---

[222] Supreme Court, *Roe v. Wade*, 22 January 1973, 410 U.S. 113, in regard to which refer above, as well as to Part II.

[223] This, showing that: 'Where government policy not only allows the biological parents to choose what to do with their embryos, but also safeguards the independence of their decision with strict conditions, the connection between injury and policy is a "purely speculative" one'.

[224] With regard to this decision, see 'Frozen Embryos Do Not Have Standing to Sue' (2011) 30 *Biotechnology Law Report* 338 et seq.

[225] As recalled above.

revoked. The two foundations held that the patents did not meet the criteria of non-obviousness, in that the fundamental elements of the invention had already been published several years before.[226]

The case required two years of consultation and the examination of a long series of documents presented by the appellants and by WARF. In a preliminary decision of 2007, the Patent Office declared the patents invalid, accepting the reconstruction of the appellants.[227] Subsequently, WARF reduced its claims, in an attempt to reinforce its position, and at the same time granted public research institutions the possibility of using the technique without the need to pay any royalty (while this remains a requirement for private facilities and pharmaceutical companies). In March 2008, evidently in light of their new 'perimeter', outlined by the changes introduced, the Patent Office confirmed the validity of the patents. In April 2010, finally, the Board of Patent Appeals and Interferences of the PTO declared WARF's patents invalid: the reasoning provided was not moral in nature, but rather centered on the application of the criteria of novelty and non-obviousness, in that the claim on patents had already been the subject of a patent in 1992.[228]

---

[226] In particular, they held that the method of use was that, going back to 1981, used to gather and cultivate embryonic stem cells of mice. See A.R. Chapman, 'The Ethics of Patenting Human Embryonic Stem Cells', cit., p. 275; C. Holden, 'Prominent Researchers Join the Attack on Stem Cell Patents (2007) 317 *Science* 187.

[227] See F. Murray, 'The Stem-Cell Market: Patents and the Pursuit of Scientific Progress' (2007) 356 *New England Journal of Medicine* 2341 et seq.

[228] With regard to the WARF case, see S.R. Donnelly, 'The Patentability of Human Embryonic Stem Cells', cit., p. 126; J.M. Golden, 'WARF's Stem Cell Patents and Tensions between Public and Private Sector Approaches to Research' (2010) 38 *Journal of Law, Medicine and Ethics* 314 et seq.; A.R. Chapman, 'The Ethics of Patenting Human Embryonic Stem Cells', cit., p. 275 et seq., who, in particular, highlights the aggressive approach adopted by WARF in protecting its own patents, which cover the main aspects of the activity of human embryonic stem cells; C. Holden, 'Wisconsin Stem Cell Patents Upheld' (2008) 319 *Science* 1602 et seq.; WARF News, 'U.S. Patent Office Issues Certificates to Uphold WARF Stem Cell Patents', 26 June 2008, accessed 17 December 2014 at http://www.warf.org/news/news.jsp?news_id=234, and 'WARF Submits Additional Language to Protect Key Stem Cell Patents', 2 October 2007, accessed 17 December 2014 at http://www.warf.org/news/news.jsp?news_id=216. The restrictive effect of the cited agreement between WARF and Geron was criticized by a part of doctrine, which showed how stem cells are essential elements for research, whose use should be allowed in a non-exclusive way in order to promote future research (A.K. Rai and R.S. Eisenberg, 'The Public Domain: Bayh-Dole Reform and the Progress of Biomedicine', cit., p. 303).

In 2009, President Barack Obama adopted an Executive Order[229] which overcame the limitation previously in force of research solely on stem cell 'lines' existing prior to August 2001, which prohibited the allocation of funds to create further lines. In July 2010, the Food and Drug Administration granted Geron Corporation authorization to carry out, for the first time in the world, clinical tests on human stem cells derived from embryos.[230] The procedures will be developed jointly by the Geron Corporation and by the University of California in some patients with spinal cord damage. The FDA had already authorized these experiments in 2009, but they were then suspended because of some complications found in test animals.[231] Subsequently, Geron developed further methods and perfected an improved technique in order to make the new cells purer. The experimentation of 'Phase 1' will be conducted in multiple centers (up to seven) on a small number of patients: the main objective is to verify the safety of the technique, before the efficacy. Patients with Grade A, complete, subacute lesions of the thoracic spinal cord, according to the ASIA (American Spinal Injury Association) scale of evaluation will be enrolled, and the treatment will consist in the injection of embryonic stem cells treated by the Geron Corporation itself and identified by the abbreviation Grnopc1, derived from oligodendrocyte progenitor cells, that is nerve cells which surround the long filaments connecting the nerve cells like a sleeve. The objective of the treatment is to restore full function of the nerve connections, repairing the lesion which inhibits it.

The decision immediately raised significant critical reactions, among which there was particular condemnation from the Holy See, through the declarations of Monsignor Elio Sgreccia, president emeritus of the Pontifical Academy for Life, who defined the authorization of the clinical tests on man with the use of stem cells from human embryos an 'unacceptable decision', in that 'the use of the human embryo receives a

[229] 'Executive Order 13505 – Removing Barriers to Responsible Scientific Research Involving Human Stem Cells', 9 March 2009, accessed 17 December at http://www.whitehouse.gov/the_press_office/Removing-Barriers-to-Responsible-Scientific-Research-Involving-Human-Stem-Cells.

[230] On this point, see A. Pollack, 'Stem Cell Trial Wins Approval of F.D.A.', in *New York Times*, 30 July 2010, accessed 17 December 2014 at http://www.nytimes.com/2010/07/31/health/research/31stem.html.

[231] Cysts had been discovered in the mice injected with the cells.

completely negative judgment not only from Catholic morality but from anyone who respects the human individual, the human being'.[232]

In August 2010, in the decision of *Shirley v. Sebelius*, the Court of Appeals for the District of Columbia Circuit interpreted the Dickey-Wicker amendment[233] in the limited sense which bans the use of federal funds to 'directly' carry out the extraction of stem cells from an embryo which gives rise to its destruction; on the contrary, according to the court the same amendment does not ban the funding of a research project which 'indirectly' uses previously extracted embryonic stem cells;[234] in this way, the path for experimentation on this matter was further opened.

*(ii) … the issue of human cloning*

American patent law does not encourage inventions regarding human cloning or transgenic human beings. Both case law and legislature are against such inventions. On the basis of the constitutional mandate of the XIII Amendment,[235] the USPTO holds that human cloning is prohibited by the constitution. Similarly, in the case *Pioneer Hi-Bred International*,[236] the Court of Federal Circuit specified how human cloning was excluded from the sphere of patentable subject matter. On the other hand, it was confirmed that some aspects of human cloning, such as the method for producing components of a human clone, may be patentable.

After 1997, the event of the birth of the cloned sheep Dolly[237] raised enormous debates on the ethical profiles involved in biotechnological research. Authoritative voices said that biotechnology was putting the

---

232 See 'Vaticano contro gli Usa: "Inaccettabili i test sull'uomo con le embrionali"', in *Corriere della Sera*, 31 July 2010, accessed 17 December 2014 at http://www.corriere.it/esteri/10_luglio_31/vaticano-usa-staminali_1ed5949a-9caf-11df-80c5-00144f02aabe.shtml.

233 In regard to which see above, as well in Part II.

234 For more details on the case, refer to Part II.

235 Cited previously.

236 United States Court of Appeals – Eighth Circuit, *Pioneer Hi-Bred International v. Holden Foundation Seeds Inc.*, 35 F.3d 1226, 31 USPQ 2d 1385 (1994).

237 On the theme of cloning, besides referring to Part II, see among others, S. Bhattacharya, 'Cloning: The Ethical Concerns', in V. Kumari (ed.), *Issues In Applied Ethics,* New Delhi: Satyam Publishing House, 2009, p. 198; C. Kuppuswamy, 'Cloning: National and International Law' (2006) 3 *Journal of International Biotechnology Law* 192 et seq. (204 et seq.); M. Damayanthi and A.V. Menon, 'Ethical and Legal Implications of Human Cloning' (2004) 2 *ICFAI Journal of Health and Law* 65 et seq.

ethics of society at risk.[238] In the meantime, cloning experiments continued on other mammals such as rabbits, monkeys and cows, causing apprehension on the future possibility of cloning human beings.

Given that patent law in the United States does not provide a limit to patentability based on ethics and morality, case law is free to extend patent protection from microorganisms to human genetic material. On the other hand, although moral boundaries are not provided at a legislative level, case law could have taken the initiative, imposing them through interpretation. The path chosen, however, was to not impose moral limits in light of the potential benefits obtained from biotechnology and the patentability of genetically modified living beings.

Further stimulus to the debate on the matter was raised by the patent application filed in 1997 on the non-human chimera, with which[239] Professor Stuart Newman and the sociologist Jeremy Rifkin requested a patent for an invention in which a human gene was incorporated into an animal, resulting in a product which was neither animal nor human.[240]

The USPTO found itself for the first time required to consider ethical standards of society, and held that transgenic humans or non-human chimeras did not constitute patentable subject matter, in that it contrasted public order and morality. In particular, the Patent Office stated that such inventions were not patentable pursuant to the XIIIth Amendment of the U.S. Constitution. Thus, the request was rejected.

In this manner, it was clear that the patentability of human beings was prohibited on moral grounds,[241] and also that the Patent Office, like U.S. case law, excluded human cloning and its patentability pursuant to the Constitution and patent law. Given that the XIIIth Amendment prohibits the reduction into slavery of human beings, the same regulation was considered pertinent in hindering human cloning and the patentability of human beings.

---

[238] See 'Human Cloning and Human Dignity: An Ethical Inquiry', July 2002, cit.

[239] As seen above.

[240] U.S. Application Serial No. 08/993,564, e divisional application U.S. Application Serial No. 10/308,135, cit.

[241] See S.R. Donnelly, 'The Patentability of Human Embryonic Stem Cells', cit., p. 125; O. Mills, *Biotechnological Inventions. Moral Restraints and Patent Law*, cit., p. 49; J. Chambers, 'Patent Eligibility of Biotechnological Inventions in the US, Europe and Japan: How Much Patent Policy is Public Policy?', cit., p. 223 et seq.

Both the Patent Office and case law, as shown in the decision of *Pioneer Hi-Bred International*,[242] rest on the same foundation, holding that 'cloning of human being violates the inherent dignity of human life and patenting of human being amounts to slavery'. The decision, in particular, confirmed that human cloning is prohibited given that it is contrary to morality and public order. This implies, clearly, that the ban extends also to inventions regarding transgenic humans.

Upon final analysis, then, moral standards on the subject of cloning, which were not expressly provided in patent law, were indirectly introduced into the decision making practice of the PTO and case law, essentially through the use of the 'open clause' of the XIIIth Amendment.

What is more, also following further doubts raised recently with respect to the position of the United States on the issue of human cloning, it has been confirmed that processes or methods for human cloning are patentable, and in particular the Patent Office has granted some patents on the techniques for cloning 'mammals'.[243] The underlying logic to the patentability of such methods, on the other hand, has been criticized, in that the use of a method of human cloning may lead in fact to the cloning of a human being. It appears, then, that the issue must be discussed further, also in relation to future developments of cloning techniques.[244]

---

242 United States Court of Appeals – Eighth Circuit, *Pioneer Hi-Bred International v. Holden Foundation Seeds Inc.*, cit.

243 See, in particular, U.S. Patent No. 6,211,429 and in particular U.S. Patent No. 6,781,030, which also included humans in the description.

244 N.S. Sreenivasulu and C.B. Raju, *Biotechnology and Patent Law. Patenting Living Beings*, cit., p. 207; *contra*, see L.M. Knedlik, 'Cloning Patents: A Patent Office and Legislative Update', (2004) *FindLaw*, accessed 17 December 2014 at http://library.findlaw.com/2004/Sep/1/231219.html.

# PART IV

# The evolution of the discipline and problematic issues in Europe

## FOREWORD

The biotechnology patent law system in Europe has followed a significantly similar path to that seen in the United States, though with legislation and case law taking on different roles.[1]

In general, the roots of intellectual property in Europe are national, and the law regulating their use is delegated to the single national authorities of responsibility. Although a form of so-called 'community patent' should be introduced as soon as possible,[2] to the state the concept of 'European

[1] In the broad sense – inclusive of the decisions of the patent office as creators of law – adopted (as mentioned) in this book.

[2] The so-called community patent should consist in a single patent title valid for the entire territory of the European Union. The institution of such a patent is provided by the Luxembourg Convention, signed on 15 December 1975 by all those who were Member states of the EC at the time, but which never entered into force due to the resistance of some countries (especially Denmark, Ireland, Italy and Spain) after the signing. Over the years, there have been several attempts to resolve the static situation, connected overall to the attribution of exclusive rights to European judges with the authority to decide on the invalidity of the patent and to the language in which the patent request should be drafted. In June 2012, a new agreement for a 'European Union patent' has been made between the EU Parliament and Council on the so-called 'EU patent package' (single patent, unified language system and patent court; see http://www.europarl.europa.eu/news/it/pressroom/content/20111219IPR34540/html/EU-patent-gets-Legal-Affairs-Committee-green-light, accessed 18 December 2014). The necessary EU legislation for enhanced co-operation was approved by the

patent', in effect composed by a bundle of national patents, is the dominant one.

At the European level, regulations have been adopted which have shaped the subject of patents, and, with the European Patent Convention and Directive 98/44/EC, have gone so far as to provide the patentability of biotechnological inventions, with specific provisions for the exclusion of living organisms from patentability.

The path was initiated, also in Europe, by the patentability of micro-organisms. Later came the patentability of plants, followed by that of animals and finally of human genetic material. This step appears to be the final one, in consideration of the shared position in excluding human beings from the sphere of patentability, which is particularly supported within the European context.

## IV.1 BIOTECHNOLOGICAL INVENTIONS IN LEGISLATION

### IV.1.1 The Evolution of Patent Regulation

In Europe, when businesses began to carry out their activity at a cross border level and to request patents outside their countries, the request for an 'international' patent system and for harmonization and centralization became pressing and led to the adoption of three fundamental regulatory acts, which represent the origins of the European discipline on the subject.

The first is the Paris Convention for the protection of intellectual property of 1883.[3] The Convention, negotiated in order to protect industrial property rights such as patents, trademarks, designs and models, and so on,[4] formed the basic principles of patent law in Europe. In the first place, they set aside the discrimination and requirement of

---

European Parliament on 11 December 2012, and entered into force in January 2013. The provisions will apply once the related Agreement on a Unified Patent Court, which was signed by all EU Member states except Croatia, Poland and Spain, enters into force.

[3] 'Paris Convention for the Protection of Industrial Property' (or 'Paris Convention'), an international convention signed on 20 March 1883, and subsequently modified in 1900 (Brussels), 1911 (Washington), 1925 (L'Aia), 1934 (London), 1958 (Lisbon), and 1967 (Stockholm) and amended in 1979, currently includes 174 party states.

[4] See art. 1(3) of the Paris Convention.

reciprocity between states, providing the criteria of 'national treatment'. Pursuant to article 2 of the Convention, with regard to criteria for patentability, every State party to the agreement is held to apply the same treatment it reserves for its own citizens to the citizens of other states. What is more, the Convention established a common minimum standard for the protection of industrial property, with the institution of a system of 'priority'. On the basis of article 4 of the Convention, an individual who correctly files an application for a patent, design or utility model, or trademark, in a Member state, enjoys the right of priority for its filing in other Member states. This means that an application filed in one of the states allows for a period of time – equal to one year[5] – in which to present the application in some of the others with the same priority (the date of the original application filing). In this way, the applicant ensures a single date of deposit, after which any other disclosure will not be part of the state of the art against which the novelty of the invention will be examined.[6]

The second important act of the European system is the Convention of Strasbourg,[7] adopted by the European Council with the objective of establishing a common European market. The Strasbourg Convention was essentially aimed at remodelling the substantial provisions of patents, without dealing with profiles of morality in any significant way. The Convention, which represented the basic regulatory text for the subsequent European Patent Convention, forged an essentially universal patent system subject to extension to all technical fields.

The Strasbourg Convention, in particular, established that the substantial requirements for patentability included the concepts of novelty, originality and industriality, which could be remodelled to satisfy new

---

[5] Following the Brussels review of 1900.

[6] The large number of states which signed the Paris Convention made it difficult to carry out further developments in the context of relative regulatory framework. In any case, art. 19 of the Paris Convention provides a flexible element, reserving the conclusion of a separate special agreement for the protection of intellectual property to the states, on the condition that they are not contrary to the provisions of the Paris Convention. The use of these special agreements has allowed the continuance of the 'internationalization' of patent law: the Convention on the European Patent itself (see below), in the preamble, shall state that it is a special agreement according to the meaning of art. 19 of the Paris Convention.

[7] 'Convention on the Unification of Certain Points of Substantive Law on Patents for Invention' (or 'Strasbourg Convention'), signed by the Member states of the European Council in Strasbourg on 27 November 1963 and entered into force on 1 August 1980.

needs. In this sense, with reference to the test of novelty, previously local and limited, the Convention adopted a completely new approach, for which the state of the art against which the novelty is to be measured is universal, including 'everything made available to the public by means of a written or oral description, by use, or in any other way, before the date of the patent application or of a foreign application, the priority of which is validly claimed'.[8] With regard, then, to the criteria of originality, although it is common opinion that patentable inventions should be distinct from the state of the art for something more than novelty, the adopted definitions at a national level were considerably different.[9] The compromise reached associated the criteria with non-obviousness (in the American and British model), providing that: 'An invention shall be considered as involving an inventive step if it is not obvious having regard to the state of the art'.[10] Regarding then the criteria of industriality, though it is common opinion that in the manufacturing industry inventions are patentable, many countries excluded medicines and food stuffs from patentability, in that it was believed that the protection of life and health are universal objectives transcending any property right.[11] What is more, various states had limitations on the patentability of procedures on living organisms. Thus, patent protection was not available for all technical fields. The text of the Convention, on this point, provided broadly that: 'An invention shall be considered as susceptible of industrial application if it can be made or used in any kind of industry including agriculture'.[12]

Another issue arising during the drafting of the Convention was connected to the doubt as to whether the sphere of protection should be limited to the words used in the claim or if the protection should be extended to its general inherent inventive concept,[13] or if a solution should be found between these two extremes. A related issue was

---

[8] The Strasbourg Convention, art. 4, para. 2. The harmonization reached in Strasbourg gave rise to an international standard adopted for the criteria of novelty in the subsequent patent regulations.

[9] In particular, German case law put the profile of technical progress alongside that of the inventive step.

[10] Strasbourg Convention, art. 5. Also for the criteria of originality, the harmonization reached in Strasbourg gave rise to an international standard adopted in the subsequent patent regulations.

[11] In this regard, see R. Moufang, 'Methods of Medical Treatment under Patent Law' (1993) 24 *IIC* 18 et seq.

[12] Strasbourg Convention, art. 3.

[13] Respectively, the solutions adopted in the United Kingdom and in Germany.

whether the claim could be used to protect an element contained in the description but which was not explicitly indicated in the claim itself. The solution of compromise adopted on these themes provided, on one hand, that: 'The extent of protection conferred by the patent shall be determined by the terms of the claims';[14] on the other, that the description could be used to interpret the claims.[15]

Despite the universal nature of the Convention of Strasbourg, some exceptions to patentability were inserted for political and economic reasons. In particular, as far as morality is concerned, two – optional – exceptions were provided, indicated in article 2 of the Convention.

The first exception allowed, fundamentally as far as this case is concerned, was in regard to the refusal of patents for inventions 'the publication or exploitation of which would be contrary to ordre public or morality, provided that the exploitation shall not be deemed to be so contrary merely because it is prohibited by a law or regulation'.[16] Some commentators, further, have shown that this provision was not included in the first draft of the Convention, and the same data is found[17] in the examination of the historical documentation regarding the drafting of article 53 of the European Patent Convention, which is based on article 2 of the Strasbourg model. Moreover, the same commentators stated that the insertion of the provision was not intended to deal with the essence of patent law, but rather was added to allow the states to exercise power, provided by national laws, to deny a patent when its granting would be unacceptable under profiles of public order or morality.[18] As a consequence, it has been shown that the provisions of the Convention of Strasbourg were aimed at the recognition of two legitimate concerns of national governments. The first was that governments should not find themselves having to publish obscene documents, and the second that

---

14 Strasbourg Convention, art. 8(3). In this regard, the compromise reached in Strasbourg laid the basis for the subsequent adoption of the 'Protocol on the Interpretation of Article 69 of the European Patent Convention', dated 5 October 1973, then modified in the 'Revision Act' of 29 November 2000.

15 So, not to add elements to them.

16 Strasbourg Convention, art. 2(a).

17 As will be seen below.

18 See, in general, E. Armitage and I. Davies, *Patents and Morality in Perspective*, London: Common Law Institute of Intellectual Property, 1994.

they should not publish instructions concerning the carrying out of acts which endanger peace or violate morality.[19]

The second exception to patentability allowed by the Convention concerned the non-patentability of animals and plants, with the explicit exclusion of microbiological processes and products,[20] to safeguard the already widespread use of microorganisms in patented processes. The exclusion from patentability of plant varieties was dictated by the conviction that they could be better protected by specific national laws, or, alternatively, by the UPOV Convention of 1961 on the protection of new plant varieties.[21] The exclusion of new animal varieties from patentability was a logical corollary to the exclusion of plants, in order to render the discipline consistent with that of the regime of higher life forms (beyond leaving space for any future *ad hoc* protection of them).

The fact that the exclusions were allowed and not obligatory was held to be an argument in favour of the statement that the Convention was limited to facilitating state practices and that in the context of patentability moral concerns were marginal. This observation, it was added, is supported by the observation that a single practice at the national level regarding patent denial on the basis of public order and morality does not exist, and in particular there is disagreement among states as to the meaning of the concept of morality.[22]

The third fundamental act is the European Patent Convention, agreed in Munich in 1973,[23] which provided a common system of patent concession at European level, establishing the European Patent Office.[24] Following the advent of the European Economic Community with the Treaty of Rome in 1957, it became necessary to harmonize the national rights of Member states. The removal of trade barriers was essential for a proper functioning of the common market. On the other hand, it was also clear that the harmonization of national patent legislation would not change the fact that granted patents would be limited to the national

---

[19] See O. Mills, *Biotechnological Inventions. Moral Restraints and Patent Law*, Farnham (UK)/Burlington (USA): Ashgate Publishing, 2010 (Revised Edition), p. 26.

[20] Strasbourg Convention, art. 2(b).

[21] 'International Convention for the Protection of New Varieties of Plants' (or 'UPOV Convention'), signed on 2 December 1961, entered into force on 10 August 1968, and most recently modified in 1991.

[22] On this point, see O. Mills, *Biotechnological Inventions. Moral Restraints and Patent Law*, cit., p. 28.

[23] 'Convention on the Grant of European Patents' (European Patent Convention) of 5 October 1973 modified several times; hereafter also 'EPC'.

[24] Hereafter 'EPO'.

territory. Therefore, a European patent regulated by community law and administrated by European institutions was needed. This led, as a result of the work of a special working group, to the preparation of a conventional text on the subject, with the aim of creating a patent system: (a) supranational, in conformity with the Convention of Paris; (b) which coexists with existing national systems (already in force for some time), with a supranational patent office; (c) which supports the economic structure of the community; (d) through which a compromise is reached among systems in which patents are granted without preventive examination and those for which it is required; and (e) which allows an equal balance between the interests of the inventor and those of the public.[25]

Upon final analysis, there were essentially two aims of the EPC: on one hand, the enforcement of the Treaty of Rome, contributing to the removal of trade barriers; on the other, the creation of a harmonized supranational regulation for a European patent, valid in a larger number of states with respect to the number of EEC members.[26] Subsequently, however, there being disagreement among Member states of the EEC on the point of whether the text must only involve themselves or should be extended to other nations, it was decided that two Conventions would be drafted, one related to a European patent, on the model of the substantial law of Strasbourg, extended to non-EEC Member states (the European Patent Convention, signed in Munich in 1973); the other in which the discipline of such patents was restricted to the EEC Member states (the Convention on the community patent, signed in Luxembourg in 1975,

---

25 See G. Oudemans, *The Draft European Patent Convention: A Commentary with English and French Texts*, London: Stevens & Sons/New York: Matthew Bender & Co., 1963, *passim*.

26 The negotiations of Strasbourg were carried out among states of the European council, which included the original six members of the EEC and EFTA members (European Free Trade Area). The European Patent Convention is broadly based on the Convention of Strasbourg, also open to non-Member states of the EEC. The legal basis of the first purpose was art. 100 of the Treaty of Rome, which mandated the Council to adopt norms aimed at making the establishment and function of the common market effective, and art. 9, which provided a customs union as the basis for the community. In regard to the second purpose, in order to render the new system attractive for as many states as possible, and avoid having to deal with the differences existing among national legislations, it was decided that the new supranational legislation should coexist with the rights of national patents.

which never entered into force because it was not ratified by enough countries).[27]

The EPC, basically, allowed for the resolution of the problem of filing several patent applications in various European nations, establishing the possibility of a unified procedure of issuance for a bundle of national patents, with preventive examination, by the EPO.

As far as concerns the contents of the EPC, first of all, it must be noted that the obligatory substantial provisions of the Strasbourg Convention were incorporated into the new Convention. The EPC, then, in compliance with the previous conventional texts, states that European patents must be granted to inventions 'which are susceptible of industrial application, which are new and which involve an inventive step'.[28]

The Convention, article 52(2), contains a list of four classes of inventions which are considered patentable, that is: '(a) discoveries, scientific theories and mathematical methods; (b) aesthetic creations; (c) schemes, rules and methods for performing mental acts, playing games or doing business, and programs for computers; (d) presentations of information'. Article 52(3), then specifies that patentability is precluded only in that the application or patent 'relates to such subject-matter or activities as such'.

With regard to the optional exceptions provided by article 2 of the Convention of Strasbourg, given that the harmonization obliged the Member states of the EEC to operate within the same regulatory system, it was decided to adopt the option of not providing patent protection to certain types of inventions, making the exceptions obligatory in article 53 of the new Convention. In this regard, from the preparatory work of the

---

27 For further details on the attempts which finally brought about the (above mentioned) enactment of the European Union patent, see http://en.wikipedia.org/wiki/European_Union_patent, accessed 18 December 2014.

28 EPC, art. 52 (1). With regard to the inventive step, the EPO is oriented towards a 'problem-solution approach', based on the following steps: (a) to identify the state of the art available in the case at hand; (b) to determine the objective technical problem that the invention resolves; (c) to verify whether the solution of problems identified by the inventor is obvious or not for the average expert in the field. This last step is based on the 'could-would approach', for which it is evaluated as to whether the state of the art in general there is some teaching which would have, and not simply may have, pushed the expert – in the face of that technical problem with the unclear characteristics from the available state of the art – to modify or adapt that state of the art taking into consideration that teaching, in order to achieve a similar result as that of the invention.

EPC[29] emerges the limited attention given to moral profiles in the debate on the text: considering that definitions of public order and morality shared at the European level did not exist, and agreeing that the interpretation of concepts should be used by the community institutions, it was held to be sufficient to mention the exception within the text without providing further details, as a sort of safeguard given to national legislatures but operating at the margins of the system.[30]

The definitive formula of the text of article 53 of the EPC, instead, makes the exceptions to patentability on moral grounds obligatory.[31] So, first and foremost the Convention provides, in article 53(a), that patents are not to be granted for 'inventions the publication or exploitation of which would be contrary to "ordre public" or morality, provided that the exploitation shall not be deemed to be so contrary merely because it is prohibited by law or regulation in some or all of the Contracting States'.[32] What is more, article 53(b) of the Convention states that patents also should not be granted for 'plant or animal varieties or essentially biological processes for the production of plants or animals; this provision does not apply to microbiological processes or the products thereof'.[33]

---

[29] 'Proceedings of the first meeting of the Patents Working Party,' Brussels, 17–28 April 1961 (Patents Working Party Document: section 5, IV/2767/61-E), available at https://www.epo.org.

[30] Thus O. Mills, *Biotechnological Inventions. Moral Restraints and Patent Law*, cit., p. 31 et seq. See also E. Armitage and I. Davies, *Patents and Morality in Perspective*, cit.

[31] In this regard, the European system has been defined as 'ask (moral) question first, then patent', in opposition to the American system, defined instead as 'patent first, then ask (moral) questions', see M.A. Bagley, 'A Global Controversy: The Role of Morality in Biotechnology Patent Law', in P. Yu (ed.), *Intellectual Property and Information Wealth: Issues and Practices in the Digital Age*, Westport (USA): Praeger Press, 2006, p. 317 et seq. (especially p. 322 and 331).

[32] The 'Guidelines for Examination in the European Patent Office' published for the first time by the EPO in 1977 stated that, in order to interpret art. 53(a) EPC, '(a) fair test to apply is to consider whether it is probable that the public in general would regard the invention as so abhorrent that the grant of patent rights would be inconceivable' (Guidelines cit., 2010, Part C, Chapter 4, 4).

[33] Naturally, inventions which fall under the aforementioned exceptions to patentability are not patentable even if they satisfy the general requirements of novelty, originality and industriality. The regulation pursuant to art. 53(b), prohibits then the granting of the patent over living beings produced through processes which are essentially biological, but not on living beings produced through processes which are not biological; this provision was inserted into the

The provisions of article 53 EPC, thus, can be held to be a veiled attempt to regulate innovation, including biotechnology, on the one hand by withdrawing the incentive to be dedicated to creating and distributing certain inventions by denying relative patents on moral grounds (which may be defined as 'non-patentability of the result/purpose'), on the other precluding the patentability of certain living organisms or their parts (which may be defined as 'non-patentability of the subject/material').[34]

In 1998, after a turbulent path lasting ten years,[35] Directive 98/44/EC for the legal protection of biotechnological inventions was approved,[36] which provided an answer to the pressing requests for adequate legal protection for biotechnological inventions in the European Union in light of the constant progress made by the continental industry of biotechnology.[37] The Directive had the fundamental objective of harmonizing the

---

Convention given that in Europe, already prior to its adoption there had been applications for patents on living beings produced through processes which were not biological. Given that the regulation prohibits the patentability of essentially biological processes and related products, the interpretive line followed (as shall be seen) by the EPO led to the conclusion that all others are patentable, in particular biotechnological products and processes – or rather 'inventions which concern a product consisting of or containing biological material or a process by means of which biological material is produced, processed or used' (EPC, ex Chapter VI, Rule 23b.2, now Chapter V, Rule 26.2), – such as transgenic microorganisms, plants or animals and purified human genetic material – and microbiological processes, that is 'any process involving or performed upon or resulting in microbiological material' (EPC, ex Chapter VI, Rule 23b.6, now Chapter V, Rule 26.6). The provisions of the EPC, then, following the adoption of Directive 98/44/EC, were modified and integrated (as just mentioned) to render them consistent with the new regulations introduced by the latter (in this regard, see below).

34 For this distinction, see that which will be said more extensively below and in Part V.

35 See, among others: D. Gitter, 'Led Astray by the Moral Compass: Incorporating Morality into European Biotechnology Patent Law' (2001) 19 *Berkeley Journal of International Law* 1 et seq.

36 Directive 98/44/EC of the European Parliament and of the Council of 6 July 1998 on the legal protection of biotechnological inventions, in [1998] OJ L 213/13.

37 Various sources highlighted that the pre-existing confusion on the matter had negatively impacted the industry of innovation and development of biotechnology, in particular in competitive comparison with the American and Japanese markets. See, in particular, European Commission, Communication 'Promoting the Competitive Environment for the Industrial Activities Based on Biotechnology within the Community', 1991, SEC (91) 629; in doctrine: N. Jones, 'The New Draft Biotechnology Directive' (1996) 6 *European Intellectual*

criteria for the patentability of genetic material, and to facilitate a uniform application of exceptions to patentability on moral grounds in the various Member states, in order to encourage investment in biotechnology.[38] To this end, further, it was held to be appropriate to preserve the right of Member states to consider moral implications in determining patentable subject matter, as already provided pursuant to article 53(a) EPC.[39]

The procedures for the adoption of the Directive began when, in 1985, the White Paper of the European Commission on the completion of the internal market[40] stated that the differences in legislation on intellectual property among Member states had a direct and negative impact on intra-community trade. Thus, a proposal was reached for a directive on the patentability of biotechnological inventions, adopted by the Commission in October 1988.[41] The procedure for the adoption of the Directive, however, was slowed due mainly to moral issues regarding the patentability of living matter, given that the proposal did not hold this profile in consideration in establishing the requirements and limits to patentability of biotechnological inventions.[42]

Negotiations, then, began between the European Commission and Parliament on the modification of the proposal to include some fundamental moral principles, in which the Parliament requested an 'integral approach', that is the inclusion of moral and social aspects within the

---

*Property Review* 363 et seq.; D.G. Scalise and D. Nugent, 'Patenting Living Matter in the European Community: Diriment of the Draft Directive (1993) 16 *Fordham International Law Journal* 990 et seq.

38 See, among others, J. Black, 'Regulation as Facilitation: Negotiating the Genetic Revolution' (1998) 61 *Modern Law Review* 621 et seq.

39 See M.A. Bagley, 'A Global Controversy: The Role of Morality in Biotechnology Patent Law', cit., p. 333 et seq.; D. Gitter, 'Led Astray by the Moral Compass: Incorporating Morality into European Union Biotechnology Patent Law', cit., p. 2 et seq.

40 'Completing the Internal Market: White Paper from the Commission to the European Council', COM(85) 310, Milan, 28–29 June 1985, accessed 18 December 2014 at http://europa.eu/documentation/official-docs/white-papers/index_it.htm.

41 COM(88) 496.

42 See, among others, B.A. Brody, 'Intellectual Property and Biotechnology: The European Debate' (2007) 17 *Kennedy Institute of Ethics Journal* 69 et seq. (73 et seq.); G. Porter *et al.*, 'The Patentability of Human Embryonic Stem Cells in Europe' (2006) 26 *Nature Biotechnology* 653 et seq.

framework of the patent discipline.[43] In particular, the Parliament proposed the introduction of principles which reflect the exclusion from patentability of the human body and its parts and inventions contrary to public order or morality, as well as an exception for farmers.[44] In 1994, the Council announced its position, and the differences between Parliament and the Council were discussed in a Conciliation Committee which produced a joint text.[45] However, in 1995 Parliament rejected the joint text.[46] So, a new proposal of the Commission was drafted,[47] reflecting the reasoning expressed by Parliament. The Commission highlighted how without a common legislative framework which was compatible with the single market, European research was at a disadvantage with respect to third party country competitors; thus the legal protection of biotechnological inventions needed to be clarified.

The previous European patent regulations, that is the Convention of Strasbourg and the EPC, had been drafted in the 1960s and 1970s, when the developments in biotechnology could not have been imagined, and all debates that had taken place with respect to the initial proposal of 1988 confirmed the necessity to further harmonize the regulations. It was widely accepted that the protection of biotechnological inventions should be based on fundamental principles of patent law, but there had been different visions regarding the extension of protection.

For the States parties to the EPC, except for the exceptions to patentability pursuant to article 53, biotechnological material was patentable according to the patent law in force. Those opposed to the protection of biotechnological inventions held that no material existing in nature could be patented. On the other hand, the directive proposal followed a balanced approach, and in principle did not impose specific restrictions to the patentability of human matter, including, however, the principles of non-marketability of the human body and the free and informed consent of the person from whom samples were obtained on the basis of which

---

43 In this regard, see S. Sterckx, 'European Patent Law and Biotechnological Inventions', in S. Sterckx (ed.), *Biotechnology, Patents and Morality*, Farnham (UK)/Burlington (USA): Ashgate Publishing, 1997, p. 20.

44 With regard to these principles, see below.

45 'Council Common Position', 26 February 1995.

46 [1995] OJ C68. This was the first time in which the European Parliament rejected a report from the Conciliation Committee.

47 COM(95) 661, 8 October 1996.

the invention was to be developed.[48] With regard to moral profiles, which remained then a central theme to the debate on the matter, the new proposal of the Commission was in any case the subject of amendments by the European Parliament.

In 1998 an agreement was finally reached,[49] and Directive 98/44/EC was adopted. However, there were still profound differences between institutional and public opinion, and the initial opposition to the proposal had been smoothed over by virtue of the commercial importance of the patenting of biotechnology.[50]

Directive 98/44/EC, then, introduced a model of protection for biotechnological inventions at the European level which takes into consideration and transposes at a legislative level, on one hand, the principles identified at the end of the aforementioned debate between community institutions, and on the other, the principles expressed in various EPO decisions and by its Technical Board regarding the granting of patents on non-natural living beings through biotechnology; this occurred, unlike in the American model, with specific attention to moral profiles.[51]

The Directive provides an obligation on Member states to protect biotechnological inventions on the basis of their respective patent laws, requiring that they be modified to make the provisions of the Directive effective. To this end, some definitions, first of all, are provided, such as 'biological material', 'microbiological procedure', 'essentially biological process' and so on, which are significant in relation to the patentability of living beings like microorganisms, plants, animals and human genetic material produced through biotechnological processes, in that they represent a starting point.

The Directive does not contain any list specifying which inventions are patentable. It states that inventions which are new, original and subject to industrial application must be considered patentable, even where they concern a product that consists in or contains biological material, or rather a procedure through which biological material is produced,

---

[48] In this regard, see M. Herdegen, 'Patenting Human Genes and Other Parts of the Human Body under EC Biotechnology Directive' (2000–2001) 3 *Bio-Science Law Review* 102 et seq.

[49] See 'Opinion of the European Parliament' 16 July 1997, [1997] OJ C286; as well as [1996] OJ C 295/11; [1996] OJ C 296/4; [1997] OJ C311/12.

[50] On this point, see R. Brownsword W.R. Cornish and M. Llewelyn, 'Editors Comment: Human Genetics and the Law: Regulating a Revolution' (1998) 61 *Modern Law Review* 593 et seq.

[51] S. Jasanoff, *Designs on Nature: Science and Democracy in Europe and the United States*, Princeton: Princeton University Press, 2005, p. 217 et seq.

manipulated or used. In this sense, according to the Directive, inventions which involve microorganisms, plants or animals are patentable. The Directive declares biological material produced or isolated from its natural environment through technical procedures patentable,[52] given that it is believed that isolation and purification involve an inventive step. At the same time, it considers a biological procedure which requires human intervention, in possession of certain non-natural characteristics which make the procedure non-biological, patentable.

What is more, the Directive stipulates that certain biotechnological inventions are not patentable, in consideration of moral and bioethical standards, as well as in compliance with the provisions of the EPC[53] and the TRIPS Agreement.[54] The Directive, also, moves a step beyond such regulations, excluding the human body and its various stages of formation and development from the sphere of patent protection, and also excludes human cloning from patentability. In particular, pursuant to articles 4, 5 and 6 of the Directive, the following are excluded from patentability:

(a) plant varieties and animal breeds;
(b) essentially biological procedures for the production of plants or animals; moreover, the provision does not concern the patentability of inventions whose subjects are microbiological procedures or other technical procedures;[55]
(c) the human body, in various stages of its constitution and development, as well as the mere discovery of one of its elements, including therein the sequencing or partial sequencing of a gene; on the other hand, an isolated element of the human body, or which is produced in another manner, through a technical procedure, including therein the sequencing or partial sequencing of a gene, may constitute a patentable invention, even if the structure of said element is identical to that of a natural element;[56]
(d) inventions whose commercial use is contrary to public order or morality, and in particular:
    (i) processes of human cloning;
    (ii) processes of modification of the germinal genetic identity of the human being;

52 Directive 98/44/EC, art. 3.
53 Art. 53(a) and (b).
54 Art. 27(2) and (3).
55 Directive 98/44/EC, art. 4(1)–(3).
56 Directive 98/44/EC, art. 5(1)–(2).

(iii) the use of human embryos for industrial or commercial use;
(iv) processes of modification of the genetic identity of animals in order to provoke their suffering without substantial medical use for man or animals, as well as animals resulting from such processes.

The use of an invention, however, cannot in and of itself be considered contrary to public order or morality for the single fact that it is prohibited by legislative or regulatory provisions.[57]

The Directive then, while, in general, encouraging and protecting biotechnological inventions, prohibits some on the basis of public order and moral and bioethical standards, attempting to provide a balance between them. Finally, with a provision of particular importance to provide the EPO with adequate support on the subject, the Directive instructs the European Group on Ethics in Science and New Technologies, to evaluate all aspects of the field of bioethics.[58]

On the basis of the Directive, then, a biotechnological invention which does not fall within its provisions of exclusion is patentable, provided that it satisfies the requirements of novelty, originality, industriality and sufficient description. In the last respect, there is no possibility for sufficient description for biotechnological inventions. Thus, the alternative possibility of filing the invention is provided.

Given the peculiar nature of biotechnological inventions, the Directive states that in the case in which the invention requires the use of or involves biological material which is not available to the public and that cannot be described in the patent application in order to allow the reproduction of the invention by an expert in the field, the description must be held to be inadequate for the purposes of the patent, unless the biological material was deposited according to the provisions of the Treaty of Budapest.[59]

---

57 Directive 98/44/EC, art. 6(1)–(2).

58 Directive 98/44/EC, art. 7 and Recital 44.

59 Budapest Treaty on the International Recognition of the Deposit of Microorganisms for the Purposes of Patent Procedure, signed on 28 April 1977, entered into force on 9 August 1980 and subsequently amended on 26 September 1980. The deposit must occur prior or together with the presentation of the patent application. The public can have access to the filed invention, and if the deposited material ceases to be available a new deposit must be made (see Directive 98/44/EC, art. 13; EPC, art. 14).

Directive 98/44/EC[60] marked a fundamental step on the path of patentability for living beings in the European Union, which had begun with the foundations laid by the EPC.

However, the implementation process for the Directive was made complex and slow by the resistance of many Member states.[61] During the Council meeting, the Netherlands had voted against the adoption of the Directive, and Belgium and Italy abstained. After its approval, the Directive was immediately appealed by some Member states before the Court of Justice: in the case of *The Kingdom of the Netherlands v. European Parliament and Council*,[62] the Netherlands, joined by Italy and Norway, requested the abrogation of the Directive in its entirety. The reasons for the appeal were essentially the following: the fact that article 100(a) of the EC Treaty was not the correct legal basis for the Directive; the violation of the principle of subsidiarity; the violation of the principle of certainty of law; the violation of the obligations of international law; the violation of fundamental law with respect to human dignity; and the violation of procedural rules in the adoption of the Commission's proposal.[63]

---

[60] Whose acceptance in the single Member states was expected by July 2000, but was postponed by many of them for some years (see below).

[61] In this regard, see, among others, S.R. Donnelly, 'The Patentability of Human Embryonic Stem Cells: Is the Inconsistent Application of the European Union Biotechnology Directive's Moral Exclusion Clause Undermining Investor Confidence in Europe, Providing a Competitive Advantage to the United States?' (2011) 20 *Dalhousie Journal of Legal Studies* 106 et seq. (113); G. Porter, 'The Drafting History of the European Biotechnology Directive', in A. Plomer and P. Torremans (eds), *Embryonic Stem Cell Patents: European Law and Ethics*, Oxford/New York: Oxford University Press, 2009, p. 3 et seq. (16); A. Hellstadius, 'A Comparative Analysis of the National Implementation of the Directive's Morality Clause', therein, p. 118.; see as well, M.A. Bagley, 'A Global Controversy: The Role of Morality in Biotechnology Patent Law', cit., p. 334, which shows that, even in Member states that implemented the Directive, the form of implementation generated controversy: for instance, the regulations adopted in France, Germany and Italy strictly limited genetic patent protection to uses specifically detailed in the application, with the consequence that the broadest exceptions on patented genes at the EPO may not be valid in those countries.

[62] Case C-377/98 [2001] ECR I-7079.

[63] In this regard, see, among others, S.R. Donnelly, 'The Patentability of Human Embryonic Stem Cells', cit., p. 113; L. Palombi, 'The Impact of TRIPS on the Validity of the European Biotechnology Directive' (2005) 2 *Journal of International Business and Law* 62 et seq.

In particular, in rejecting the complaints of the appellants that article 5, para. 1, of the Directive, providing for the patentability of isolated elements of the human body, was not respectful of the principle of human dignity, the Court of Justice, having recognized its obligation to ensure the protection of fundamental rights of human dignity and integrity, held that article 5(1), was laid out in strict enough terms to guarantee that the human body should not be patentable and remains inviolable, and thus safeguards human dignity. According to the Court, it was necessary to make a distinction between the discovery of a DNA sequence or of an element of the human body, none of which would have been patentable pursuant to the Directive as such, and the inventions that combined one of these natural elements with a technical procedure or application, which would instead make them patentable.

Further, with regard to the relationship between the principle of human dignity and the patentability of life, the Court revealed that human dignity was protected in the Directive through various limits, and that the reference to this fundamental principle 'is, however, clearly misplaced as against a directive which concerns only the grant of patents and whose scope does not therefore extend to activities before and after that grant, whether they involve research or the use of the patented products'. The Court also added that the recognition of a patent 'does not preclude legal limitations or prohibitions applying to research into patentable products or the exploitation of patented products, as the 14th recital of the preamble to the Directive points out. The purpose of the Directive is not to replace the restrictive provisions which guarantee, outside the scope of the Directive, compliance with certain ethical rules'. Basically, the Court reiterated to the appellant states that they could simultaneously adopt the Directive and reach many of their ethical objectives through the imposition of the specific restrictions on research and use of patented products.[64]

As far as concerns the violation of the principle of legal certainty in article 6(1) of the Directive, given that it does not provide specifications regarding the probability of violating public order or morality, it was held that the application of such criteria by national authorities, left to their broad discretion,[65] was in any case, as a last resort, subject to the control

---

64 See, in particular, B.A. Brody, 'Intellectual Property and Biotechnology: The European Debate', cit., p. 85.

65 Given that, as shown by Attorney General Jacobs: 'It is common ground that this provision allows the administrative authorities and courts of the Member state a wide scope for maneuver in applying this exclusion. However that scope for maneuver is necessary to take account of the particular difficulties to which

of the Court of Justice.[66] With reference to article 6(2), finally, the Court reiterated that 'as regards living matter of human origin, the Directive frames the law on patents in a manner sufficiently rigorous to ensure that the human body effectively remains unavailable and inalienable and that human dignity is thus safeguarded'.

The Court of Justice, in conclusion, confirmed the validity of the Directive, but the opposition to it was so strong that, despite the outcome of the controversy, eight of the then 15 Member states did not implement the Directive in their own national legal systems until 2003,[67] and four until 2005.[68] In this respect, in the decision relative to the action instituted by the European Commission against Italy, *Commission v. Italy*,[69] the Court of Justice specified, with regard to article 6 of the Directive, that: 'Unlike Article 6(1) of the Directive, which allows the administrative authorities and courts of Member states a wide discretion in applying the general moral exclusion on inventions whose commercial exploitation would be contrary to ordre public and morality, Article 6(2) allows Member states no discretion in the implementation of the specific exclusions, since the very purpose of this provision is to give definition to the exclusion laid down in Article 6(1)'. On the basis of the decision, in the reading and interpretation of specific exclusions pursuant to article 6(2), the words should be given their literal meaning, without the possibility of extensive or restrictive interpretations.[70]

---

the use of certain patents may give rise in the social and cultural context of each Member state, a context which the national legislative, administrative and court authorities are better placed to understand than the Community authorities'.

[66] See J. Kokott and T. Diehn, 'The Kingdom of the Netherlands v European Parliament and Council' (2002) 96 *American Journal of International Law* 950 et seq.

[67] Austria, Belgium, France, Germany, Italy, Luxembourg, Netherlands and Sweden, which were referred to the Court of Justice in July 2003.

[68] Italy, Luxembourg, Latvia, and Lithuania. The implementation of the Directive in all Member states did not occur until 2007.

[69] Case C-456/03, [2005] ECR I-5335.

[70] In this regard, see S.R. Donnelly, 'The Patentability of Human Embryonic Stem Cells', cit., p. 114 et seq.; M. Rowlandson, 'WARF/Stem Cells (G2/06): The Ordre Public and Morality Exception and its Impact on the Patentability of Human Embryonic Stem Cells' (2010) 32 *European Intellectual Property Review* 67 et seq. (72); A. Plomer, 'Towards Systemic Legal Conflict: Article 6(2)(c) of the EU Directive on Biotechnological Inventions', in Plomer and Torremans, *Embryonic Stem Cell Patents: European Law and Ethics*, cit., p. 173 et seq. (187 et seq.). In particular, refer for instance to the exclusion of 'uses of human embryos for industrial or commercial purposes', stating that this does not exclude patents for inventions which give rise to the destruction of embryos. In

With the adoption of the Directive, in order to coordinate the EPC provisions with those of the new Directive, in 1999 the EPO[71] inserted a new Chapter,[72] entitled 'Biotechnological inventions', in Part II of the 'Implementing Regulations' of the Convention,[73] which basically reproduces the rules of the Directive in the corpus of the Convention, harmonizing the provisions of each.

### IV.1.2 Bioethical Profiles of Patentability

Unlike what has occurred in the United States, bioethical issues in Europe have received acknowledgement at the legislative level. In particular, patent legislation specifically prohibits the patentability of certain inventions on these grounds.

The European Patent Convention[74] provides that inventions contrary to public order or morality must not be patented. The EPC does not allow for the patentability of plants, animals or essentially biological processes used to produce them. What is more, in light of the bioethical standards of society, methods for surgical or therapeutic treatment of the human or animal body and diagnostic methods are excluded from the sphere of patentability. Similarly, the patenting of living beings found in nature in their original form is also considered unethical in accordance with the Convention.[75]

---

this sense, further, the statement of the Directive *rapporteur* Rothley is given, according to whom: 'In relation to the use of embryos, the Council has set some limitations: they are not to be used for industrial or commercial purposes. But I would only ask you to remember that this was done with the UK in mind. We cannot as European legislators decree that something which does not contravene the underlying legal principles of all Member states is a contravention of public order, and we cannot brand something that we do not jointly regard as abhorrent as a contravention to common decency. That is not acceptable.' ('Report on the Proposal for a European Parliament and Council Directive on the Legal Protection of Biotechnological Inventions', COM (95) OJ C286/7, 1997).

[71] In the current European patent regulation framework, in fact, the EPO functions as both a policy-maker and enforcer of the current discipline, exercising in this way an amount of control with respect to patentable material which was not originally provided by the EPC.

[72] Then 'Chapter VI', currently 'Chapter V'.

[73] 'Implementing Regulations to the Convention on the Grant of European Patents', as most recently amended by the Administrative Council of the EPO on 26 October 2010.

[74] As seen above.

[75] EPC, art. 53.

The TRIPS Agreement incorporates, basically, the same possibilities of exclusion from patentability as are provided in the EPC.[76]

Directive 98/44/EC,[77] thus, deals with bioethical issues associated with the progress of biotechnology, prohibiting the patentability of certain biotechnological inventions. The Directive attempts to balance bioethical standards with the most recent developments on the subject, such as research on the human genome, the embryo, stem cells, and cloning. In this view, further, the task of evaluating the ethical profiles of biotechnology is delegated to the European Group on Ethics in Science and New Technologies.[78]

In order to understand the meaning of the provisions introduced by the Directive, considering that they continue on the same path taken by the provisions in article 53 of the EPC, it is opportune to consider them in detail and compare them with the aforementioned provisions, in order to identify the range of innovations introduced with respect to it.

Both the EPC and the Directive contain a 'morality clause' which prohibits the patentability of inventions in certain cases. Based on article 53(a) EPC, inventions whose use would be contrary to public order or morality are excluded from patentability. The rules of the Directive which are equivalent to this provision are inserted in article 6(1), according to which inventions whose commercial use is contrary to public order or morality are excluded from patentability; and the use of an invention in and of itself may not be considered to be contrary to public order or morality solely given that it is prohibited by a legislative or regulatory provision. The issue arises from the fact that the Directive is binding exclusively in regard to single Member states, with the consequence that single patent offices may take a national approach on the interpretation of the rules. In addition, considering that the EPO is also used as a referee for violations of the provision, further issues are raised, in that the EPO is not a body of the European Union and thus is not legally bound to follow the Directive. Thus, in order to avoid two different patent legislations coexisting in both EU and EPC Member states (the EPC and the Directive), the EPC was adapted to the provisions of the Directive. However, regardless of this adaptation, it is possible for, at respectively European and national levels, the concepts of public order and morality to be interpreted differently, giving rise to differences in practice among Member states.

---

[76] See TRIPS Agreement, art. 27, in regard to which refer above, to Part I.

[77] Entered into force in 2000.

[78] Directive 98/44/EC, art. 7.

While the text of the EPC does not specify the meaning of public order and morality, the Directive offers a guide to what is not patentable by virtue of these ideas in article 6(2), including in particular: '(a) processes for cloning human beings; (b) processes for modifying the germ line genetic identity of human beings; (c) uses of human embryos for industrial or commercial purposes; (d) processes for modifying the genetic identity of animals which are likely to cause them suffering without any substantial medical benefit to man or animal, and also animals resulting from such processes'.

Further, the formulation of the dictates of the Directive regarding this list provoked some observations. In the first place, considering that the list is clearly exemplary and not definitive,[79] it allows national patent offices to exclude a large range of inventions from patentability based on public order and morality, thus expanding the jurisdiction of national patent offices beyond that which was provided for by the EPO in article 53(a) EPC. In this manner, concerning EPO case law, with regard to article 53(a), new doubts appear to have been added regarding the limits of patentable subject matter. In the second place, article 6(2) of the Directive does not directly reference inventions concerning plants, whose exclusion from patentability seems to be based on the exclusion provided for animals; this, on the one hand, drawing attention to the fact that it missed the opportunity to clarify this point concerning the difficulties in interpreting article 53(b) EPC, which were registered in the practice of the EPO; on the other, showing how the assimilation of plants and animals, thus, does not seem to be shared, given that through the exclusion from patentability of procedures or products related to transgenic animals 'to provoke their suffering without substantial medical utility' European legislature seems to have clarified the difference between animals, higher life forms capable of suffering, and plants, lower life forms for which such a high bioethical standard is not, then,

---

[79] Besides the expression 'in particular' in the text of the provision, refer to Recital 38 of the Directive, in which it is specified 'that it is just as important to insert into the provision of this directive itself a list indicating inventions excluded from patentability, to provide judges and national patent offices general guidelines for the interpretation of the reference to public order and morality; that this list cannot certainly be considered exhaustive; that the procedures whose application produces prejudice to human dignity, such as for instance procedures for the production of hybrid beings resulting from germinal or totipotent human or animal cells, must obviously also be excluded from patentability'.

provided.[80] Finally, it is critically pointed out that the Directive does not include a reference to environmental protection as being among the factors of evaluation for the contents of the list.[81]

The Directive provides, then, indications on what is or is not considered contrary to bioethics. Articles 5 and 6 of the Directive, further, were incorporated within the EPC by virtue of the current Rules 28 and 29 of the 'Implementing Regulations to the Convention on the Grant of European Patents'.[82] In particular, besides the coordination of the provisions of the Directive referenced above, in June 1999 a regulation was introduced in the EPC, Rule 23d (now Rule 28), pursuant to which:

> Under Article 53(a), European patents shall not be granted in respect of biotechnological inventions which, in particular, concern the following: (a) processes for cloning human beings; (b) processes for modifying the germ line genetic identity of human beings; (c) uses of human embryos for industrial or commercial purposes; (d) processes for modifying the genetic identity of animals which are likely to cause them suffering without any substantial medical benefit to man or animal, and also animals resulting from such processes.[83]

---

[80] On the contrary, EPO case law suggested that animals and plants should be equal on the basis of the principle of equal treatment. On this point, see for instance the case of *Plant Genetic Systems/Glutamine Synthetase Inhibitors*, 1995, EPOR 357, as well as what shall be said below.

[81] A factor which was considered important for moral issues by the Technical Board of the EPO in the case *Harvard Onco-Mouse*, in regard to which see below. For observations recalled here, see in particular, Mills, *Biotechnological Inventions. Moral Restraints and Patent Law*, cit., p. 136.

[82] 'Implementing Regulations to the Convention on the Grant of European Patents of 5 October 1973', adopted on 7 December 2006 and most recently modified on 26 October 2010.

[83] Rule 28(c) was the subject of an interesting ruling of the EPO Enlarged Board of Appeal, in the context of a patent application for stem cells presented in 1995 by the Wisconsin Alumni Research Foundation (Patent application no. EP 0 770 125). The WARF case concerned an application regarding a method for obtaining cultures of embryonic stem cells from primates. The application did not include examples of human embryonic stem cells, but indicated the use of pre-implanted human embryos as the starting material. The Enlarged Board of Appeal in November 2008 (T 1374/04; G 2/06), confirmed the presentation of the Examining Division of the EPO, stating in particular that the rule prohibits the patenting of inventions of products which, on the date of filing, may be prepared only through methods which necessarily require the destruction of human embryos. In this regard, refer below, to para. IV.2.

As far as the patentability of the human body or its elements is concerned, then, in Rule 23e (now Rule 29), it was provided that: 'The human body, at the various stages of its formation and development, and the simple discovery of one of its elements, including the sequence or partial sequence of a gene, cannot constitute patentable inventions'.[84] On the contrary, it is expressly allowed that: 'An element isolated from the human body or otherwise produced by means of a technical process, including the sequence or partial sequence of a gene, may constitute a patentable invention, even if the structure of that element is identical to that of a natural element'.[85] It is also specified that: 'The industrial application of a sequence or a partial sequence of a gene must be disclosed in the patent application'.[86]

What is more, in November 2000 a letter (c) was added to article 53 of the Convention, in order to include methods of treatment or diagnosis of the human or animal body among the exceptions to patentability:[87] for public policy reasons, then, and no longer[88] for deficiency regarding the requirement of industriality.[89]

Basically, through the specific indication of certain inventions as being hypothetical in violation of the provisions on public order and morality of the EPC, Directive 98/44/EC eliminated the need to refer to the 'balancing of interest', 'public abhorrence' and 'public acceptability' tests previously adopted by the decisional practices of the EPO.[90] On the other

---

[84] Rule 29(1).

[85] Rule 29(2).

[86] Rule 29(3).

[87] Based on art. 53(c), then, patents should not be granted for 'methods for treatment of the human or animal body by surgery or therapy and diagnostic methods practised on the human or animal body; this provision shall not apply to products, in particular substances or compositions, for use in any of these methods'. The reasoning for the exception is 'to deny patent protection to methods which serve medical purposes, so that no one could be hampered in the practice of medicine by patent legislation' (EPO Board of Appeal Case Law 2006, Special edition 6/2007, p. 17 et seq., accessed 18 December 2014 at http://archive.epo.org/epo/pubs/oj007/08_07/special_edition_6_case_law_boards_of_appeal.pdf).

[88] As in the previous formula of art. 52(4) EPC.

[89] For a framework of the EPC revision and of the potential impact regarding the European patent practices, see S.J. Farmer and M. Grund, 'An Overview of the New European Patent Convention (EPC 2000) and Its Potential Impact on European Patent Practice' (2006/2007) 9 *Bio-Science Law Review* 53 et seq.; B.A. Brody, 'Intellectual Property and Biotechnology: The European Debate', cit., p. 83 et seq.

[90] In this regard, refer below, to para. IV.2.

hand, the patentability of morally controversial inventions not falling within the sphere of specific exclusions has remained the subject of evaluation.

Some commentators criticized the Directive for its continued inclusion of moral considerations, holding that they may contrast with its aims, given its vagueness and the contrasting interpretations of it in various Member states, while the judgment on patentability should have remained separate from such considerations. Again, in particular it was stated that, while moral issues concerning particular developments in biotechnology were resolved simply through their inclusion in the list of exclusions from patentability, there is no criterion dictated which indicates the reasons for which those cases should be excluded from patentability; thus no guide was provided for the evaluation of new cases of morally controversial inventions, such as for instance those concerning elements of the human body, in particular gene sequences, isolated or otherwise produced through a technical process.[91] Other commentators observed, however, that the Directive is a regulatory tool to be praised for its attempt, courageous though contradictory, to provide a detailed guide to patent offices and the courts regarding that which, from the point of view of the legislature, is not patentable subject matter.[92]

Thus, the exceptions included in the Convention now appear to be consistent with those provided by the Directive. The latter seems, then, to offer national courts and patent offices a general guide for interpreting the concepts of public order and morality,[93] outlined more thoroughly than had been done in the past.[94] However,[95] unless these concepts are interpreted in a uniform way by all the Member states, and consistently with EPO case law, the result may be a divergence between patent law and practice. In fact, in national legal systems, the concepts of public

---

[91] In this sense, among others, see B.A. Brody, 'Intellectual Property and Biotechnology: The European Debate', cit., p. 90.

[92] See, in particular, Bagley, 'A Global Controversy: The Role of Morality in Biotechnology Patent Law', cit., p. 335.

[93] Recital 38, cit.

[94] Some commentators concluded, then, stating that the Directive created a new and important source of law, distinct from national and EPO practices, with the result of a clarification of the legislative framework and a greater predictability and certainty of law regarding patentable subject matter and parameters of protection. Thus T. Howard, 'The Legal Framework Surrounding Patenting for Living Materials', in J. Gibson (ed.), *Patenting Lives: Life Patents, Culture and Development*, Farnham (UK)/Burlington (USA): Ashgate Publishing, 2008, p. 9 et seq.

[95] As hinted above.

order and morality are not consistent. Based on the EPC, further, these concepts would be a subject for European institutions, in particular for the EPO.

Article 6 of the Directive, then, evoking the provision in article 53(a) EPC, prohibits the patentability of inventions whose commercial use would be contrary to public order or morality. In this regard, what is provided by Recital 39 of the Directive is particularly noteworthy, where it is specified that (a) 'public order and morality correspond in particular to ethical or moral principles recognized in a Member state and whose observance is essential in particular on the matter of biotechnology, given the potential range of inventions in this sector and their intrinsic connection with living matter', and also that (b) 'these ethical or moral principles complete the normal legal tests provided by patent law, regardless of the technical sector of the invention'.

Moreover, difficulties arise when determining the context of such ethical or moral principles, an issue that gives rise to significant differences of opinion.[96]

In the first place,[97] there are those who believe that the establishment of bioethical limits on technology should be the task of legislation through specific regulations, and not of the patent discipline.[98]

Similarly, it is stated that the examiners of the EPO would not be sufficiently prepared to face ethical and moral profiles of patents.[99]

Additionally, it is shown that the Directive attempts to join the discipline of the European Union with legal obligations derived from a

---

[96] See D. Harms, 'Drafting Claims around Morality', (1996) 7 *European Intellectual Property Review* 424 et seq.

[97] As mentioned above.

[98] In this regard, see for instance, the speech given in 1994 by Prof. Gerald Dworkin at the House of Lords Select Committee on the European Communities, where he stated that: 'Few would deny that there are major ethical issues relating to developments in biotechnology and genetic engineering; that there is a need to ensure that such ethical issues are properly addressed; that there should be adequate controls and monitoring of undesirable or questionable developments … (t)he real question, though, is whether such control should be exercised in any significant way through the patent system. A rational answer must be "no"' (HL Paper 28, HMSO, March 1, 1994); in doctrine, see P. Spada, 'Liceità dell'invenzione brevettabile ed esorcismo dell'innovazione' (2000) *Rivista di diritto privato* 6 et seq., and P. Spada, 'Etica dell'innovazione tecnologica ed etica del brevetto' (1996) *Rivista di diritto privato* 217 et seq.

[99] See therein, para 46, cit. in O. Mills, *Biotechnological Inventions. Moral Restraints and Patent Law*, cit., p. 139.

broader international treaty of the EU.[100] In fact, Recital 16 states that patent law must be applied in respect of fundamental principles which guarantee the dignity and integrity of man,[101] and then Recital 43 recognizes that European Union law[102] respects the fundamental rights which are guaranteed by the European Convention on Human Rights,[103] and which result from common constitutional traditions of Member states, given that they are general principles of EU law.[104] In reference to

[100] See R. Ford, 'The Morality of Biotech Patents: Differing Legal Obligations in Europe?' (1997) 6 *European Intellectual Property Review* 317.

[101] Specifying, as well, 'that it is necessary to reinforce the principle according to which the human body, in every stage of its constitution and development, including germinal cells, the simple discovery of one of its elements or one of its products, as well as the sequence or partial sequence of a human gene, are not patentable; that such principles are consistent with patentability criteria provided by patent law, according to which a simple discovery cannot constitute the subject of a patent'.

[102] Pursuant to art. 6 of the Treaty of the European Union.

[103] Or 'ECHR', signed in Rome on 4 November 1950 and entered into force on 3 September 1953. Among the fundamental rights sanctioned in the Convention, the right to life (art. 2) is the most noteworthy.

[104] Among these, in European law and in general in international law following the Second World War, the principle of human dignity took on particular importance. In this regard, see P. Capps, *Human Dignity and the Foundations of International Law*, Oxford/Portland (USA): Hart Publishing, 2009, p. 106 et seq.; R. Andorno, 'Human Dignity and Human Rights as a Common Ground for a Global Bioethics' (2009) 34 *Journal of Medicine and Philosophy* 223 et seq., in which human dignity is defined as 'the overarching principle of international biolaw'; R. van der Graaf and J.J. van Delden, 'Clarifying Appeals to Dignity in Medical Ethics from an Historical Perspective' (2009) 23 *Bioethics* 151 et seq.; C. McCrudden, 'Human Dignity and Judicial Interpretation of Human Rights' (2008) 19 *European Journal of International Law* 655 et seq. For specific analysis of the provisions relative to the principle in the documents on the subject of bioethics, besides that which will be said below, see D. Beyleveld and R. Brownsword, 'Human Dignity in Bioethics and Biolaw', Oxford: Oxford University Press, 2002, p. 11 et seq., where it is shown that, with respect to the use of the same in international declarations of rights and in the post-war constitutionalism with prevalent aims of individual empowerment, in – more recent – documents on the theme of bioethics human dignity has taken on a function of constraint, that is limits to individual autonomy (rather than expansion of freedom); G. Bognetti, 'The Concept of Human Dignity in European and U.S. Constitutionalism', in G. Nolte (ed.), *European and U.S. Constitutionalism*, Cambridge: Cambridge University Press, 2005, p. 85 et seq.; J.Q. Whitman, '"Human dignity" in Europe and the United States: The Social Foundations', therein, p. 108 et seq., according to which the 'weak' approach in regard to human dignity in the American legal system arose from the absence in

this last point, today, principles dictated by the Charter of Fundamental Rights of the European Union are of central importance,[105] which, after the Treaty of Lisbon assumed the same legal power as EU Treaties.[106] Among these, in particular, it is necessary to recall the fundamental principles of human dignity and the right to life and the integrity of the individual;[107] in the sphere of medicine and biology, in particular, the following criteria must be respected:

- 'free and informed consent of the person of interest, according to the methods defined by law;
- the ban of eugenic practices, in particular of those whose aim is the selection of the person;
- the ban on making the human body and its parts, as such, a source of profit;
- the ban on the reproductive cloning of human beings'.[108]

---

that model of a reaction to the fascist and Nazi experiences, at the basis, instead, of the 'strong' European approach. As far as concerns the 'objective' function of constraint of the principle of dignity with respect to the evolution of science and technology, in particular with respect to the fundamental ban on the reduction of the individual to a thing (included in the Universal Declaration of Human Rights of the UN dated 10 December 1948, arts 1 and 4, according to which no individual can be held in a state of slavery or service, and provided also in the American legal system, in the XIIIth Amendment to the U.S. Constitution, that prohibits slavery or involuntary service), see J. Habermas, *The Future of Human Nature*, Cambridge: Polity Press, 2003, p. 1 et seq.; M.C. Nussbaum, 'Objectification' (1995) 24 *Philosophy and Public Affairs* 249 et seq. (264 et seq.); J.P. Fletcher, 'Human Dignity as a Constitutional Value' (1984) 22 *University of Western Ontario Law Review* 171 et seq. (180 et seq.). The latter function, according to a widely believed reconstruction, may work also to the benefit of people 'in formation', future generations, as well as the deceased; see P. Häberle, 'Die Menschenwürde als Grundlage der staatlichen Gemeinschaft', in J. Isensee and P. Kirchhof (eds), *Handbuch des Staatsrechts der Bundesrepublick Deutschland, II, Verfassungsstaat*, Heidelberg: Müller Verlag, 2004 (3rd edn), para. 55.

105 Proclaimed on 7 December 2000 in Nice, and subsequently, in an adapted version, on 12 December 2007 in Strasbourg.

106 With regard to the relevance of the principles sanctioned in the Charter on the subject of biotechnology, see P. Šturma, 'The EU Charter of Fundamental Rights and the Governance of Biotechnology in the European Union', in F. Francioni and T. Scovazzi (eds), *Biotechnology and International Law*, Oxford/Portland (USA): Hart Publishing, 2006, p. 369 et seq.; N. Lenoir, 'Biotechnology, Bioethics and Law: Europe 21st Challenge' (2006) 69 *Modern Law Review* 1 et seq.

107 Charter of Fundamental Rights of the European Union, arts 1–3.

108 Charter of Fundamental Rights, art. 3. In this regard, it was shown that these provisions most clearly and specifically express the close link among

The question arising, then, is if the evaluation of bioethical profiles of patent applications should be based on the ECHR and on the Charter of Nice.

With reference to the ECHR, it has been shown that a positive response to this query may lead to the belief that all European regulations subsequent to the Convention should be interpreted with respect to the human rights it includes. Thus, even where there are no moral regulations in the Directive, the resolution of relative issues should be based on the provisions of the ECHR. A similar approach based on rights would be focused on the interests of the individual and would be aimed at protecting him from claims according to which the damages caused to him are justified in the interest of society or third parties.[109] What is more, it has also been held that, on the one hand, where the Directive is interpreted according to moral criteria based on the Convention, the result may be in conflict with existing case law, not only among national patent offices, but also within the EPO;[110] on the other, an evaluation based only on the respect for human rights provided in the ECHR may lead to the result of not considering other instances which are relevant within the sphere of biotechnology, such as its economic and social impact.[111] On the other hand, given that the Convention is binding (also) in regard to national patent offices, the failure to apply rigorous moral evaluations gives rise to its violation.

The issue seems to have found specific conclusive criteria following the creation, and in particular the subsequent 'constitutionalization', of the Charter of Nice. The latter, currently,[112] has legal value equal to that of European Union Treaties: thus, it seems to be believed that EU

---

personal dignity, independence and identity, found already in art. 2 of the UNESCO Universal Declaration on the Human Genome and Human Rights of 1997, according to which 'dignity makes it imperative not to reduce individuals to their genetic characteristics and to respect their uniqueness and diversity', and art. 1 of the European Convention on Human Rights and Biomedicine of the European Council of the same year, which provides the commitment of the states 'to protect the dignity and identity of all human beings and guarantee everyone, without discrimination, respect for their integrity and other rights and fundamental freedoms with regard to the application of biology and medicine'.

[109] In this regard, see J. Herring, *Medical Law and Ethics*, Oxford: Oxford University Press, 2008 (2nd edn).

[110] On the other hand, in this regard refer to the evolution of the same examined below, at para. IV.2.

[111] See R. Ford, 'The Morality of Biotech Patents: Differing Legal Obligations in Europe?', cit., p. 318.

[112] Pursuant, as said, to art. 6 of the EU Treaty.

regulations must, without a doubt, comply and be interpreted on the basis of the fundamental principles included within it. In this sense, it must be noted that the Charter itself, in article 52(1), provides specifically that: 'Any limitations to the use of rights and freedoms recognized by this Charter must be provided by the law and respect the essential content of the aforementioned rights and freedoms. In regards to the principle of proportionality, limitations may be made only where they are necessary and answer effectively to the purposes of general interest recognized by the Union or to the need to protect the rights and freedoms of others'. What is more, regarding the relationship between the Charter and the ECHR, in article 52(3), it is provided that: 'Where this Charter contains rights corresponding to those guaranteed by the European Convention for the safeguarding of the Rights of Man and fundamental freedoms, the meaning and range of these are equal to those conferred by the aforementioned convention'.[113] Through this provision, then, the regulations of substantial law of the ECHR are 'incorporated' into the Charter,[114] and thus inserted together with it at the head of the European Union legal system, as indispensable parameters of reference and interpretation for every European regulation.

Returning to the provision of article 6 of Directive 98/44/EC, in any case, it must be read in light of Recital 38 of the same, which states[115] that the list inserted in the aforementioned regulation is aimed at providing judges and national patent offices with general guidelines for the interpretation of the reference to public order or morality, but that in any case this list cannot be considered exhaustive. As a consequence, procedures whose application creates prejudice to human dignity, such as for instance procedures for the production of hybrid beings resulting from human or animal, germinal or totipotent cells, must also be excluded from patentability.

The reasoning of article 6 of the Directive is to provide a 'catalogue' of inventions whose patentability would be contrary to public order or morality, in order to 'substantiate' these concepts. Thus, for the limited aim of the subject of the Directive on biotechnological inventions, the European legislature intervened to identify common concepts of public

---

[113] Specifying, also, that: 'This provision does not preclude that Union law grant more extensive protection'.

[114] In this regard, see in particular P. Šturma, 'The EU Charter of Fundamental Rights and the Governance of Biotechnology in the European Union', cit., p. 382 et seq.

[115] As mentioned above.

order and morality, though merely through a list of some 'specific violations' and in any case in a non-clarifying manner.[116]

This provision raises important issues to be resolved, which show the relevance of bioethical profiles in the application of the Directive. In this sense, it is useful to consider the example of substantial medical utility for man or animal contemplated in article 6(2)(d) on the theme of the patentability of transgenic animals, the hypothesis presented in the list of exceptions to patentability which appears logically nearer to the provisions of the EPC. This provision of the Directive also should be interpreted on the basis of Recital 45 of the same, which limits itself to listing the fields in which such utility may be found, or rather research, prevention, diagnosis or treatment. Given that the Recital is expressed in such broad terms, this suggests that in the interpretation of the aforementioned regulations the meaning attributed to the concepts of public order and morality must consequently be extended. With regard to article 53 EPC, then, it can without a doubt be stated that the provisions of the Directive pursuant to article 6 and the related Recital require more extensive bioethical control.

Similarly, article 5(1) of the Directive includes bioethical considerations, in that it essentially prohibits the patentability of the human body establishing that: 'The human body, in its various stages of constitution and development, as well as the mere discovery of one of its elements, including therein the sequence or partial sequence of a gene, may not be considered patentable inventions.'

The reasoning of the rule is to specify that the human body, in every stage of its formation and development, does not constitute patentable subject matter; this is in order to specifically clarify the status of the non-appropriability and inalienability of the human body.[117] What is

---

[116] In this regard, besides referring above to para. IV.1.1., see also S.J.R. Bostyn, 'A Decade After the Birth of the Biotech Directive: Was it Worth the Trouble?', in E. Arezzo and G. Ghidini (eds), *Biotechnology and Software Patent Law: A Comparative Review of New Developments*, Cheltenham, UK and Northampton, MA, USA, Edward Elgar, 2011, p. 221 et seq. (236).

[117] The aforementioned rule, then, functions also as a limit to the criteria of extension for the protection of material in which the product is incorporated, provided (besides that for biological material in art. 8) in particular in art. 9 of the Directive, according to which in fact: 'Except for article 5, paragraph 1, the protection granted by a patent to a product containing or consisting in genetic information is extended to any material in which the product is incorporated and in which the genetic information is contained and carries out its function' (in this regard, see in particular S.J.R. Bostyn, 'A Decade After the Birth of the Biotech Directive: Was it Worth the Trouble?', cit., p. 243 et seq.).

more, the protection of human dignity, a principle which inspired the legislature when laying out the provisions of the Directive,[118] is confirmed. In particular, it was stated that the words 'in various stages of its constitution and development' refer also to the initial stages, implying that, for instance, a blastocyst[119] could not be the subject of patent protection.[120]

The provision, fundamentally, is aimed at the preclusion of granting patents for speculative applications, which present experimental data based entirely on DNA sequences; thus, DNA sequences which codify proteins with unknown functions are not patentable. In this sense, article 5 should be read in light of the statement in Recital 23, or that a simple DNA sequence without indication of a function does not contain any technical teaching and may not then be considered a patentable invention.[121] The provision, clearly, considers the difficulty in identifying the moment in which it may be held that the sequence of DNA does not contain any technical teaching.

Article 5(3), then, provides that the industrial application of a gene sequence or partial sequence must be specifically indicated in the patent

---

[118] In particular, such provisions are based on that shown in Recital 16 of the Directive, according to which 'patent law must be exercised in respect to fundamental principles which guarantee the dignity and integrity of man; for which it is necessary to reaffirm the principle according to which the human body, in every stage of its constitution and development, including germinal cells, the simple discovery of one of its elements or one of its products, as well as the sequence or partial sequence of a human gene, are not patentable; that such principles are not in compliance with the criteria of patentability provided by patent law, according to which a simple discovery cannot constitute the subject of a patent'.

[119] The blastocyst is an embryonic phase that in mammals is formed in the initial phases of the process of embryogenesis, with the formation of blastocele; generally blastocysts are formed from the fourth to 14th day after fertilization.

[120] In this sense, see S.J.R. Bostyn, A Decade After the Birth of the Biotech Directive: Was it Worth the Trouble?', cit., p. 223. In this regard, refer also below, to para. IV.2.

[121] The proposal to introduce a 'genetic sequence right' was proposed, which would not be under the control of a single person or body. The proposal recognizes that beyond the fact that a sequence of genes represents an invention or not, the identification of its function represents an important contribution and should be encouraged, allowing the researcher to receive a fee in exchange for disclosure, rather than the granting of exclusive rights as in the patent system. In this regard, see L. Palombi, 'The Genetic Sequence Right: A Sui Generis Alternative to the Patenting of Biological Material', in J. Gibson, *Patenting Lives: Life Patents, Culture and Development*, cit., p. 75 et seq.

application.[122] The debate on the rule is centred on two possible interpretations of it. According to the first, it is nothing more than a clarification of the requirement of industriality. According to another interpretation, instead, with this provision protection limited to the purposes of the product would be introduced for inventions concerning genes.[123] What is more, the latter thesis would seem to be strengthened by what is stated in Recital 24 of the Directive, on the basis of which, in order to ensure respect for the criterion for industrial application, in the case of partial gene sequences used to produce a protein or partial protein, it is necessary to specify which protein or partial protein is produced, or what function it serves.[124] The Court of Justice seems to have accepted the thesis of non-absolute protection, but limited to specific purposes, in the ruling on the case *Monsanto v. Cefetra*,[125] and

---

[122] The provision was inspired by patent applications presented in 1995 by the NIH, which concerned partial sequences of DNA without any noted function (see S.J.R. Bostyn, 'Enabling Biotechnological Inventions in Europe and the United States. A study of the patentability of proteins and DNA sequences with special emphasis on the disclosure requirement', Eposcript Series, no. 4, EPO, Munich, 2001, p. 134 et seq.).

[123] See, among others, S.J.R. Bostyn, 'A Decade After the Birth of the Biotech Directive: Was it Worth the Trouble?', cit., p. 225 et seq.; L. Van Raden and D. Von Renesse, '"Überbelohnung" – Anmerkungen zum Stoffschutz für biotechnologische Erfindungen' (2002) *GRUR* 393 et seq.; in favour of absolute protection, R.S. Crespi, 'Gene and Compound Claims – Another View' (2001/2002) 5 *Bio-Science Law Review* 3 et seq. (as well as other references cited therein).

[124] In general, according to the principle accepted in European law and in all those of the single Member states, the patent provides absolute protection of the product, that is it covers all the uses of it and the methods of its production. What is more, the possibility of obtaining protection for new uses of patented products or new procedures for the production of them is provided; these latter patents shall have a sphere of protection limited to the specific medical aim pursued, and shall be dependent on the patent of the product (art. 54(4) and (5), EPC; on this theme, see V. Falce, *Profili pro-concorrenziali dell'istituto brevettuale*, Milan: Giuffrè, 2008, p. 242 et seq., who, regarding the distinction between a product patent and new use patent, clarifies that the first is necessary when from the state of the technique it was not possible to predict the creation of a product dedicated to a certain single use, while the second is used when the teaching to reach a product was already available previously and only the particular function found was not obviously accessible).

[125] *Monsanto Technology LLC v. Cefetra BV, Cefetra Feed Service BV, Cefetra Futures BV, Alfred C. Toepfer International GmbH*, decision of 6 July 2010, case C-428/08, [2010] E.C.R. I-6765.

the EPO frames it in the context[126] of specification of the requirement of industriality.[127] Similarly, some Member states, such as France, Germany[128] and Italy,[129] introduced protection of a product for limited purposes for genetic inventions. On this point it seems, then, that clarification is necessary.[130] In its current state, by virtue of the meaning of the regulation pursuant to article 5(2), and while awaiting further clarification on the point, it seems possible to say that the preferable interpretation of the rule is that based on which the test of industriality in cases regarding gene sequences has a higher standard than in those which concern other inventions, and in particular that the protection of genetic inventions would be limited to the purposes of the product.

Article 5(1) of the Directive, then, states that the simple discovery of an element of the human body is not patentable, but does not indicate a reason. The exclusion appears justified, *prima facie*, by traditional patentability criteria, for which discoveries are not in and of themselves patentable. It is natural to ask, further, if the exception can also be justified on a bioethical basis.

---

[126] Inspired, as said, by Recital 24 of the Directive.

[127] See S.J.R. Bostyn, 'A Decade After the Birth of the Biotech Directive: Was it Worth the Trouble?', cit., p. 227 et seq.

[128] *Patentgesetz*, para. 1, regarding which see C. Kilger, J. Feldges and H.R. Jaenichen, 'The Erosion of Compound Protection in Germany: Implementation of the EU Directive on the Legal Protection of Biotechnological Inventions. The German Way' (2005) 87 *Journal of Patent & Trademark Office Society* 569 et seq.; see also M.A. Bagley, 'A Global Controversy: The Role of Morality in Biotechnology Patent Law', cit., p. 334.

[129] 'Codice della proprietà industriale' (legislative decree dated 10 February 2005, no. 30, as modified by legislative decree dated 13 August 2010, no. 131), art. 81 *quater* (d), and art. 81 *quinquies*. In doctrine, see O. Capasso, 'La brevettabilità degli elementi naturali (art. 81 quater, D.Lgs. 10.2.2005, n. 30)', and Capasso, 'Le ulteriori categorie di trovati biologici esclusi dalla brevettabilità e la limitazione della protezione alla funzione indicata per i brevetti aventi ad oggetto sequenze o sequenze parziali di un gene (art. 81 quinquies, D.Lgs. 10.2.2005, n. 30)', in C. Galli and A.M. Gambino, with the collaboration of V. Falce (eds), *Codice commentato della proprietà industriale e intellettuale*, Turin: Utet, 2011, p. 821 et seq.; V. Di Cataldo and E. Arezzo, 'Scope of the Patent and Uses of the Product in the European Biotechnology Directive' (2007) *Italian Intellectual Property* 11 et seq.

[130] Thus S.J.R. Bostyn, 'A Decade After the Birth of the Biotech Directive: Was it Worth the Trouble?', cit., p. 234.

Article 5(2)[131] specifies that an element which is isolated from the human body or otherwise produced, through a technical process, including the sequencing or partial sequencing of a gene, may constitute a patentable invention, even if the structure of this element is identical to that of a natural element.[132] In this regard, it is clear that it cannot be held that a gene is no longer a gene simply because it has been isolated from the human body. The identification of a gene sequence and the protein it codifies is merely a discovery as long as the gene remains in the human body, given that at this stage man cannot produce the protein. However, the isolation of the gene and its transfer into a new environment in which the protein can be produced constitutes an invention.[133] Moreover, as far as concerns the reference to elements produced through a technical procedure, it has been shown that in genetic engineering the cDNA is always produced as a result of such a procedure, and thus in this regard specification is not necessary. Similarly, regarding the particular requirement of novelty provided for isolated elements of the human body in the final part of the rule, according to which it is not specified that the structure of the identified element be identical to that of a natural element, the test does not concern the existence of the substance, but rather the fact that it is known in terms of its structure,[134] and also in this case with regard to cDNA the formula risks being misleading, given that cDNA does not exist in nature.[135]

Unlike the United States, then, in Europe bioethical profiles are specifically outlined within patent regulations. Both the EPC and Directive 98/44/EC, in fact, deal with the moral issues raised by biotechnological inventions. As a consequence, though the provision of a general ban on human cloning is lacking, certain inventions in the sphere of human genetics are expressly excluded on a bioethical basis.

---

131 Confirming the previous decision making practice of the EPO.

132 On this point, see S.J.R. Bostyn, 'Patenting DNA sequences (polynucleotides) and scope of protection in the European Union: an evaluation – Background study for the European Commission within the framework of the Expert Group on Biotechnological Inventions', Brussels, European Commission, 2004 (accessed 18 December 2014 at http://ec.europa.eu/internal_market/indprop/docs/invent/patentingdna_en.pdf), p. 37 et seq.

133 See U. Schatz, 'Patents and Morality', in Sterckx, *Biotechnology, Patents and Morality*, cit., p. 159 et seq.

134 See S. Crespi, 'Biotechnology Patenting: The Wicked Animal Must Defend Itself' (1995) 9 *European Intellectual Property Review* 431 et seq.

135 See O. Mills, *Biotechnological Inventions. Moral Restraints and Patent Law*, cit., p. 143.

In this regard, as mentioned, the provisions of some human rights conventions, which have significant influence at a European level, are also specifically interesting. In particular, the ECHR, in line with the Universal Declaration on Human Rights[136] and other similar human rights conventions,[137] calls for respect for human dignity. On the basis of such conventions, fundamental rights are guaranteed: the dignity, life and liberty of man, as well as, among others, the right to enjoy the benefits of scientific progress. Moreover, the UNESCO Universal Declaration on the Human Genome and Human Rights[138] states that research on the human genome and human genetics must respect ethical standards of society.[139] The Declaration promotes the development of ethical studies in view of scientific and technological progress in the fields of biology and genetics.[140] The basic principle of the Declaration is that no biological or genetic research should prevail over the respect for human rights and human dignity.[141] Freedom of research is recognized as being a part of freedom of thought and is encouraged, especially in the fields of biology, genetics and medicine, as a necessary means for guaranteeing the right to health and the relief of human suffering.[142] On the other hand, the Declaration pushes Member states to establish ethics committees in order to evaluate ethical, legal and social issues raised by research on the human genome and its applications. These ethics committees should also take the initiative in promoting education in bioethics.[143] What is more, the Declaration aims to identify the practices which may be contrary to human dignity and ethics in particular, such as, actions affecting germinal lines.[144] Finally, Member states were mandated to adopt appropriate legal

[136] Approved on 10 December 1948 in Paris by the General Assembly of the United Nations.

[137] See International Covenant on Civil and Political Rights, adopted by the General Assembly of the United Nations on 16 December 1966 and in force since 23 March 1976, and International Covenant on Economic, Social and Cultural Rights, adopted by the General Assembly of the United Nations on 16 December 1966 and in force since 3 January 1976.

[138] 'Universal Declaration on the Human Genome and Human Rights', adopted by the General Conference of UNESCO on 11 November 1997 and acknowledged by the General Assembly of the United Nations in 1998.

[139] Declaration, arts 13 and 14.

[140] Declaration, preamble.

[141] Declaration, art. 10.

[142] Declaration, art. 20.

[143] Declaration, arts 16 and 20.

[144] Declaration cit., art. 26.

measures to guarantee respect for ethics and human rights in light of the progress made in the field of research on human genetics.[145]

A further action of significant interest on this subject is represented, then, by the Convention on Human Rights and Biomedicine.[146] The Convention, aimed at protecting the dignity of human beings who are the subjects of research in the biomedical sciences,[147] provides that research on human beings, tissues, organs or the human genome may be executed only after having received the informed consent of the individual involved, who must be aware of the possible related risks.[148] The interest and well-being of the individual must prevail over the exclusive interest of the firm or science.[149] The use of the human body and its parts for purposes of profit is prohibited by the Convention,[150] which, given recent developments in research regarding human embryos in particular, attempts to ensure it adequate protection. In this sense, the creation of human embryos for purposes of research is held to be contrary to ethics and thus is prohibited.[151] From this perspective, then, the Convention encourages greater debate on ethics regarding developments in the field of biomedicine.[152]

In order to evaluate the ethical and social effects of biomedical research, finally, advisory groups were established in Europe. In particular, some positions expressed by these groups are interesting. In this sense, the European Group on Ethics in Science and New Technologies has intervened at various times in support of the European Commission, on the main themes raised by the development of biotechnological research. In its Opinions, further, in general, it was recommended that human dignity be guaranteed in light of developments in biomedical research, and that research on human genetics, especially on the human body, be carried out solely for therapeutic purposes and after having obtained the informed consent of the person involved, and it was further

---

[145] In this regard, see G. Kamstra *et al.*, *Patents on Biotechnological Inventions: The EC Directive*, London: Sweet and Maxwell, 2002, p. 21.

[146] European Council, 'Convention for the protection of Human Rights and dignity of the human being with regard to the application of biology and medicine: Convention on Human Rights and Biomedicine', CETS No. 164, Oviedo, 4 April 1997, entered into force on 1 December 1999.

[147] Convention, arts 1, 15 and 17.

[148] Convention, arts 13, 21 and 22.

[149] Convention, art. 5.

[150] Convention, arts 2, 21 and 22.

[151] Convention, art. 18.

[152] Convention, art. 28. On this point, see C. Campbell, 'A Commission for the 21st Century' (1998) 61 *Modern Law Review* 598 et seq.

shown that in the context of procedures before the EPO and national patent offices, it may be necessary to proceed with ethical evaluations, with the help of independent experts. In particular, then, it was stated that patents should not be granted for the human body or its elements at various levels of its constitution and development, that the use of human tissues in biomedical and genetic research should ensure respect for human dignity and integrity, that a restrictive application of the principle of non-commercialization of human embryos is appropriate (in line with that of non-commercialization of the human body), and that the ethical implications of reproductive cloning lead to the belief that it should be prohibited.[153]

## IV.2 BIOTECHNOLOGICAL INVENTIONS IN CASE LAW

### IV.2.1 The Evolution of Patent Protection

Case law has played an extremely important role in Europe as well, though less decisively than American case law, on the path to patentability of living beings. The EPO and the decisions of its Technical Board have contributed significantly to the evolution of biotechnological patent law, though in Europe[154] this has found its basis in the legislative provisions of the EPC and Directive 98/44/EC.

#### IV.2.1.1 The patentability of microorganisms

As early as 1969, a patent application in Germany was filed for a method for breeding doves with red feathers. The German patent office denied the application on the grounds that the method was not replicable, and the decision was confirmed by the *Bundespatentgericht*.[155] This was the

---

[153] See, in particular, European Group on Ethics in Science and New Technologies, 'Opinion n°16 – Ethical aspects of patenting inventions involving human stem cells', 7 May 2002; 'Opinion n° 15 – Ethical aspects of human stem cell research and use', 11 November 2000; 'Opinion n° 11 – Ethical aspects of human tissue banking', 21 July 1998; 'Opinion n° 9 – Ethical aspects of cloning techniques', 28 May 1997; and 'Opinion n° 8 – Ethical aspects of patenting inventions involving elements of human origin', 25 September 1996. In doctrine, see G. Kamstra *et al.*, *Patents on Biotechnological Inventions: The EC Directive*, cit., p. 20; C. Campbell, 'A Commission for the 21st Century', cit., p. 598 et seq.

[154] As seen above.

[155] Federal patent court. With regard to the case, see M. Adelman, R.R. Rader, J.R. Thomas and H.C. Wegner, *Cases and Materials on Patent Law*,

first case in which a procedure for the production of a living organism was the subject of a patent application in a country of today's European Union, and it was the first case which opened the door to the patentability of biotechnological inventions.

At the beginning of the 1970s, then, five years prior to the U.S. Supreme Court ruling on the *Chakrabarty* case, the *Bundespatentgericht* declared the patentability of new microorganisms, under the condition that the inventor could demonstrate the possibility of replicating the method used for their production.[156]

Meanwhile, the EPC was adopted, which excluded the patentability of products obtained from essentially biological processes, but not those which are not biological.

In the case of *Genentech-I/Polypeptide expression,*[157] for the first time after the adoption of the Convention, a patent application was filed for a living being. The invention was a plasmid, a microorganism adapted to transform host bacteria, which included a control sequence of the expression, or 'regulon', which would allow the expression of foreign DNA as a recoverable polypeptide (protein). The applicants for the patent invented a general principle which would allow the plasmids to control the expression of the polypeptides in the bacteria. The Board of Appeal of the EPO, considering that the Convention excluded the patentability of processes and relative, essentially biological (and not those which are not biological), products, held that the invention in question, not being a product of such a process, was patentable.

The decision of the Board, then, allowed the patentability of microorganisms at the European level, giving rise to the granting of patents for procedures and relative biotechnological products, also in regard to other life forms, such as plants, animals and human genetic material.[158]

---

Eagan (USA): West Group, 1998, p. 156; F. Lederer, 'A Perspective on Patenting Microorganisms under the European Convention: Prospects and Considerations' (1979) 7 *APLA Quarterly Journal* 288 et seq.

156 *Bundespatentgericht*, decision *Bäckerhofe*, BGHZ 64, GRUR 430 (1975). In this regard, see N.S. Sreenivasulu and C.B. Raju, *Biotechnology and Patent Law. Patenting Living Beings*, Noida (India): Manupatra, 2008, p. 39; F. Lederer, 'A Perspective on Patenting Microorganisms under the European Convention: Prospects and Considerations', cit., p. 288 et seq.

157 T 292/85 (1989), EPO Official Journal, 1989, 275.

158 See M. Adcock and M. Llewelyn, 'TRIPS and the Patentability of Microorganisms' (2000/2001) 4 *Bio-Science Law Review* 91 et seq. As far as plants are concerned, in Europe the EPC, in art. 53, excludes the patentability of plant varieties and essentially biological procedures for the production of plants (but not for microbiological procedures and relative products). Plant varieties, on the other

### IV.2.1.2 The patentability of animals

The evolution of biotechnological patent law at the European level took another fundamental step with the patentability of animals.[159]

Such a patent application was registered, for the first time in European history, in the case of the *Harvard Onco-Mouse*.[160] The invention concerned a transgenic mammal, a mouse with foreign genes (oncogenes)[161] which would make it more predisposed to developing cancer, and thus useful in related research. The inventor presented[162] two patent applications for a method to produce a non-human transgenic mammal predisposed to cancer and containing oncogenes and for the relative product.[163]

---

hand, are protected on the basis of EC Regulation 2100/94 (inspired by the UPOV Convention). The theme, further, as noted, goes beyond the subject of this book (in this regard, among others, see C.B. Onwuekwe, 'Plant Genetic Resources and the Associated Traditional Knowledge: Does the Distinction between Higher and Lower Life Forms Matter?', in J. Gibson (ed.), *Patenting Lives: Life Patents, Culture and Development*, Farnham (UK)/Burlington (USA): Ashgate Publishing, 2008, p. 139 et seq.; M.R.P. Temmerman, 'The Patentability of Plant Genetic Inventions', NCCR *Trade Regulation Working Paper* 2007/04, Berne, 2007, accessed 18 December 2014 at http://phase1.nccr-trade.org/images/stories/publications/IP9/The%20Patentability%20of%20Plant%20Genetic%20Inventions,%20Final%20version.pdf; G. Van Overwalle, 'Patent Protection for Plants: A Comparison of American and European Approaches' (1999) 39 *Journal of Law & Technology* 143 et seq.)

[159] On this theme, see among others, N. Peace and A. Christie, 'Intellectual Property Protection for the Products of Animal Breeding' (1996) 4 *European Intellectual Property Review* 213 et seq.; R. Moufang, 'Patentability of Genetic Inventions in Animals' (1989) 20 *International Review of Industrial Property and Copyright Law* 823 et seq.; R.S. Crespi, 'Prospects for International Cooperation', in W. Lesser (ed.), *Animal Patents: The Legal, Economic and Social Issues*, London: Macmillan, 1989, p. 30 et seq.; B. Hoffmaster, 'The Ethics of Patenting Higher Life Forms' (1988) 4 *Intellectual Property Journal* 1 et seq.

[160] Technical Board of Appeal, T 19/90 (1990) EPO Official Journal, 1990, 476.

[161] The sequence of oncogenes was introduced into the animal in the embryonic stage.

[162] Following the application filed in the United States, with respect to which the USPTO had granted the patent in 1988 (in this regard, refer above, to Part III).

[163] The method provided for the introduction of a sequence of activated oncogenes into the mammal (with the explicit exclusion of its applicability to human beings).

Initially, the patent applications were denied by the Examining Division of the EPO on the basis of the fact that the invention was contrary to public order and morality and also of the fact that the invention concerned an animal breed, excluded from the sphere of patentability pursuant to the EPC.[164] What is more, the application for a non-human mammal was considered too broad with regard to the description of the invention, which concerned only a method for a transgenic mouse.

On appeal, the EPO Technical Board of Appeal reformed the interpretation of article 53(a) and (b) of the Convention which excluded the animal invention in question from patentability. The Board held that the Examining Division was wrong in denying the patent application on the basis of such exceptions to patentability. On the contrary, the Board, specifying that every invention which does not fall within the sphere of exclusions or exceptions to patentability must be considered patentable,[165] held that on the basis of the Convention text animal inventions are not excluded from patentability.

Thus, the issue raised before the Court was to decide whether or not the invention fell under one of the exceptions provided by the Convention. Under this framework, the Board came to analyze and interpret the concepts of public order, morality, animal breed, essentially biological processes and microbiological processes, and so on.

With reference to the exception for public order and morality pursuant to article 53(a) EPC, it was shown that biotechnological inventions have different implications that require them to be evaluated carefully. These include the suffering of animals, as well as possible risks to the environment, which must be weighed against the utility of the invention for humanity. The Board, then, considered that where the risk associated with the invention was greater than and could not be balanced with the positive aspects, issues of public order and morality may arise and the patent application could be denied. Where the new technology involves higher forms of life, not only should the risk be evaluated, but also the possible threat which may be caused by such a life form.

In this case, various interests were involved which required balance: one was the basic interest of humanity in curing widespread and dangerous illnesses; another was the protection of the environment from the uncontrolled distribution of unwanted genes; a third was against

---

[164] The EPC, basically, in art. 53(a), excludes patent protection for inventions whose use is contrary to public order and morality, and in art. 53(b), excludes the patentability of animal varieties.

[165] With a restrictive interpretation of the latter (according to the precedents in case law; see *Lubrizol/Hybrid Plants*, T320/87, 1990 EPOR 173).

cruelty to animals. If the invention caused some damage to the environment or suffering to animals the denial of the patent on the basis of public order and morality would have been justified.[166]

With regard to the invention in question, the Board showed that its utility for humanity could not be denied. Cancer is one of the most frequent causes of death in many countries and the cause of great suffering. Every contribution to the development of new anti-cancer treatments represented, thus, a significant benefit for humanity and was to be considered worthy of incentive. Given that there were no alternatives to experimentation on animals, the models of animal testing were indispensable for cancer research.[167]

The invention in question, further, gave rise to the need for a smaller number of animals compared to traditional experiments. Thus, the invention contributed to reducing the overall extent of animal suffering. Therefore, in the interest of humanity, the invention could be granted, despite its causing animal suffering.[168]

In consideration of the possible risks to the environment caused by the spread of genetically modified inventions, the Board evaluated the aim of the invention, which was to provide subjects for animal testing, to be used exclusively in the laboratory, in controlled conditions, by qualified staff. As a consequence, there would be no risk of uncontrolled distribution, except in the case of any improper use by the laboratory personnel where the test would take place, a hypothesis which did not appear to be an adequate basis for denial of the patent.

In conclusion, then, it was held that the invention which was the subject of the application was not contrary to public order or immoral, in consideration of the fact that it was of great use to cancer research, a social utility which was more important than animal suffering.[169]

As for the evaluation regarding the nature of animal breeds of the invention pursuant to article 53(b) EPC, it was shown that the invention

---

[166] See W.R. Cornish, *Cases and Materials on Intellectual Property*, London: Sweet and Maxwell, 1999 (3rd edn), p. 62 et seq.

[167] On this point, see A. Berns, 'Separating the Wheat from the Chaff' (1991) 1 *Current Biology* 28 et seq.

[168] In this regard, see, among others, A. Ottolia, 'Moral Limits to Biotech Patents in Europe: A Quest for Higher Harmonization', in E. Arezzo and G. Ghidini (eds), *Biotechnology and Software Patent Law: A Comparative Review of New Developments*, cit., p. 309 et seq. (314 et seq.); G. Porter, 'The Drafting History of the European Biotechnology Directive', cit., p. 3 et seq. (12).

[169] See B.A. Brody, 'Intellectual Property and Biotechnology: The European Debate', cit., p. 91 et seq.

gave rise to a transgenic animal, and did not constitute, instead, an animal breed.

In regard to the ban on the patenting of essentially biological processes, the Board highlighted that the process in the application required the insertion of an oncogene into a vector (a plasmid), which had to be micro-injected into the ovum in an early stage of the embryo, and this would not be possible through essentially biological processes. The Board evaluated the invention as being the result of a microbiological process, which is not excluded from patentability.

Finally, the Board took into consideration the issue regarding the fact that the patent application for the production of a non-human mammal was too broad, in that the description of the invention only identified a method for producing a transgenic mouse. The Board held that for the biological invention it is sufficient that at least one method be clearly indicated so that an expert in the field is able to recreate it, while it is not necessary to indicate all manners of putting the invention into practice on various mammals. As a consequence, it was decided that the disclosure of one non-natural method of producing transgenic mice was sufficient to put it into practice on other non-human mammals, and the application for the method of producing non-human transgenic mammals was valid.

In conclusion, then, the Examining Division intervened to review its previous decision,[170] stating in particular that the invention could not be considered immoral pursuant to article 53(a) EPC, and was therefore patentable and highlighting, however, that the evaluations carried out were applicable only to this case.

The test adopted for the decision, then, did not indicate a definition of the concept of morality, nor a rational basis for the choice of those particular factors for the balancing in opposition to other possible circumstances.[171] On the other hand, it did provide the EPO with criteria for the evaluation of the patentability of morally controversial inventions pursuant to article 53(a).[172]

---

[170] Examining Division (1991) EPO, R. 525.

[171] See O. Mills, *Biotechnological Inventions. Moral Restraints and Patent Law*, cit., p. 59; R. Nott, 'The Biotech Directive: Does Europe Need a New Draft?' (1995) 17(12) *European Intellectual Property Review* 563 et seq.; A. Abbott, 'Oncomouse Hearing Ends Up in Confusion' (1995) 378 *Nature* 427.

[172] See S.R. Donnelly, 'The Patentability of Human Embryonic Stem Cells', cit., p. 110; M.A. Bagley, 'A Global Controversy: The Role of Morality in Biotechnology Patent Law', cit., p. 331 et seq.; H. Jaenichen and A. Schrell, 'The "Harvard Onco-Mouse" in the Opposition Proceedings Before the European Patent Office' (1993) 15 *European Intellectual Property Review* 345 et seq. In

What is more, the decision was subsequently the subject of much opposition before the EPO,[173] which only came to a final decision in 2004,[174] confirming the validity of the patent but restricting the field of application only to mice.[175]

With the *Harvard Onco-Mouse* decision, then, a patent was granted for the first time at a European level for a transgenic animal, opening the door to the patentability of similar biotechnological products.

---

this sense, for instance, in the case of *Upjohn's Application* the EPO denied the patent application for a transgenic genetically modified, with an oncogene, animal, so that it would lose its hair and would thus be useful for studies on human baldness. In this regard, the EPO held that the suffering of the animal was not justified by the less relevant aim pursued, that is the cure for baldness (in regard to the case, for which refer, extensively, below to para. IV.2.2, see M.A. Bagley, 'A Global Controversy: The Role of Morality in Biotechnology Patent Law', cit., p. 332; R. Nott, 'The Biotech Directive: Does Europe Need a New Draft?', cit., p. 563 et seq. (565 et seq.); S. Conner, 'Patent Ban on Baldness "Cure" Mouse', *The Independent*, 2 February 1992, p. 5). This so-called 'balancing of interest' test was not followed, instead for the decision regarding the patentability of transgenic plants, *Greenpeace v. Plant Genetic Systems* (1995 EPOR 357, 373), in which the Technical Board of the EPO judged a patent on transgenic plants developed to be resistant to a certain class of herbicide valid, holding that the invention did not contravene to requirements of non-contrariety to public order and morality pursuant to art. 53(a), EPC. Denying specifically, also, the utility of the test used in the decision of Harvard Onco-Mouse, in that it was not only a method for the evaluation of patentability, but rather 'just one possible way, perhaps useful in situations in which an actual damage (e.g., suffering of animals) … exists'. This circumstance in this case was not relevant, so it was deemed appropriate instead to concentrate on the 'public acceptability' of the categories of patentable subject matter, which did not appear to violate the provisions of art. 53(a) given that they concerned 'activities (production of plants and seeds, protection of plants from weeds or fungal diseases) and products (plant cells, plants, seeds) which cannot be considered to be wrong as such in the light of conventionally accepted standards of conduct of European culture' (see S.R. Donnelly, 'The Patentability of Human Embryonic Stem Cells', cit., p. 111; M.A. Bagley, 'A Global Controversy: The Role of Morality in Biotechnology Patent Law', cit., p. 332 et seq.; A. Ottolia, 'Moral Limits to Biotech Patents in Europe: A Quest for Higher Harmonization', cit., p. 316).

[173] See H.R. Jaenichen and A. Schrell, 'The Harvard Oncomouse in the Opposition Proceedings Before the European Patent Office', cit., p. 35 et seq.

[174] EPO Board of Appeals, case T 0315/03/3.3.8, 6 July 2004.

[175] In regard to this decision, see B.A. Brody, 'Intellectual Property and Biotechnology: The European Debate', cit., p. 69 et seq. (97 et seq.).

### IV.2.1.3 The patentability of genetic material and human cells

After transgenic animals it was the turn of human genetic material, to be the subject of patent applications in Europe. With the development of biotechnology, human genetic material, such as cells, genes and gene sequences, DNA and sequences and fragments of DNA began to be the subject of patent applications, together with the products of such genetic material, like proteins, antibodies, and so on.

In the case of *Howard Florey/Relaxin*,[176] the EPO, for the first time in Europe, found itself having to decide whether human genetic material was patentable or not.[177] The application concerned a fragment of DNA which codified a hormone called relaxin, capable of provoking light muscle contractions in the human body. In particular, the hormone is found in a pregnant woman's body during labour to relax the pelvic structure in order to facilitate the delivery of the baby during childbirth. The DNA sequence which codified relaxin for the invention was taken from a pregnant woman and then the DNA was integrated into a host bacterium and afterwards used for the industrial production of relaxin.

The application was the subject of opposition based on the fact that human genetic material does not constitute patentable subject matter, pursuant to article 53(a) EPC. What is more, it was held that the exploitation of a pregnant woman was immoral and constituted a violation of human dignity. Finally, it was stated that the patentability of human genes was equal to ownership of rights for the human body, contrary to the right of man to self-determination, putting it at the same level as slavery.

The Opposition Division of the EPO rejected the objections and the patent was granted. The Division stated, in particular, that:

> The function of the Article has to be seen as a measure to ensure that patents would not be granted for inventions universally regarded as outrageous. This interpretation reflects the 'fair test' provision in the relevant passages of the Guidelines for Examination,[178] which state that Article 53(a) is likely to be invoked only in rare and exceptional cases.[179]

In this manner, it supported the 'light' bioethical system with regard to the application of article 53(a), but on the other hand, with respect to the

---

[176] *Howard Florey/Relaxin v. Fraktion der Grünen im Europäischen Parlament*, EPO Official Journal, 1995, 388; 1995 EPOR 541.

[177] See D. Beyleveld and R. Brownsword, 'Human Dignity, Human Rights, and Human Genetics' (1998) 61 *Modern Law Review* 661 et seq.

[178] 'Guidelines for Examination in the European Patent Office', cit.

[179] Decision cit., 6.2.1.

previous decision on the Harvard Onco-Mouse, they went from the 'balancing of interest' test to that of 'public abhorrence',[180] specifying in this sense that:

> A fair test to apply is to consider whether it is probable that the public in general would regard the invention as so abhorrent that the grant of patent rights would be inconceivable. If it is clear that this is the case, objection should be raised under Article 53(a); otherwise not.[181]

According to the Division, in particular, in this case it would have been repugnant for the great majority of the public if the invention had given rise to the patentability of human life, to the exploitation of a pregnant woman, or to a return to slavery with rights of ownership over women.

Contrarily, the Division held that in general the patentability of human genes is not against bioethics, given that patenting genes is not the same as patenting a human being, dealing rather with chemical substances which carry genetic information, potentially of use in medicine.[182] Given that the patentability of other human substances was not a subject of dispute,[183] there could not be moral distinction between the patenting of genes and of other substances.[184]

With respect to the objection regarding the exploitation of a pregnant woman as an offence to human dignity, the Division replied, on the one hand, that the woman who donated the tissue had expressed her consent; on the other, that the removal of human tissue was acceptable in that there were the appropriate procedures of informed consent, according to the framework of the then draft of the Convention on Human Rights and

---

180 See, in this regard S.R. Donnelly, 'The Patentability of Human Embryonic Stem Cells', cit., p. 111 et seq.; M.A. Bagley, 'A Global Controversy: The Role of Morality in Biotechnology Patent Law', cit., p. 333.

181 In this respect, among others, see A. Ottolia, 'Moral Limits to Biotech Patents in Europe: A Quest for Higher Harmonization', cit., p. 316.

182 See W. Cornish and D. Llewelyn, *Intellectual Property: Patents, Copyright, Trade Marks and Allied Rights*, London: Thomson Sweet and Maxwell, 2007, p. 884.

183 Such as for instance the interferons and erythropoietin, or blood and bones, which were widely used for many years as a source of useful products.

184 The Opposition Division even went so far as to state that: 'Even if every gene in the human genome were cloned (and possibly patented), it would be impossible to reconstitute a human being from the sum of its genes' (however, the statement appears inconsistent with the ban of art. 6, para. 2, of the Directive regarding the patentability of procedures for cloning human beings; see O. Mills, *Biotechnological Inventions. Moral Restraints and Patent Law*, cit., p. 145).

Biomedicine of the European Council.[185] Finally, to the objection regarding the ownership of rights over the human body they replied that in general a patent grants rights in order to exclude third parties from the use of the invention without consent, and not rights over individuals.[186] The only stage of the invention in which the woman was involved was the initial stage, as a voluntary source of relaxin, and after the granting of the patent the donor was free to live her life as she wished and enjoyed the same right to self-determination as always.[187]

With the decision, then, the Opposition Division opened the door to the patentability of human genes pursuant to the EPC, furnishing also guidelines for future cases regarding what constitutes an abhorrent invention to be excluded from patentability.[188]

Following the start of patentability of human genetic material registered in the *Relaxin* case, subsequently in *Biogen v. Medeva*[189] the subjects of the patent application were DNA molecules. The invention consisted in genetically modified DNA molecules which contained an insert necessary for the production of an essential protein for the vaccine

---

[185] 'Convention for the Protection of Human Rights and Dignity of the Human Being with regard to the Application of Biology and Medicine: Convention on Human Rights and Biomedicine', cit.

[186] In this regard, see L. Bently and B. Sherman, *Intellectual Property Law*, Oxford/New York: Oxford University Press, 2009 (3rd edn), p. 456.

[187] The subsequent appeal against the decision was rejected by the Technical Board of Appeal in 2002 (*Relaxin/Howard Florey Institute*, T 0272/95, 23.10.2002).

[188] See O. Mills, *Biotechnological Inventions. Moral Restraints and Patent Law*, cit., pp. 63 and 147. The test of 'public abhorrence', then, represents a lesser obstacle for morally controversial inventions than that of 'public acceptability', given that fewer inventions are likely to be considered 'abhorrent' rather than 'unacceptable' for society. None of the three tests described in EPO case law ('balancing of interest', 'public abhorrence' and 'public acceptability'), though, require the exclusion patentability to be connected to a ban on commercial use of the invention. In consideration of the confusing EPO case law on the matter, then, the European Commission held it necessary to intervene with the aim of further harmonizing patent law for biotechnological inventions, which led to the adoption of Directive 98/44/EC (see S.R. Donnelly, 'The Patentability of Human Embryonic Stem Cells', cit., p. 112; M.A. Bagley, 'A Global Controversy: The Role of Morality in Biotechnology Patent Law', cit., p. 333; G. Porter, 'The Drafting History of the European Biotechnology Directive', cit., p. 7 and 349).

[189] 1997 RPC 1, HL.

against the Hepatitis B virus. Thus, Biogen obtained the patent.[190] Subsequently, Biogen deposited a complaint in the United Kingdom against the alleged violation of the patent by Medeva, which objected to the validity of the patent on the basis of the lack of the inventive step in that the state of the art implied the resulting object of the invention and Biogen had in reality used known procedures and known resulting products. In support of these arguments, Medeva made use of a publication of an expert in the field for the creation of the direct expression of a eukaryotic protein in a prokaryotic cell.

While the UK Patent Court deemed the patent valid, noting the difference to the state of the art in the particular execution of the idea to produce genetically modified DNA, the Court of Appeal overturned the decision stating that, for the invention, Biogen used known and available techniques and materials, without having developed new processes or products with respect to the state of the art. The House of Lords, then, confirmed the decision of the appeal.

Meanwhile, in the same period in which the House of Lords declared the invalidity of Biogen's invention, the EPO Technical Board of Appeal established that the patent was valid. Thus, the position of the biotechnology industry was upheld, according to which the increased level of purity or quality of known products requires particular research effort, and thus is worthy of patent incentive. Further, the Board held that Biogen's invention represented an important result in the fight against Hepatitis B, in that many researchers were working towards the same goal. The decision of the EPO Board confirmed, then, that genetic material such as DNA and genes falls under the sphere of patentable material.[191]

Thus, currently in Europe, microorganisms, plants and relative inventions, animals and related inventions, and human genetic material such as cells, genes and DNA and their relative parts, and biotechnological products are considered patentable.

---

190 See W. Cornish, *Cases and Materials on Intellectual Property*, cit., p. 89 et seq.

191 Subsequently, in the case *Novartis* the same EPO Board held that genetic material such as cells and their parts are considered microorganisms, stating that as such they were patentable from the moment in which the patentability of microorganisms was recognized in Europe. In this regard, see R. Nott, 'The *Novartis* case in the EPO' (1999) 21 *European Intellectual Property Review* 33 et seq.

## IV.2.2 Bioethical Profiles of Protection

### IV.2.2.1 Bioethics in the patentability of microorganisms

In Europe, bioethical issues regarding the patentability of biotechnological inventions were first taken into consideration in the case of *Genentech-I/ Polypeptide expression*,[192] which dealt with a genetically modified microorganism (a plasmid). The EPO Technical Board of Appeal, in evaluating the morality of the patenting of a living subject like a microorganism, held that its patentability did not violate public order or morality. The patentability of microorganisms, in any case, did not raise serious ethical concerns; thus, with respect to these no particular objection was made.[193]

The decision to grant a patent for microorganisms, further, stimulated patent applications for other, more advanced life forms.

### IV.2.2.2 Bioethics in the patentability of animals

The granting of patents for microorganisms was followed by patent applications for animals. Unlike the former hypothesis, the latter raised significant ethical doubts.

The ethical issue regarding the patentability of animals was brought for the first time before European courts, and in particular before the EPO Technical Board of Appeal, in the case of *Harvard Onco-Mouse*.[194] The application met with strong opposition, which focused mainly on the profiles of suffering caused to the animal and the possible risks associated with the release of a genetically modified organism into the wild. Many animal rights associations, together with the opponents to the patentability of life, stated that it was not acceptable to subject animals to risks and suffering in the name of social benefit. They claimed that animals are living beings created by God, and man cannot manipulate them according to his whim, nor own them as a commercial good, given that this would be contrary to public order and morality. What is more, it was held that the spread of genetically modified organisms could lead to unexpected and irreversible risks for the environment.[195] Recalling the

---

192 T 292/85 (1989).

193 See N.S. Sreenivasulu and C.B. Raju, *Biotechnology and Patent Law. Patenting Living Beings*, cit., p. 215.

194 Decision T 19/90, EPO Official Journal, 1990, 476.

195 See L. Bently and B. Sherman, 'The Question of Patenting Life', in L. Bently and S.M. Maniatis (eds), *Intellectual Property and Ethics*, London: Sweet and Maxwell, 1998, p. 111 et seq. (112 et seq.).

provision of the EPC according to which every invention whose use is contrary to public order or to morality cannot be patented,[196] it was argued in fact that the application could not be accepted given that the invention violated these principles.

The patent applicants, however, were able to convince the EPO that the potential benefit for cancer research could be considered superior and prevail over ethical elements. What is more, the EPO Board took the intention at the basis of the manipulation of animals into consideration, which consisted in not wanting to go against ethical standards or to cause risk or suffering to animals, but rather in conducting research on a chronic illness which affects millions of people throughout the world. Thus, it was held that there was nothing disgraceful in the patenting of an animal in view of the potential benefit it might produce.[197] In this evaluation, the Board attempted to balance the bioethical issue and the social benefit, stating that the benefit which humanity would have gained from cancer research using the invention had to be considered prevalent for the purposes of granting the patent. The EPO, then, concluded that the patenting of an invention with such a clear use is not immoral,[198] and thus granted the patent.

Subsequently, in the case of *Upjohn's Application*,[199] the EPO Board found itself again having to balance bioethical standards with the benefits of a biotechnological invention regarding a patent application for the invention of a transgenically modified mouse useful for a model aimed at favouring research on methods to stimulate hair growth. The moral objection was in regard to the fact that one among a group of alternative genes proposed for insertion into the mouse was an oncogene.

The applicants stated that the invention was useful for research on human hair loss and baldness. Based on the *Harvard Onco-Mouse* decision, they highlighted that the benefit to society from the invention had to be considered superior and prevailing over the moral issues.[200]

Contrarily, moving away from its decision on *Harvard Onco-Mouse*, the Board held that the potential social benefits produced by the invention could not be balanced with the possible risks and suffering of

---

196 EPC, art. 53.

197 See W. Cornish, *Cases and Materials on Intellectual Property*, cit., p. 62 et seq.

198 See J. Chambers, 'Patent Eligibility of Biotechnological Inventions in the US, Europe and Japan: How Much Patent Policy is Public Policy?' (2002) 34 *George Washington International Law Review* 223 et seq.

199 Case No. 89913146.0.

200 In this regard, see L. Bently and B. Sherman, 'The Question of Patenting Life', cit., p. 113.

the animals. This case was different from the previous one, in that the possibility for research on hair loss and balding could not be compared to cancer research. Thus, the Board held that the patent for a genetically modified mouse which is susceptible to hair loss would be immoral and contrary to public order.[201] At that point, however, the applicant filed the application again, eliminating reference to the oncogene, and so the patent was granted.[202]

Another case worthy of attention, finally, is that concerning the patent application filed by the University of Edinburgh on transgenic animal stem cells. The patent was granted by the EPO a short time after the adoption of Directive 98/44/EC.[203]

Due to the international reaction to the patent, and in particular from the opposition of the German, Italian and Dutch governments, the University modified the claims specifying that they were limited to non-human cells. The Opposition Division held then that, while the original application violated article 53(a) EPC, the amended application was legitimate.[204]

### IV.2.2.3 Bioethics in the patentability of genetic material and human cells: The issue of embryonic stem cells

There have been patent applications proposed for human genetic material in Europe as well, asking the EPO then to decide on the morality of such requests.

---

201 See M.A. Bagley, 'A Global Controversy: The Role of Morality in Biotechnology Patent Law', cit., p. 332; R. Nott, 'The Biotech Directive: Does Europe Need a New Draft?', cit., p. 563 et seq. (565 et seq.); S. Conner, 'Patent Ban on Baldness "Cure" Mouse', cit. Further, it was also shown that the concepts of public order and morality are not defined, and as a consequence remain entrusted to a case by case evaluation (N.S. Sreenivasulu and C.B. Raju, *Biotechnology and Patent Law. Patenting Living Beings*, cit., p. 216).

202 See G. Laurie, 'Biotechnology and Intellectual Property: A Marriage of Inconvenience?', in S.A.M. McLean (ed.), *Contemporary Issues in Law, Medicine and Ethics*, Aldershot (UK)/Brookfield (USA): Dartmouth Publishing, 1996, p. 237 et seq. (252 et seq.); R. Nott, 'Plants and Animals: Why They Should Be Protected by Patents and Variety Rights' (1993) *Patent World* July/August 45 et seq.

203 European Patent No. 0695 351, issued on 8 December 1999.

204 EPO Press Release, 'Edinburgh Patent Limited After EPO Opposition Hearing', 24 July 2002. In regard to the case, see extensively below, in the subsequent paragraph.

In the *Relaxin* case,[205] the EPO found itself for the first time having to decide on bioethical profiles of patenting for human genetic material.[206] The application concerned a gene which codified the hormone called relaxin, present in the body of women during labour to relax the body undergoing fatigue and pain. The patent request was the subject of opposition on the basis of the considerations for which the patentability of human genes: (a) offends morality or public order and is contrary to human dignity, given that it requires the extraction of tissues from the human body; and (b) is equal to a modern form of slavery, given that it entails the dislocation of women and their gradual subservience to commercial businesses.

The Opposition Division, however, held that the extraction of tissue from a human body was not immoral, in that it represented a standard medical practice and that there was the consent of the person involved. With respect to the objection that the patenting of genes was equivalent to slavery, it was decided to exclude it due to the fact that the patent for genes does not grant rights over the individual. Finally, it was concluded that the patenting of genes does not give rise to a patent for human life and thus it is not contrary to morality.[207]

The decision of the Opposition Division, then, provided a guide for future cases with regard to what criteria should be adopted in considering an invention repugnant and thus excluding it from patentability. In particular, on the point of the intrinsic immorality of patenting human genes, the Division claimed that this was based on the assumption of an overwhelming consensus among Member states on the fact that such patenting is repugnant and thus prohibited by article 53(a), an assumption that, according to the Division, was instead false. With reference to the application of article 53(a), the criteria of 'overwhelming consensus' adopted by the Opposition Division in this case limited the hypothesis of refusal of the patent by the EPO, with respect to the 'balancing of interests' test adopted by the Examining Division in the case of *Harvard Onco-Mouse*.

The decision supported the proposal that the consent of the Member states was a determining factor, and that the EPO or another European institution did not have the power to decide what represented a repugnant invention. The decision confirmed, then, the opinion according to which

---

205 EPO Official Journal, 1995, 388; 1995 EPOR 541.

206 See D. Beyleveld and R. Brownsword, 'Human Dignity, Human Rights, and Human Genetics', cit., p. 675.

207 See again L. Bently and B. Sherman, 'The Question of Patenting Life', cit., p. 114 et seq.

the underlying intention of the design of article 53(a), was to allow the state practice of denying patentability on bioethical grounds to continue.

Given that the position of the Opposition Division in the *Relaxin* case appeared completely different from that of the Examining Division in the *Harvard Onco-Mouse* decision, clearly there was confusion within the EPO regarding the application of article 53(a).[208]

In three more recent cases regarding the opposition to patents on genes for the predisposition of ovarian and breast cancers, BRCA1[209] and BRCA2,[210] granted by the EPO to Myriad Genetics and subsequently transferred to the University of Utah Research Foundation, the application of article 53(a) EPC, was again brought up. All three cases concerned the decision of the Opposition Division regarding patents, subject to appeal before the Technical Board of Appeal.[211]

In the first case, *Breast and ovarian cancer/University of Utah*,[212] the arguments for appeal concerned, also, the fact that the patented material was excluded from patentability according to article 53(a) EPC. In the first place, the appellants[213] stated that there was no proof that the donors of the cells used to identify the breast cancer gene had given their prior informed consent. In the absence of such proof, it was held that the initial obtainment of research results had raised significant bioethical issues, and thus was a violation of public order and morality pursuant to article 53(a). The Technical Board responded by showing that the EPC does not contain provisions which establish a requirement for which the applicant

---

[208] See O. Mills, *Biotechnological Inventions. Moral Restraints and Patent Law*, cit., p. 63.

[209] EP0699754, EP0705903 and EP0705902.

[210] EP0785216.

[211] In the *Myriad* case, and in particular regarding the motives of the opposition, besides referring (for what occurred in the US legal system) to Part III, see, in particular, among others, A.S. Kesselheim and M.M. Mello, 'Gene Patenting. Is the Pendulum Swinging Back?' (2010) 362 *New England Journal of Medicine* 1855 et seq.; G. Matthijs, 'The European Opposition against the BRCA Gene Patents' (2006) 5 *Familial Cancer* 95 et seq.; J. Paradise, 'Lessons from the European Union: The Need for a Post-Grant Mechanism for Third-Party Challenge to U.S. Patents' (2005) 7 *Minnesota Journal of Law, Science & Technology* 315 et seq., and J. Paradise, 'European Opposition to Exclusive Control Over Predictive Breast Cancer Testing and the Inherent Implications for United States Patent Law and Public Policy: A Case Study of the Myriad Genetics' BRCA Patent Controversy' (2004) 59 *Food & Drug Law Journal* 133 et seq.; see also B.A. Brody, 'Intellectual Property and Biotechnology: The European Debate', cit., p. 104 et seq.

[212] Technical Board of Appeal, decision 27 September 2007, T 1213/05.

[213] *Sozialdemokratische Partei der Schweiz, Greenpeace e.V., et al.*

for a patent must provide proof of prior informed consent. The second argument of the appellants was that the patented matter gave rise to significant social and economic consequences, in that not only would there be an increased cost to patients, but it would influence the way in which diagnosis and research was organized at a European level. The Board replied that the text of article 53(a), excludes from patentability when the 'use of the invention', and not the 'use of the patent', is contrary to public order or morality. Based on this, then, the Board dismissed the appeal.

In the second case, *Mutation/University of Utah*,[214] the appellants[215] argued that the methods for diagnosing a predisposition for cancer of the mammary gland and ovaries should not be considered an invention pursuant to article 53(c), which prohibits the patentability of methods for treatment or diagnosis of the human or animal body. The reply of the Technical Board was that the regulation cited excluded patentability for those diagnostic methods only if the technical interventions the method is based on, which make diagnosis possible, are carried out on living human or animal bodies.[216] Another reason for the appeal was that the patented material gave rise to moral issues, in violation of the provisions of article 53(a). The board also rejected this argument, confirming its decision on the previous case T 1213/05.

In the third case, *Method of diagnosis/University of Utah*,[217] the appellants[218] stated that the granting of the patent would not have allowed patients to have their genetic information read and examined by the company they preferred, nor could it be guaranteed that the criminal and medical genetic databases would remain completely separate, an ethical principle endorsed by the EPO. Once again the Board preferred to follow its own decision in the case T 1213/05, and rejected the appeal.

The cases referenced up to this point seem to demonstrate the need for a uniform approach by the EPO Examining Division, Opposition Division and Technical Board of Appeal, with reference to the extension to the meaning and interpretation of article 53(a) EPC.[219]

---

214 Technical Board of Appeal, decision 13 November 2008, T 0666/05.

215 *Institut Curie, Assistance Publique-Hôpitaux de Paris, et al.*

216 Thus: Enlarged Board of Appeal, opinion G 1/04, EPO Official Journal, 2006, 334.

217 Technical Board of Appeal, decision dated 19 November 2008, T 0080/05.

218 *Institut Curie, Assistance Publique-Hôpitaux de Paris, et al.*

219 See O. Mills, *Biotechnological Inventions. Moral Restraints and Patent Law*, cit., p. 64.

*(i) The issue of embryonic stem cells* In recent years, the most debated and central issue to the main legal decisions on the subject of biotechnological inventions at a European level was that relative to the patentability of embryonic stem cells. The controversy arose with the first isolation of these cells by Doctor James Thomson, biologist and inventor from the University of Wisconsin,[220] in November of 1998,[221] immediately focusing attention on the ban on using human embryos for industrial or commercial use as per article 6(2)(c) of Directive 98/44/EC.[222]

The first formal instance on the theme was presented to the European Commission in December 1998 by Doeke Eisma, a member of the European Parliament.[223] The written investigation required that the Commission clarify whether article 5 of the Directive prohibited the patentability of the invention of a method for the development of embryonic stem cells in a laboratory. The Commission replied that, though the patentability of such cells may concern articles 5 and 6(2)(c), it did not have jurisdiction over the specific issue, and the decision was in the hands rather of national patent offices and judges. In fact, however, it was these latter bodies, rather than the patent applications presented to the EPO, which sparked debate on the subject.[224]

An early case concerned the application related to the so-called *Edinburgh Patent,*[225] whose subject was the extraction of human embryonic stem cells, the genetic manipulation of these cells and the raising of genetically manipulated embryos from them.[226] The Examining Division granted the patent. Germany, Italy and the Netherlands opposed this decision. On the contrary, a report of the European Group on Ethics

---

[220] See J. Thomson *et al.*, 'Embryonic Stem Cell Lines Derived from Human Blastocysts' (1998) 282 *Science* 1145. In this regard refer, *amplius*, to Chapter 3.

[221] Four months after the adoption of Directive 98/44/EC.

[222] See G. Porter, 'The Drafting History of the European Biotechnology Directive', cit., p. 25 et seq.

[223] Question No. 3668/98 'Applicability of the Directive on the Legal Protection of Biotechnological Inventions to Human Research with Embryonic Tissue', 1998 OJ C325/024.

[224] In this regard, see S.R. Donnelly, 'The Patentability of Human Embryonic Stem Cells', cit., p. 115.

[225] EP0695351, cit.

[226] See B.A. Brody, 'Intellectual Property and Biotechnology: The European Debate', cit., p. 99 et seq.; L. Bently and B. Sherman, *Intellectual Property Law*, cit., p. 116.

concluded that there were no bioethical obstacles to the patentability of processes related to embryonic stem cells, regardless of their source.[227]

The Opposition Division recognized that there was no shared approach on the theme of embryonic stem cells, in legislation or in other standards of conduct generally accepted by European culture, and decided to declare the patent valid with changes in the claims, excluding from the sphere of protection human or animal embryonic stem cells. In particular, the Opposition Division observed that the exclusion in Rule 23d(c)[228] EPC could be interpreted either restrictively or broadly.[229] According to a restrictive interpretation, only commercial uses of embryos as such were excluded from patentability. The broad interpretation meant that patenting was precluded not only for commercial uses of human embryos, but also for stem cells derived from the destruction of embryos, regardless of whether or not the application claimed the direct use of the embryo. The Division accepted the second interpretation, showing that otherwise the provision according to Rule 23d(c), would result in a repetition of guaranteed protection for the embryo in Rule 23e.[230]

In 2005, the EPO suspended the granting of new patents regarding embryonic stem cells, holding that there were 'too many ethical aspects that have not been resolved at the political level'.[231] In the same year, the European Commission[232] first intervened regarding this theme, followed by the European Parliament,[233] who in particular reinforced that 'the production of embryonic stem cells involves the destruction of human embryos and thus the patent for the process which involves embryonic

[227] European Group on Ethics in Science and New Technologies, 'Opinion n°16 – 07/05/2002 – Ethical aspects of patenting inventions involving human stem cells', cit. In doctrine, see W. Cornish and D. Llewelyn, *Intellectual Property: Patents, Copyright, Trade Marks and Allied Rights*, cit., p. 885.

[228] Now Rule 28.

[229] See S.R. Donnelly, 'The Patentability of Human Embryonic Stem Cells', cit., p. 116; G. Porter, 'The Drafting History of the European Biotechnology Directive', cit., p. 25.

[230] Now Rule 29; the equivalent of art. 5(1) of Directive 98/44/EC.

[231] In this regard, see S. Schubert, 'Europe Halts Decisions on Stem Cell Patents' (2005) 435 *Nature* 720 et seq.

[232] Report of the Commission to the European Council and Parliament: Developments and implications of patent rights in the sector of biotechnology and genetic engineering, COM(2005) 312, Brussels, 14 July 2005.

[233] Resolution of the European Parliament on patents relative to biotechnological inventions, Strasbourg, 26 October 2005.

stem cells or cells developed from stem cells from human embryos is contrary to article 6, par. 2, letter c) of the Directive'.[234]

Despite the change to the claims, the University of Edinburgh presented an appeal. After a decade, also following the aforementioned institutional interventions; during an oral proceeding before the Board of Appeal in 2007, the appeal was withdrawn.[235]

The decision on the *Edinburgh Patent* was a leading case on the subject of the limits to the bioethical order of patentability in the European context, in that it represented an inversion of the tendency compared with the previous restrictive interpretation of the 'morality ruling' by the EPO, influencing subsequent decisions.

Another interesting case on the subject concerned the application presented by the *Wisconsin Alumni Research Foundation*[236] in relation to the research carried out by Doctor James Thomson for a method to obtain cultures of embryonic stem cells from a primate embryo, as well as for the cultures themselves.[237]

In 2004, the Examining Division held it opportune to reject the claims that may be given in regard to embryonic stem cells, holding them to be contrary to morality, and showing in particular that, though isolated embryonic stem cells are not in and of themselves 'embryos', they could

---

234 Resolution, point 14.

235 See EPO Press Release, 'Edinburg Patent appeal Withdrawn', November 2007. In doctrine, see A. Agovic, 'Stem Cell Patents on a Knife Edge' (2008) 3 *Journal of Intellectual Property Law and Practice* 718 et seq.

236 Hereafter 'WARF'.

237 With regard to the case, besides referring above, to Part III, see S. Sterckx and J. Cockbain, 'Assessing the Morality of the Commercial Exploitation of Inventions Concerning Uses of Human Embryos and the Relevance of Moral Complicity: Comments on the EPO's WARF Decision' (2010) 7 *SCRIPTed* 83 et seq., accessed 18 December 2014 at http://www.law.ed.ac.uk/ahrc/script-ed/vol7-1/sterckx.asp; P.L.C. Torremans, 'Patentability of Human Stem Cell or Synthetic Biology Based Inventions', in E. Arezzo and G. Ghidini, *Biotechnology and Software Patent Law: A Comparative Review of New Developments*, cit., p. 288 et seq. (302 et seq.); S.J.R. Bostyn, 'A Decade after the Birth of the Biotech Directive: Was it Worth the Trouble?', cit., p. 237 et seq., and S.J.R. Bostyn, 'Patenting Human Embryonic Stem Cells in Peril: The Decision of the Enlarged Board of Appeal in G 2/06' (2009) 10 *BioScience Law Review* 13 et seq.; R. Fitt, 'New Guidance on the Patentability of Embryonic Stem Cell Patents in Europe' (2009) 27 *Nature Biotechnology* 338 et seq.; R. Plaggenborg and F.J. Zimmer, 'The EPO's Decision G 2/06 on the Patentability of Human Embryonic Stem Cells: Sounding the Bell for the Next Round?' (2009) 28 *Biotechnology Law Report* 332 et seq.; B.A. Brody, 'Intellectual Property and Biotechnology: The European Debate', cit., p. 100 et seq.

not be obtained without the previous destruction of embryos.[238] The WARF presented an appeal against the decision to the Technical Board of Appeal and requested also that the Enlarged Board of Appeal put the issue to the interpretation of the Court of Justice, given that it concerned the interpretation of European Union law.

The Technical board of Appeal, first, held the hypothesis of using the 'balancing of interest' test adopted in the *Harvard Onco-Mouse* decision unconvincing for the evaluation of inventions related to embryonic stem cells pursuant to article 53(a) EPC.[239]

Consequently, it put a series of issues to the interpretation of the Enlarged Board of Appeal. The latter reached a decision in November 2008,[240] dismissing the appeal of the WARF.

Further, in the decision of the Enlarged Board of Appeal, no response was given regarding the request for referral to the Court of Justice, stating that there was no institutional or legal connection between it and the EPO. The EPO, which deems itself an international organization, showed how it was impossible to assume that the Member states of the EPC, some of which are not members of the EU, could give jurisdiction to the Court of Justice. In the decision, in particular, it was shown that, due to the fact that the invention described in the WARF's application could be made only through the destruction of human embryos, and was to the benefit of industrial and/or commercial use, it was clearly within the sphere of the ban on the use of human embryos for such purposes. This ban, according to the Board, had to be broadly interpreted, and the exception for inventions for therapeutic or diagnostic purposes which are useful for the embryo was not applicable, given that it required that they be beneficial to the embryo itself, while in this case the embryos were destroyed.[241]

---

238 See G. Porter, 'The Drafting History of the European Biotechnology Directive', cit., p. 26.

239 In particular, the Board put the doubt of 'whether, when it comes to human life, it would be ethically acceptable to make a decision by weighing the interests of human beings which could potentially benefit from the exploitations of the technology against a right, if any, of human embryos' at the center of their argument. In doctrine, see M. Rowlandson, 'WARF/Stem Cells (G2/06): The Ordre Public and Morality Exception and its Impact on the Patentability of Human Embryonic Stem Cells', cit., p. 67 et seq. (70).

240 G 2/06, *Use of embryos/WARF*, 25 November 2008.

241 See M. Rowlandson, 'WARF/Stem Cells (G2/06): The Ordre Public and Morality Exception and its Impact on the Patentability of Human Embryonic Stem Cells', cit., p. 71; R. Fitt, 'New Guidance on the Patentability of Embryonic Stem Cell Patents in Europe', cit., p. 338 et seq.

The decision may be read in two ways. According to a restrictive approach, based on the concluding statement of the Enlarged Board of Appeal for which it had a limited impact on the case and did not concern the patentability of embryonic stem cells in general, the exclusion would be limited to inventions which for their creation involve the 'direct' destruction of human embryos. In this sense, patent protection regarding the methods for the extraction of human stem cells and the latter, where they are based on previously existing cell lines, would not be excluded.[242] Nor, clearly, may the decision be reasonably extended to technological progress regarding 'induced pluripotency' stem cells, or to methods for the obtainment of embryonic stem cells which do not involve the destruction of the embryo.[243]

According to the broader interpretation, the ban given in the decision extends to all inventions derived from an original destructive use of human embryos.[244] In this sense, it is shown that the board stated within the decision that the exclusion refers to industrial or commercial uses, holding that the creation of the product constitutes an industrial or commercial use of the invention, even when there is the intention to use the product for further research. The logical consequence to such an approach is that there should be alignment between moral regulations applied in the sphere of patent regulations and those applied beyond it, such as is provided in Recital 39 of Directive 98/44/EC.[245] Also in this

---

[242] See A. Plomer, 'Towards Systemic Legal Conflict: Article 6(2)(c) of the EU Directive on Biotechnological Inventions', cit., p. 173 et seq. (177); R. Fitt, 'New Guidance on the Patentability of Embryonic Stem Cell Patents in Europe', cit., p. 339.

[243] See G. Bahadur and M. Morrison, 'Patenting Human Pluripotent Cells: Balancing Commercial, Academic and Ethical Interests' (2010) 25 *Human Reproductions* 14 et seq. (18).

[244] Also 'indirect', then, as for instance in the case of cell lines which were previously extracted from embryos and deposited in a stem cell bank. In this sense, see M. Burda and M. Grund, 'G2/06 WARF EPO Upholds a Policy of Refusing European Patents for Inventions Covering Products that "Can Only Be Obtained" via the Destruction of Human Embryos' (2006/2007) 9(6) *Bioscience Law Review* 239 et seq.

[245] Where it is stated that 'public order and morality correspond in particular to ethical or moral principles recognized in a Member state and whose observance is indispensable in particular on the subject of biotechnology, given the potential range of inventions in this sector and there intrinsic link with the living matter', and that 'these ethical or moral principles complete the normal legal examinations provided by patent law, regardless of the technical sector of the invention'. In doctrine, see A. Plomer, 'Towards Systemic Legal Conflict: Article 6(2)(c) of the EU Directive on Biotechnological Inventions', cit., p. 178.

view, however, it seems reasonable to exclude the extension of the ban to technological progress in regard to 'induced pluripotency' stem cells, or to methods of obtaining embryonic stem cells which do not involve the destruction of the embryo.

The decision of the Enlarged Board of Appeal is based on sources of EU law, such as the evolution of negotiation for the Directive[246] and the policy of the Commission within the Framework Programs,[247] which excluded funding of research which involves the destructive use of human embryos. On the other hand, it was shown[248] how this may be contrary to the European non-patent regulations which involve the use of human embryos, in particular to Directive 2004/23/EC on the use of human tissues and cells[249] and to EC Regulation 1394/2007 on medicines for advanced treatments.[250] Based on these, the industrial and commercial use of

---

246 On this point, among others, see (*contra*): S.J.R. Bostyn, 'A Decade After the Birth of the Biotech Directive: Was it Worth the Trouble?', cit., p. 237 et seq.; P.L.C. Torremans, 'Patentability of Human Stem Cell or Synthetic Biology Based Inventions', cit., p. 303.

247 In particular, recently, Regulation (EU) no. 1291/2013 of the European Parliament and of the Council of 11 December 2013 establishing Horizon 2020 – the Framework Programme for Research and Innovation (2014–2020) and repealing Decision No 1982/2006/EC, and the previous Decision 1982/2006/EC of the European Parliament and of the Council of 18 December 2006 concerning the Seventh Framework Programme of the European Community for research, technological development and demonstration activities (2007–2013) (see Part II).

248 See S.R. Donnelly, 'The Patentability of Human Embryonic Stem Cells', cit., p. 118 et seq.

249 Directive 2004/23/EC of the European Parliament and of the Council of 31 March 2004 on setting standards of quality and safety for the donation, procurement, testing, processing, preservation, storage and distribution of human tissues and cells.

250 Regulation (EC) no. 1394/2007 of the European Parliament and of the Council of 13 November 2007 on advanced therapy medicinal products and amending Directive 2001/83/EC and Regulation (EC) No 726/2004. See as well, in argument: 'Communication from the Commission to the European Parliament, the Council, the European Economic and Social Committee and the Committee of Regions – Safe, Innovative and Accessible Medicines: a Renewed Vision for the Pharmaceutical Sector', COM(2008) 666, Brussels, 10 December 2008. In this regard, further, Recital 17 of Directive 98/44/EC, given 'significant progress in the treatment of diseases has already been made thanks to the existence of medicinal products derived from elements isolated from the human body and/or otherwise produced, such medicinal products resulting from technical processes aimed at obtaining elements similar in structure to those existing naturally in the

embryonic stem cells is considered capable of providing important benefits for the health and economy of European citizens.[251]

On the other hand, these same regulations contain explicit caveats on the theme of the use of embryonic stem cells (and germinal cells). In particular, Directive 2004/23/EC, in Recital 12, provides that its governance:

> should not interfere with the decisions of Member states relative to the use or non-use of particular types of human cells, including germinal cells and embryonic stem cells. If, however, a Member state authorizes a particular use for such cells, this directive shall apply all provisions necessary for the protection of public health, in consideration of the specific risks of such cells based on scientific knowledge and their particular nature, and shall guarantee the respect of the fundamental rights. What is more, this directive should not interfere with the provisions of the Member states which define the legal term of 'person' or 'individual'.

In an extremely similar way, Regulation 1394/2007, in Recital 7, provides that its rules:

> should not interfere with the decisions made by the Member states in regards to the opportunity to authorize this or that type of human cell, for instance embryonic stem cells or animal cells. It should also not influence the application of national legislation which prohibits or limits the sale, supply or use of medicines containing, consisting in or derived from such cells.

From these provisions, aimed essentially at delegating the decision on the authorization of use of embryonic stem cells to Member states, in effect, seems to give rise to potential risks of systematic inconsistency between such broad discretion of evaluation left to the Member states[252] and the general ban laid out in patent regulations, where the use of stem cells leads to the destruction of embryos.[253]

---

human body', provides specifically that 'consequently, research aimed at obtaining and isolating such elements valuable to medicinal production should be encouraged by means of the patent system'.

[251] See S.R. Donnelly, 'The Patentability of Human Embryonic Stem Cells', cit., p. 119; A. Plomer, 'Towards Systemic Legal Conflict: Article 6(2)(c) of the EU Directive on Biotechnological Inventions', cit., p. 186.

[252] Except, in the case of Directive 2004/23/EC, where a Member state authorizes a particular use of such cells, the provision of the application of all provisions necessary for the protection of public health and in respect of fundamental rights.

[253] In this regard, it was shown that the restrictive interpretations provided in the *Edinburgh* and *WARF* cases would have made the context of art. 6, para. 2(c),

Recently, another fundamental declaration on this subject, which clarified the even more clearly restrictive approach of the EU Court of Justice with respect to the patentability of embryonic stem cells, was made in regard to the case of *Oliver Brüstle v. Greenpeace eV*.[254] The cause concerned a patent, filed in December 1997 by Doctor Oliver Brüstle, concerning isolated and purified neural progenitor cells,[255] obtained from human embryonic stem cells used to treat neurological illnesses.[256]

Based on the request made by Greenpeace eV, the *Bundespatentgericht*[257] declared the patent of Dr. Brüstle null and void, in that its subject was a procedure that allowed for the obtainment of progenitor cells from human embryonic stem cells. The *Bundesgerichtshof*,[258] on the appeal of Dr. Brüstle, consulted the EU Court of Justice with regard to the interpretation of article 6(2)(c), of Directive 98/44/EC, and in particular of the notion of 'human embryo', not defined within the Directive. In particular, clarification was requested regarding whether the patentability of the human embryo concerned all stages of life, from the fertilization of the egg, or if other conditions need to be met, for instance that a certain stage of development be reached.[259]

Upon examination of the notion of 'human embryo', first of all, the Court of Justice recalled that, according to consistent case law, the uniform application as much for European Union law as for the principle of equality requires that a provision of law of the Union which does not

---

of Directive 98/44/EC uncertain, and pushed to attempt different strategies of expression in patent applications (see G. Porter *et al.*, 'The Patentability of Human Embryonic Stem Cells in Europe', cit., p. 653 et seq.). This uncertainty on the bioethical exclusions from patentability, it is added, would produce the result of jeopardizing research and investment on this matter (A. Ottolia, 'Moral Limits to Biotech Patents in Europe: A Quest for Higher Harmonization', cit., p. 324).

254 Decision of 18 October 2011, case C-34/10, [2011] ECR I-9821.

255 Progenitor cells are intended as immature body cells which are still capable of multiplying, developing and differentiating themselves into specific mature body cells. Neural progenitor cells have the capability of forming mature cells of the nervous system, neurons for instance.

256 According to the indications provided by Dr. Brüstle, there were already clinical applications for the invention, in particular for patients affected by Parkinson's disease.

257 German federal court on the subject of patents.

258 German federal court of cassation.

259 With regard to the evolution of the case and the issues raised by the *Bundesgerichtshof*, see in particular: S.J.R. Bostyn, 'A Decade After the Birth of the Biotech Directive: Was it Worth the Trouble?', cit., p. 240 et seq.

contain any specific reference to the law of Member states as regards the determination of its meaning and its range must give rise, throughout the Union, to an independent and uniform interpretation.[260] Thus, it highlighted that:

> The lack of a uniform definition of the concept of human embryo would create a risk of the authors of certain biotechnological inventions being tempted to seek their patentability in the Member states which have the narrowest concept of human embryo and are accordingly the most liberal as regards possible patentability, because those inventions would not be patentable in the other Member states. Such a situation would adversely affect the smooth functioning of the internal market which is the aim of the Directive.[261]

However, as for the meaning attributed to the notion of 'human embryo' referred to in article 6(2)(c), of the Directive, the Court highlighted that, although the definition of human embryo constitutes a particularly delicate social issue in several Member states, due to the diversity of their values and traditions, it has not been called, by a preliminary ruling, to deal with issues of a medical or ethical nature, but rather to a legal interpretation of pertinent provisions of the Directive.[262]

Thus, the Court proceeded to highlight that from the preamble of the Directive it emerges that, though it is true that it is aimed at the encouragement of investments in the biotechnology sector, the use of biological material of human origin must occur in respect of fundamental rights and, in particular, of human dignity. In this sense, Recital 16 of the Directive provides that 'patent law must be exercised in respect of fundamental principles which guarantee the dignity and integrity of man'.[263]

In the same sense, the Court showed that the context and aims of the Directive demonstrate that the Union legislature intended to exclude any possibility of the obtaining of a patent where respect for human dignity may be compromised.

---

[260] EU Court of Justice, decision *Oliver Brüstle v. Greenpeace eV*, cit., point 25, which refers also, in this sense, to decisions dated 18 January 1984, case 327/82, *Ekro* [1984] ECR I-107, point 11; 19 September 2000, case C 287/98, *Linster* [2000] ECR I-6917, point 43; 16 July 2009, case C 5/08, *Infopaq International* [2009] ECR I-6569, point 27, and 21 October 2010, case C 467/08, *Padawan* [2010] ECR I-10055, point 32.

[261] Court of Justice, decision cit., point 28.

[262] Court of Justice, decision cit., point 30, where there is reference in the same sense to the decision of 26 February 2008, case C 506/06, *Mayr* [2008] ECR I-1017, point 38.

[263] Court of Justice, decision cit., point 32.

As a consequence, according to the Court, the notion of the human embryo must be intended broadly: a human egg must be considered as such from the phase of fertilization, since fertilization is the event which initiates the development process of a human being. What is more, the qualification of human embryo must also be recognized to the unfertilized human egg in which the nucleus of a mature human cell was implanted, and to the unfertilized human egg induced to divide and develop through parthenogenesis. In fact, though such organisms may not have been fertilized in the proper sense, they, due to the technique used to create them, are capable of initiating the process of development of a human being similar to the embryo derived from a fertilized egg.[264]

As far as concerns, then, stem cells obtained from a human embryo in the blastocyst stage – to which the invention that is the subject of Dr. Brüstle's patent refers – the Court stated that it is the task of the national judge to establish, in consideration of scientific development, if they are such as to initiate the process of development of a human being and, as a consequence, fall under the notion of 'human embryo'.[265]

The Court of Justice proceeded, then, to evaluate whether the notion of 'use of human embryos for industrial or commercial purposes', not patentable pursuant to article 6(2)(c), of the Directive, also includes the use for purposes of scientific research. In this regard, provided that the Directive is not intended to regulate the use of human embryos in the sphere of scientific research, rather its subject is exclusively the patentability of biotechnological inventions, the Court observed that the fact of granting patent protection for an invention implies, in principle, its industrial and commercial use.[266] As a consequence, scientific research

---

[264] Court of Justice, decision cit., points 34–36.

[265] Court of Justice, decision cit., point 37. More specifically, in point (1) of the ruling the Court stated that: 'Article 6(2)(c) of Directive 98/44/EC of the European Parliament and of the Council of 6 July 1998 on the legal protection of biotechnological inventions must be interpreted as meaning that: – any human ovum after fertilisation, any non-fertilised human ovum into which the cell nucleus from a mature human cell has been transplanted, and any non-fertilised human ovum whose division and further development have been stimulated by parthenogenesis constitute a "human embryo"; – it is for the referring court to ascertain, in the light of scientific developments, whether a stem cell obtained from a human embryo at the blastocyst stage constitutes a "human embryo" within the meaning of Article 6(2)(c) of Directive 98/44'.

[266] According to the Court: 'Although the aim of scientific research must be distinguished from industrial or commercial purposes, the use of human embryos for the purposes of research which constitutes the subject-matter of a patent application cannot be separated from the patent itself and the rights attaching to it';

which involves the use of human embryos may not obtain protection from patent law. Contrarily, the Court underlined that the patentability of the uses of human embryos for industrial or commercial uses is not prohibited, pursuant to the Directive, where it concerns uses for therapeutic or diagnostic purposes which are applied to and are useful for the human embryo.[267]

Finally, with reference to the specific issue of the patentability of an invention related to the production of neural progenitor cells, which involves the use of stem cells extracted from a human embryo in its blastocyst phase, the Court highlighted that such extraction leads to the destruction of the embryo. In this regard, the Court stated that the patentability of the invention must be excluded in all cases in which its execution requires the destruction of human embryos or their use as a starting material;[268] this is so even if the patent claims (as in the case under discussion) do not involve their use.[269]

---

in fact, 'the use of human embryos for purposes of scientific research which is the subject-matter of a patent application cannot be distinguished from industrial and commercial use and, thus, avoid exclusion from patentability' (decision cit., points 43–45, where it is also specified that this interpretation coincides with that adopted by the EPO in the decision G 2/06, *Use of embryos/WARF*, cit., points 25–27, in that case with respect to the provision pursuant to Rule 28(c), EPC, identical to that referred to in art. 6, para 2(c), of the Directive).

[267] For instance, in order to correct a deformity and improve its future outlook on life. See point (2) of the ruling, on the basis of which: 'The exclusion from patentability concerning the use of human embryos for industrial or commercial purposes set out in Article 6(2)(c) of Directive 98/44 also covers the use of human embryos for purposes of scientific research, only use for therapeutic or diagnostic purposes which is applied to the human embryo and is useful to it being patentable'.

[268] In this regard, the Court specified that: 'The fact that destruction may occur at a stage long before the implementation of the invention, as in the case of the production of embryonic stem cells from a lineage of stem cells the mere production of which implied the destruction of human embryos is, in that regard, irrelevant' (decision cit., point 49; also in this regard the Court highlights, in point 51, the concurrence of this interpretation with that adopted by the EPO in decision G 2/06, *Use of embryos/WARF*, cit., point 22).

[269] See point (3) of the ruling, according to which: 'Article 6(2)(c) of Directive 98/44 excludes an invention from patentability where the technical teaching which is the subject-matter of the patent application requires the prior destruction of human embryos or their use as base material, whatever the stage at which that takes place and even if the description of the technical teaching claimed does not refer to the use of human embryos'. The lack of exclusion from patentability of the latter hypothesis, in fact, would have the consequence of removing the useful effect of the regulation as per art. 6, para. 2(c), of the Directive, allowing the applicant of the patent to avoid the prohibition of

The declaration of the Court of Justice, then, provides a definition of human embryo which is extremely broad and which specifically expresses the dignity and integrity of human life from the initial stage of its development.

On the other hand, national courts are left to ensure that stem cells obtained from human embryos in the blastocyst phase, in light of scientific developments, may give rise to the process of development of a human being and, thus, fall under the notion of human embryo. This margin of discretion left to national judges seems meant not to interfere with national provisions (for instance, in the Spanish legal system) which are more permissive regarding so-called pre-embryo research.[270]

A predictable consequence of the decision – extremely critical in that it could reduce interest in obtaining patents on the results of research on the subject at a European level[271] – could be to concentrate research on instruments which are more respectful to bioethics, in order to obtain patents regarding stem cells obtained from adult 'induced pluripotency' stem cells, or from embryonic stem cells not involving the destruction of the embryo.[272]

---

patentability through a clever drafting of the claim (decision cit., point 50). Such a 'whole content approach' adopted by the Court of Justice – as well as by the EPO in the *WARF* case (see above) – then, has been restated by the European Parliament in the Resolution of 10 May 2012 on the patenting of essential biological processes (2012/2623(RSP), which at point 6 '(w)elcomes the recent decision of the European Patent Office in the WARF case and of the European Court of Justice in the Brüstle case, as they appropriately interpret Directive 98/44/EC and give important indications on the so-called whole content approach', and 'calls on the European Commission to draw the appropriate consequences from these decisions also in other relevant policy areas in order to bring EU policy in line with these decisions'.

[270] On this point, refer to Part II.

[271] Thus, according to some, reducing the interest to develop such research; while, according to others – with an observation which seems shared – pushing the patenting of such results in legal systems with disciplines which are less geared towards bioethical profiles (for instance, as was seen, in the American system). For comments on the decision, see M.I. Schuster, 'The Court of Justice of the European Union's Ruling on the Patentability of Human Embryonic Stem-Cell-Related Inventions (Case C-34/10)' (2012) 43 *IIC* 626 et seq.; S. Burke, 'Interpretive Clarification of the Concept of 'Human Embryo' in the Context of the Biotechnology Directive and the Implications for Patentability: *Brüstle v Greenpeace eV* (C-34/10)' (2012) 34 *European Intellectual Property Review* 346 et seq.; A. Plomer, 'After *Brüstle*: EU accession to the ECHR and the Future of European Patent Law' (2012) 2 *Queen Mary Journal of Intellectual Property* 110 et seq.

[272] See, in particular: European Centre for Law and Justice, '*Brüstle* case: The ECLJ welcomes the decision of the Court of Justice of the European Union on the status of the human embryo', 18 October 2011, accessed 18 December 2014 at http://eclj.org/Releases/Read.aspx?GUID=dbe0cfa9-b5b6-4c48-a419-60115f434a

However, in the latest decision on the *International Stem Cell* case,[273] the Court of Justice of the EU was asked by the UK High Court to consider whether the *Brüstle* decision was correct insofar as it appeared to exclude parthenotes[274] from patentability.

In *Brüstle*, the Court of Justice assumed that both parthenotes and non-fertilised ova after somatic cell nuclear transfer were capable of developing into a human being and should therefore be excluded from patentability.

In *International Stem Cell*, the UK High Court explained that, according to current scientific knowledge, a mammalian parthenote can never develop to term because it does not contain the requisite paternal DNA that would permit it to do so, in contrast to a fertilised ovum. Parthenotes can develop only to the blastocyst stage, over about five days.

The Court of Justice, therefore, recognised that the information before the Court in *Brüstle* was unclear on the development capabilities of parthenotes. It acknowledged that the evidence before it, which was undisputed, indicated that parthenotes could not themselves, without additional genetic manipulation, develop to produce an actual embryo.

Thus, the Court concluded that an unfertilised human ovum whose division and further development has been stimulated by parthenogenesis does not constitute a 'human embryo' within the meaning of art. (6)(2)(c) of the Directive – and so it is not excluded from patentability – if, in light of current scientific knowledge, that ovum does not, in itself, have the inherent capacity of developing into a human being, that being a matter for the national court of determine.[275]

---

d4&s=eur. Further, the development of methods for the reprogramming of cells appears currently to be the 'new frontier' of biotechnological research, with important results already being reached also on behalf of Dr. Brüstle's team itself (see A. Abbott, 'Stem Cells: The Cell Division' (2011) 480 *Nature* 310 et seq.; P. Koch and O. Brüstle *et al.*, 'Excitation-induced Ataxin-3 Aggregation in Neurons from Patients with Machado–Joseph Disease' (2011) 480 *Nature* 543 et seq.).

273 *International Stem Cell Corporation v Comptroller General of Patents, Designs and Trade Marks*: C-364/13 [2014].

274 Which are created by the activation of an oocyte, in the absence of sperm, by a variety of chemical and electrical techniques such that they are capable of dividing and developing further.

275 For a first comment on the decision, see E. O'Sullivan, *International Stem Cell Corp v Comptroller General of Patents: The Debate Regarding the Definition of the Human Embryo Continues*, in 36 *European Intellectual Property Review*, 2014, p. 155 ss.

# PART V

# Conclusions: Biotechnological inventions and the patentability of living beings in the United States and European models between legal issues and bioethical profiles – The utility of the comparative method, the perspective of the legal process and relations with other sciences and humanistic disciplines (in particular bioethics)

## V.1 COMPARATIVE LAW OF BIOTECHNOLOGICAL PATENTS IN THE UNITED STATES AND EUROPEAN MODELS BETWEEN 'COMPETITIVE ANTIBODIES' AND 'BIOETHICAL COUNTERLIMITS'

In the western legal tradition, both in the United States and European models, patent law finds its justifying rationale mainly in the provision of incentives for innovation, in exchange for the so-called disclosure of the technology on which the invention is based to the public. Directly, this is functional also for the development of subsequent innovations, as well as for the pursuit of pro-competitive aims. This is in virtue of some 'competitive antibodies' which characterize the patent discipline, in particular: (a) the evaluation regarding the fulfillment of patent requirements; (b) the so-called *numerus clausus* of exclusive rights; and (c) the provision of specific limits to ownership.[1]

---

[1] In this regard, see *ex multis*, A. Musso, 'Grounds of Protection: How Far Does the Incentive Paradigm Carry?', in A. Ohly (ed.), *Common Principles of European Intellectual Property Law*, Tübingen (Germany): Mohr Siebeck, 2012, p. 33 et seq.; V. Denicolò and L.A. Franzoni, 'Weak Intellectual Property Rights, Research Spillovers, and the Incentive to Innovate' (2012) 14 *American Law and Economics Review* 111 et seq.; WIPO, '2011 World Intellectual Property Report. The Changing Face of Innovation', Geneva, WIPO, p. 75 et seq.; W. Kingston, *Beyond Intellectual Property: Matching Information Protection to Innovation*, Cheltenham, UK and Northampton, MA, USA, Edward Elgar, 2010, *passim*; G. Ghidini, *Innovation, Competition and Consumer Welfare in Intellectual Property Law*, Cheltenham, UK and Northampton, MA, USA, Edward Elgar, 2010, p. 33 et seq.; A. Peukert, 'Güterzuordnung und Freiheitsschutz', in R.M. Hilty, T. Jaeger and V. Kitz (eds), *Geistiges Eigentum. Herausforderung Durchsetzung*, Berlin/Heidelberg: Springer, 2008, p. 47 et seq.; P. Menell and S. Scotchmer, 'Intellectual Property', in A.M. Polinsky and S. Shavell, (eds), *Handbook of Law & Economics*, Vol. 2, Amsterdam: North Holland, 2007, p. 1475 et seq.; P.E. Geller, 'Die Auflösung des Geistigen Eigentums' (2006) *GRUR International* 273 et seq.; W.M. Landes and R.A. Posner, *The Economic Structure of Intellectual Property Law*, Cambridge (USA): Harvard University Press, 2003, p. 1 et seq.; J.H. Reichman and P.F. Uhlir, 'A Contractually Reconstructed Research Commons for Scientific Data in a Highly Protectionist Intellectual Property Environment' (2003) 66 *Law & Contemporary Problems* 315 et seq.; W. Gordon, 'Intellectual Property', in P. Cane and M. Tushnet (eds), *The Oxford Handbook of Legal Studies*, Oxford: Oxford University Press, 2003, p. 617 et seq.; R.C. Dreyfuss, D.L. Zimmermann and H. First (eds), *Expanding the Boundaries of Intellectual Property: Innovation Policy for the Knowledge Society*, Oxford/New York: Oxford University Press, 2001, *passim*; W. Cornish, 'The Expansion of Intellectual Property Rights', in G. Schricker, T. Dreier and A. Kur (eds), *Geistiges Eigentum im Dienste der Innovation*, Baden-Baden (Germany): Nomos, 2001, p. 9 et seq.; E. Kaufer, 'The Incentives to Innovate under Alternative Property Rights Assignments with Special Reference to the Patent System' (1986) 142 *Zeitschrift für die gesamte Staatswissenschaft – Journal of Institutional and Theoretical Economics* 210 et seq.; F. Machlup, *An Economic Review of the Patent System – Study No 15 of the Subcommittee on Patents, Trademarks and Copyrights of the Committee on the Judiciary,* U.S. Senate, 85th Congress, Second Session, Washington DC, 1958, accessed 19 December 2014 at http://library.mises.org/books/Fritz%20Machlup/An%20Economic%20Review%20of%20the%20Patent%20System_Vol_3.pdf. In particular in Anglo-Saxon doctrine, on both sides of the Atlantic, further proposals have been made aimed at the substitution of the exclusive right – 'property' – with a credit right – 'liability' – upon payment of compulsory licensing fees; in this sense, see especially W. Kingston, 'Compulsory Licensing with Capital Payments as an

Through the patent system, then, the 'differential price' of the innovation is paid only by the first generation of those who actually make use of the innovation itself: which shall then circulate, and be used, in the system of free competition – and at the basically subsequent level of price and variety/quality of production.

In particular, with regard to the aforementioned internal corrective measures to the patent system itself, the evaluation of the requirements for patentability, at least according to the traditional setting of the patent system, is aimed at the identification of findings which are actually worthy of exclusive rights, in that they are the expression of a 'flash of genius' which gives rise to a substantial innovation with respect to the previous technical state.

Further, over time – following the advent of industrial design and thus of software and in particular genetic engineering,[2] and the relative decline of the secret as a typical instrument for the protection of

---

Alternative to Monopoly Grants for Intellectual Property' (1994) *Research Policy* 1275 et seq.; as well as: J.H. Reichman, 'Of Green Tulips and Legal Kudzu: Repackaging Rights in Subpatentable Innovation' (2000) *Vanderbilt Law Review* 1743 et seq., which limited the proposal to lower level innovations. For a different critical approach of the view of over-protectionism of intellectual property in favour of greater emphasis of the 'public domain', see J. Boyle, *Public Domain: Enclosing the Commons of the Mind*, New Haven (USA): Yale University Press, 2008, *passim*; B. Hugeholtz and L. Guibault (eds), *The Future of the Public Domain*, The Hague: Kluwer Law International, 2006, *passim*. More extreme positions were taken, then, to criticize *tout court* the granting of intellectual property rights, disregarding their function of stimulating innovation and holding rather that they are obstacles to the diffusion of new ideas, in this sense, see M. Boldrin and D.K. Levine, *Against Intellectual Monopoly*, Cambridge: Cambridge University Press, 2008, *passim*.

[2] See, among others: S. Aymé, G. Matthijs and S. Soini, 'Patenting and Licensing in Genetic Testing – Recommendations of the European Society of Human Genetics' (2008) 16 *European Journal of Human Genetics*, S3 et seq. (S6), where in particular it is highlighted that 'due to the properties of genetic material, many claims that have been granted are overly broad. Broad claims are intrinsic in gene patents, due to the universal nature of gene translation and transcription and protein synthesis', and Aymé, Matthijs and Soini, 'Patenting and Licensing in Genetic Testing: Ethical, Legal and Social Issues', therein, S10 et seq. (S19 et seq.); J.P. Walsh, W.M. Cohen and A. Arora, 'Research Tool Patenting and Licensing and Biomedical Innovation', in W.M. Cohen and S.A. Merrill, (eds), *Patents and the Knowledge-Based Economy*, Washington: National Academic Press, 2003, p. 285 et seq.; M.A. Heller and R.S. Eisenberg, 'Can Patents Deter Innovation? The Anticommons in Biomedical Research' (1998) 280 *Science* 698 et seq.

innovation – legislators and case law,[3] on both sides of the Atlantic, gave rise to the proliferation of exclusive rights, coming to recognize them also with respect to innovations characterized by a reduced, so-called incremental, level of originality, that is produced through cumulative and sequential work which proceeds 'by trial and error'.[4]

This phenomena had a significant impact also with regard to the *numerus clausus* of exclusive rights,[5] in that *sui generis* exclusive systems were introduced in order to regulate 'technology-specific' applications of know-how which appeared likely to evade even the broadest paradigms of patent and copyright laws.[6] Subsequently, these *sui generis* systems were extended in view of the protection for investors with regard to subsequent unauthorized applications. Thus, hybrid systems evolved into a sort of 'miniature patent discipline', or 'patent-like hybrids', that

---

[3] According to the broad acceptance of case law – including the decisions of patent offices in that they are the creators of law – adopted in this book.

[4] See, in particular J.H. Reichman, 'Saving the Patent Law from Itself: Informal Remarks Concerning the Systemic Problems Afflicting Developed Intellectual Property Regimes', in F. Scott Kieff (ed.), *Perspectives on Properties of the Human Genome Project*, Waltham (USA): Academic Press, 2003, p. 289 et seq. (295 et seq.); S. Scotchmer, 'Standing on the Shoulders of Giants: Cumulative Research and the Patent Law' (1991) 5 *Journal of Economic Perspectives* 29 et seq.

[5] At the foundation of the *numerus clausus*, as is known, the recognition of exclusive rights on intangible goods is based on a basically typical system, in which the interests of entities other than things and without legislative recognition (direct or analog) enjoy limited protection and are characterized by the absence of exclusivity; see V. Zeno-Zencovich, entry *Cosa*, in *Digesto delle discipline privatistiche. Sezione civile*, IV, Turin: Utet, 1989, p. 438 et seq. With regard to the acceptance of the principle in European legal systems, see for instance, F. Zenati-Castaing and T. Revet, *Les Biens*, Paris: PUF, 2008, p. 114; A. Ohly, 'Gibt es einen Numerus clausus der Immaterialgüterrecht?', in A. Ohly, T. Bodewig and T. Dreier (eds), *Perspektiven des geistigen Eigentums und Wettbewerbsrecht. Festschrift für Gehrard Schricker zum 70. Geburstag*, Munich: Beck, 2005, p. 105 et seq. With regard to the debate on the theme of the American model, see N.M. Davidson, 'Standardization and Pluralism in Property Law' (2008) 61 *Vanderbilt Law Review* 1597 et seq. (especially 1658 et seq.); H. Hansmann and R. Kraakman, 'Property, Contract, and Verification: The *Numerus Clausus* Problem and the Divisibility of Rights' (2002) 31 *Journal of Legal Studies* 373 et seq. (esp. 408 et seq.); T.W. Merrill and W.E. Smith, 'Optimal Standardization in the Law of Property: The *Numerus Clausus* Principle' (2000) 110 *Yale Law Journal* 1 et seq. (esp. 19 et seq.).

[6] Such as for instance plant varieties databases, software, industrial design, etc.

hindered access to information, limited subsequent innovation and obstructed competition.[7]

On the theme of *numerus clausus*, further, criticisms found through a traditional reading of the same were recently highlighted in doctrine, in particular in the intellectual property sector: these consist essentially, on the one hand, in view of the supervening complication of the system of sources with respect to the dominance of the state law paradigm, in the underestimation of the dynamic dimension of the law and the role of case law as a 'creator of law'; on the other, in light of the internationalization of the sources themselves and the always greater importance, in the related system, of the rules of soft law, of technical regulations, private codes of conduct, the *lex mercatoria*, and so on, in inadequate consideration of the ease with which the phenomena of 'regulatory capture' can occur in various sectors. Thus, the admissibility of the recognition of new intellectual property rights through case law was confirmed, where it is not in contrast with 'ordering' provisions dictated in specific legal rules (such as for instance those relative to the exclusion from patentability).[8]

In regard to the limits of ownership, it is necessary to show, in particular, that the provisional acts of exclusive rights coming from the patent discipline, such as licenses, can be used in order to extend the *ius excludendi* coming from the patent, up to the possible obtainment of actual market power. Thus, beyond any intervention pursuant to the antitrust discipline, the patent system itself is characterized by a dual *fil rouge* of protection for innovation, whether it is: (a) created by the inventor, and as such also encourages further innovations by her/him; or

---

[7] See J.H. Reichman, 'Saving the Patent Law from Itself: Informal Remarks Concerning the Systemic Problems Afflicting Developed Intellectual Property Regimes', cit., p. 296 et seq., who highlighted that: 'This expansion of exclusive rights into the realm of routine applications of know-how to industry converts the collective knowledge available to the technical community as a whole into artificial private preserves, which have to be negotiated and combined in order to support investment in research and development. It generates the anticommons effects – or at least the perceived threat of them – currently experienced in many quarters, and it fosters proposals for reform, some of them quite radical, which supporters of the patent system view with alarm'; see also J.H. Reichman, 'Of Green Tulips and Legal Kudzu: Repackaging Rights in Subpatentable Innovation', cit., p. 1749 et seq., and 'Legal Hybrids Between the Patent and Copyright Paradigms' (1994) 94 *Columbia Law Review* 2432 et seq.

[8] See S. van Erp, 'A *Numerus Quasi-Clausus* of Property Rights as a Constitutive Element of a Future European Property Law?' (2003) 7 *Electronic Journal of Comparative Law* 1 et seq.); A. Ohly, 'Gibt es einen *Numerus clausus* der Immaterialgüterrecht?', cit., p. 118.

(b) subsequently by third parties, with clear pro-competitive effects (indirect, in that they are not connected to antitrust criteria).[9]

As a consequence, both in the United States and European models, limitations of ownership are included, in fact, in order to keep the possibility for third parties to enter the market of innovation open and to provide further contributions to scientific progress.

In this regard, further, it has been shown that in the European context these limitations are provided more clearly in the discipline on the matter, and thus are more easily applied, with respect to that of the U.S. system, for instance in regard to the exception for experimental uses, to compulsory licenses, and to the specific identification of a function which is actually carried out by the molecule of DNA in the patent application.[10]

In regard to the scenario laid out up to this point, the advent of biotechnology introduced new issues and specific cases for protection, also not within the patent system,[11] which lead to 'broadening the

---

[9] See, among others, G. Ghidini, *Innovation, Competition and Consumer Welfare in Intellectual Property Law*, cit., p. 209 et seq., who concludes from the reconciliation of the two disciplines (patent and antitrust) in two guiding principles of the legal system, and in particular to that (first of all) of freedom of competition.

[10] On these aspects, refer also to that which will be further specified below.

[11] Giving rise, as seen, to a broad and complex debate in regard to their patentability; in this regard, besides referring to that which is noted above in the previous Parts as well as below, see *ex multis*, M.M. Karlan, 'Patent Policy, Natural Products, and the Gene Patent Debate: Seeking the Proper Judicial Mode of Analysis' (2011) 67 *New York University Annual Survey of American Law* 95 et seq.; E. Kica and N. Groenendijk, 'The European Patent System: Dealing with Emerging Technologies' (2011) 24 *Innovation – The European Journal of Social Science Research* 85 et seq.; T. Caulfield, 'Human Gene Patents: Proof of Problems?' (2010) 84 *Chicago-Kent Law Review* 133 et seq.; O. Mills, *Biotechnological Inventions. Moral Restraints and Patent Law*, Farnham (UK)/ Burlington (USA): Ashgate Publishing, 2010 (Revised Edition), p. 10 et seq.; E.R. Gold, 'The Ethics of Biotechnological Intellectual Property', in E.R. Gold and B.M. Knoppers (eds), *Biotechnology IP & Ethics*, Markham (Canada): LexisNexis, 2009, p. 15 et seq. (28 et seq.); M.A. Majumder *et al.*, 'Ethical Challenges of Patenting 'Nature': Legal and Economic Accounts of Altered Nature as Property', in B.A. Lustig, B.A. Brody and G.P. McKenny (eds), *Altering Nature – Volume I: Concepts of 'Nature' and 'The Natural' in Biotechnology Debates*, Berlin: Springer, 2008, p. 199 et seq.; J. Johnston and A.A. Wasunna, 'Patents, Biomedical Research, and Treatments: Examining Concerns, Canvassing Solutions', Special Report (2007) 37 *Hastings Center Report* S1 et seq.; S. Jasanoff, 'In the Democracies of DNA: Ontological Uncertainty and

perspective' of analysis carried out by the jurist, through the use of comparative methods.

Patent protection for biotechnological inventions, as known, is justified in order to guarantee adequate incentive and return on the huge investments which are necessary to do research in the field.[12] In fact, the subject matter protected by patents is basically information, as such not rivalled or easy to copy: so, without a system of protection which enables innovators to charge a price for innovative products above the marginal cost, they would not be effectively motivated to either bear the research and development expenses and to innovate or disclose innovation, to the detriment of the public welfare.[13]

However, as known, patent protection also gives rise to relevant social costs, which are due basically to: (a) the static welfare losses due to the above mentioned mark up on the marginal cost of producing the result of the invention; (b) the possible waste of resources originating from the patent race and related litigation; and (c) the increased cost of secondary innovation, that is especially relevant in the biotechnology framework, where innovative activity is to a large extent a cumulative process, with present innovations which are mostly incremental, depending on past innovations.[14]

---

Political Order in Three States' (2005) 24 *New Genetics and Society* 139 et seq.; R.S. Crespi, 'Ethico-Legal Issues in Biomedicine Patenting: A Patent Professional Viewpoint' (2005) 11 *Science and Engineering Ethics* 117 et seq. (120 et seq.); R.M. Cook-Deegan and S.J. McCormack, 'Patents, Secrecy, & DNA' (2001) 293 *Science* 217; R. Eisenberg, 'Re-Examining the Role of Patents in Appropriating the Value of DNA Sequences' (2000) 49 *Emory Law Journal* 783 et seq.

[12] For an overview of the economic studies in this regard, and the related acknowledgment of the patent's incentive function, at least in the biotechnological, pharmaceutical and chemical sectors (as characterized by high risk research projects), see especially B. Hall and D. Harhoff, 'Recent Research on the Economics of Patents' (2012) 4 *Annual Review of Economics* 541 et seq.; E. Mansfield, 'Patents and Innovation: An Empirical Study' (1986) 132 *Management Science* 173 et seq.

[13] See, among others, T. Eger, P. Ebermann and P. Ramanujam, 'Incremental Innovation and Patent Protection for Pharmaceutical Products in India', in P.G. Babu, T. Eger, A.V. Raja, H.B. Schäfer and T.S. Somashekar (eds), *Economic Analysis of Law in India*, New Delhi: Oxford University Press, 2010, p. 128 et seq. (esp. 146).

[14] In this regard, see V. Denicolò, 'Do Patents Over-compensate Innovators?' (2007) 22 *Economic Policy* 679 et seq.; S. Scotchmer, *Innovation and Incentives*, Cambridge (USA): MIT Press, 2004, *passim*; for an heterodox point

What is more, the application of patent protection to the biotechnology sector has raised complex issues, both within the patent system and without. On the one hand, the extension of the patent tool from the mechanical, chemical and pharmaceutical sectors to the context of living, human, animal and plant matter has and must face the clear intercurrent differences between inorganic material and that of living and self-replicating matter, starting from the possible multiple functions carried out and its ability to be reproduced in subsequent generations, which are characteristics of biotechnological inventions rather than mechanical ones.

As a consequence, in the field of genetic engineering there is serious danger – which has already been noted where patent protection has been extended to chemical and pharmaceutical inventions – in asking and granting 'barrier' patents, which reserve a field of monopoly to the inventor which is much greater than the effective contribution of knowledge made by the inventor to the collective.[15] What is more, biotechnology has given rise to the problem of interference between a mechanism of regulation and stimulation of the market like patent and general interests often considered more important than that which is at the foundation of patent protection (think, for example, of the right of the individual to not have his genes manipulated, or to the safeguarding of the genetic diversity present in nature), which may be effected by the granting or use of exclusive patents.[16] In this regard, as has been seen, bioethical profiles which arise with respect to the patentability of living beings are particularly relevant.

In reference to biotechnological inventions, then, considering that the aforementioned proliferation of exclusive rights in this case concerns goods which have the particular characteristic of being, at the same time, both the input and output of the production process, and in particular its relevance for the subject of the so-called research tool,[17] it is fundamental

---

of view, see also M. Boldrin and D.K. Levine, 'The Case against Intellectual Property' (2002) 92 *American Economic Review* 209 et seq.

[15] Thus, M. Ricolfi, 'La brevettabilità della materia vivente: fra mercato e nuovi diritti' (1993) *Giurisprudenza italiana* 292 et seq. (296).

[16] See again, among others: M. Ricolfi, 'Bioethics Market and Morals: The Case of Biotechnological Patents', in C.M. Mazzoni (ed.), *A Legal Framework for Bioethics*, London: Kluwer Law International, 1998, p. 131 et seq.

[17] Defined by the National Institutes of Health as 'the full range of resources that scientists use in the laboratory, while recognizing that from other perspectives the same resources may be viewed as "end products". ... the term may thus include cell lines, monoclonal antibodies, reagents, animal models, growth

to consider the political and economic issues concerning the extent to which restrictions on access to findings connected to exclusive patents are capable of limiting the use of the inventions by other researchers, and thus – basically – may possibly hinder subsequent innovation.

In a sector such as that of biotechnology, characterized by the cumulative nature of the process of development of innovations, the existence of a collective patrimony of information, knowledge and ideas which are freely accessible to anyone and withdrawn from ownership by private persons, and in particular the absence of exclusive rights for products of basic research, is an essential prerequisite to allow for subsequent innovation, besides competition; thus, it appears to be an essential resource, to conserve in a common form in order to guarantee both the achievement of inherent objectives in the patent system, and the coordination of it with the principles and further disciplines dictated by the legal system.[18]

In this regard, it was highlighted that, both in American and European doctrine, there has been emphasis on the negative consequences of the so-called over-protection of intellectual property, and its consequent risks of foreclosure with respect to subsequent innovation and competition (up until the so-called 'tragedy of anticommons').[19] This, in particular, may

---

factors, combinatorial chemistry libraries, drugs and drug targets, clones and cloning tools (such as PCR), methods, laboratory equipment and machines, databases and computer software' (*Report of the National Institutes of Health (NIH) Working Group on Research Tools*, June 1998, accessed 19 December 2014 at http://biotech.law.lsu.edu/research/fed/NIH/researchtools/Report98.htm). On this theme, see also: C. Clift, 'Patenting and Licensing Research Tools', in A. Krattiger *et al.* (eds), *Intellectual Property Management in Health and Agricultural Innovation: A Handbook of Best Practices*, Oxford (UK)/Davis (USA): MIHR – PIPRA, 2007, accessed 19 December 2014 at http://www.iphandbook.org, p. 79 et seq.; M.D. Walker, 'The Patent Research Tool Problem After *Merck v. Integra*' (2005) 14 *Texas Intellectual Property Law Journal* 1 et seq.; M.S. Mireles, 'An Examination of Patents, Licensing, Research Tools, and the Tragedy of the Anticommons in Biotechnology Innovation' (2004) 38 *University of Michigan Journal of Law Reform* 141 et seq.; J.P. Walsh, W.M. Cohen and A. Arora, 'Research Tool Patenting and Licensing and Biomedical Innovation', cit., p. 285 et seq.

18 In this regard, besides referring below, see among others, J. Boyle, *Public Domain: Enclosing the Commons of the Mind*, cit., *passim*; C. Hess and E. Ostrom, *Understanding Knowledge as a Commons. From Theory to Practice*, Cambridge (USA): MIT Press, 2007, *passim*; B. Hugeholtz and L. Guibault, *The Future of the Public Domain*, cit., *passim*.

19 In regard to which, besides referring to that which is said in Chapter 1 and to that which shall be said below, see M. Heller, *The Gridlock Economy.*

occur through: (a) the patenting of natural elements or information, consisting in the mere reformulation of products or natural phenomena, that are in any case already accessible to the public;[20] (b) the utility of patents – should they, as was said, be granted for innovations possessing little originality – in hindering the development of products or processes by current or potential competitors; the greater risk where such use is carried out together by more patent owners on various basic elements needed for the development of a product (a so-called patent thicket);[21] or (c) the possibility of obtaining subsequent registrations for the invention making only slight improvements, in order to surreptitiously prolong the duration of the exclusive.[22]

On the other hand, it should also be noted that the effective incidence of patent protection for biotechnological inventions on the development of scientific research is still a controversial issue, with opinions of scholars and survey results among researchers not always being unanimous with respect to the preclusive effect of the patents 'upstream' on research 'downstream'.[23]

---

*How Too Much Ownership Wrecks Markets, Stops Innovation, and Costs Lives*, New York: Basic Books, 2008, *passim*, and 'The Tragedy of Anticommons' (1998) 111 *Harvard Law Review* 621 et seq.; M. Heller and R. Eisenberg, 'Can Patent Deter Innovation? The Anticommons in Biomedical Research', cit., p. 698 et seq.

[20] This, as a consequence of the aforementioned phenomena of progressive 'degradation' of the evaluation of requirements of patentability by legislators and case law.

[21] On this theme, besides referring to that which will be said extensively below, see recently: WIPO, '2011 World Intellectual Property Report. The Changing Face of Innovation', cit., p. 78 et seq.; I. Hargreaves, *Digital Opportunity. A review of Intellectual Property and Growth. An independent report by Professor Ian Hargreaves*, 2011, accessed 19 December 2014 at https://www.gov.uk/government/uploads/system/uploads/attachment_data/file/32563/ipreview-finalreport.pdf, p. 56 et seq.

[22] In this regard, see among others: A.B. Jaffe and J. Lerner, *Innovation and Its Discontents: How Our Broken Patent System is Endangering Innovation and Progress, and What to Do About It*, Princeton: Princeton University Press, 2004, *passim*.

[23] See C.J. Shiu, Of Mice and Men: Why an Anticommons Has Not Emerged in the Biotechnological Realm' (2009) 17 *Texas Intellectual Property Law Journal* 413 et seq.; Z. Lei, R. Juneja and B.D. Wright, 'Patents versus Patenting: Implications of Intellectual Property Protection for Biological Research' (2009) 27 *Nature Biotechnology* 36 et seq.; Gold, 'The Ethics of Biotechnological Intellectual Property', cit., p. 32 et seq.; S. Aymé, G. Matthijs and S. Soini, 'Patenting and Licensing in Genetic Testing: Ethical, Legal and

More recently, in American doctrine, a shared proposal was registered for an approach to the aforementioned issues posed by biotechnological inventions which could be defined as 'technology specific', based on: (a) the accurate definition of the context of the extension for innovations 'upstream'; and (b) the cost-benefit analysis with specific regard to the various circumstances in which they may arise – in particular, to the actual state, therapeutic proteins, diagnostic methods and research tools – with different evaluations depending on their particular respective characteristics. In this manner, the rules on the matter could be interpreted, or detailed, in favour of or against the patentability of the findings, according to the specific category of scientific research in question.[24]

Biotechnological inventions, then, according to the preferred reconstruction[25] appear to introduce further profiles to the regulation of the

---

Social Issues', cit., p. S11 et seq.; G.E. Marchant, 'Genomics, Ethics and Intellectual Property', in A. Krattiger *et al.* (eds), *Intellectual Property Management in Health and Agricultural Innovation: A Handbook of Best Practices*, cit., p. 29 et seq. (35 et seq.); L.B. Andrews, 'Genes and Patent Policy: Rethinking Intellectual Property Rights' (2002) 3 *Nature Reviews Genetics* 803 et seq.

[24] In this sense, see A. Lauer, 'The Disparate Effects of Gene Patents on Different Categories of Scientific Research' (2011) 25 *Harvard Journal of Law & Technology* 179 et seq. (183 et seq.); C.J. Shiu, 'Of Mice and Men: Why an Anticommons Has Not Emerged in the Biotechnological Realm', cit., p. 442 et seq.

[25] In regard to the debate on the theme, besides referring to the previous Chapters and below, see E. Siew-Kuan, 'Immoral Inventions: Interaction between Ethics and Biotechnology Patent Law' (2010) 22 *Singapore Academy of Law Journal* 931 et seq. (935 et seq., where, in particular, it is clearly highlighted that: 'Opponents to moral scrutiny of patentable subject matter contend, inter alia, that: a) The patent system is not the proper forum to deliberate issues of ethics and morality. b) Patent adjudicators are ill-equipped, and are inappropriate persons, to decide issues that call for complex interplay of ethical, moral, social religious and policy judgments. c) There is no universal benchmark for determining what constitutes unethical or immoral invention. d) Legislative intervention will unnecessarily derail private enterprise support and cause stagnation in scientific innovation. … Proponents maintain, inter alia, that: a) The patents system should subject inventions to ethical-cum-moral scrutiny, so as to sift out inventions that are improper subject matter for legal protection in order to maintain the moral standards of society. b) States, in granting patents, cannot disclaim responsibility for inventions that they grant and should not hide behind the negative character of the patent right to avoid deciding whether an invention is inherently against the public interest. The grant of broad biotechnology patents, such as those on human genes, may block research, hamper innovation, create obstacles to life-saving treatments and run counter to the goal of the patent system'); O. Mills, *Biotechnological Inventions. Moral Restraints and Patent*

'perimeter' of the patent discipline, essentially bioethical in nature: changing, in this way, the 'playing field' of comparison, given that 'bioethical counter limits' are added to competition corrective measures.

These appear hierarchically subordinate to the others in the European model – in which their further application by the EPO and the courts is not always consistent[26] – while they seem to be all but ignored in the United States, where the tendency is to concentrate on the technicality of the patent material, especially through the evaluation of the fulfilment of

---

*Law*, cit., p. 10 et seq.; S. Aymé, G. Matthijs and S. Soini, 'Patenting and Licensing in Genetic Testing: Ethical, Legal and Social Issues', cit., p. S16 et seq.; M. Blakeney, 'Biotechnological Patenting. Law, Ethics and Science', in G. Ghidini and G. Cavani (eds), *Brevetti e biotecnologie*, Rome: Luiss University Press, 2007, p. 21 et seq.

[26] It was shown, in particular, that the EPO did not offer clear definitions and unique criteria for the application of the provision contained within the EPC on exclusions to patentability for contrasting public order or morality, with the result of a case-by-case approach in fact which is not always consistent (see Nuffield Council on Bioethics, '*The ethics of patenting DNA: a discussion paper*', July 2002, accessed 19 December 2014 at http://www.nuffield bioethics.org/patenting-dna, p. 74). On this point, the European Group on Ethics, which pursuant to art. 7 and Recital 44 of Directive 98/44/EC may be consulted 'also in regards to patent law, only at the level of evaluation of biotechnology concerning fundamental ethical principles', highlighted the necessity for ethical evaluations during the examination of patent applications, where they involve specific issues of this nature, suggesting the institution of a specific independent advisory body (European Group on Ethics in Science and New Technologies, 'Opinion n°16 – Ethical aspects of patenting inventions involving human stem cells', 7 May 2002, p. 18). In this sense, see also Kica and Groenendijk, 'The European Patent System: Dealing with Emerging Technologies', cit., p. 101, which highlighted the importance 'to establish a pre-grant patent advisory body that would control and reduce the probability that patent examiners would issue low-quality patents that harm the innovation process and have undesirable ethical and social consequences'; S. Aymé, G. Matthijs and S. Soini, 'Patenting and Licensing in Genetic Testing – Recommendations of the European Society of Human Genetics', cit., p. S7, and 'Patenting and Licensing in Genetic Testing: Ethical, Legal and Social Issues', cit., p. S19. In general, then, with regard to the criticism coming from multiple disciplines and case law existing on the subject of patents in Europe, and the relative proposals for reform, see, among others, H. Ullrich, 'Harmonizing patent law: The untamable Union Patent', in M.C. Janssens and G. Van Overwalle (eds), *Harmonisation of European IP Law: From European Rules to Belgian Law and Practice – Contributions in honour of Frank Gotzen*, Bruxelles: Bruylant, 2012, p. 243 et seq.; T. Jaeger, 'Back to Square One? An Assessment of the Latest Proposals for a Patent and Court for the Internal Market and Possible Alternatives' (2012) 43 *IIC* 286 et seq.; O. Mills, *Biotechnological Inventions. Moral Restraints and Patent Law*, cit., p. 190 et seq.

the patentability requirements, tendentially 'absorbing' here also any moral issues which may emerge on the subject.[27]

According to this U.S. approach, the criteria for identification of the patentability requirements of non-obviousness and industriality are particularly significant. The former, according to the restricted reading provided by the Court of Appeals for the Federal Circuit in the decision *In re Kubin*,[28] must be held to be excluded wherever there is a wide knowledge about the protein a target gene encodes plus general knowledge of the techniques for isolating and sequencing the same gene. The requirement of non-obviousness needs the identification of specific methods of use for the finding, and so allows the limitation of the patent exclusive only for a specific application of the invention.[29]

This 'merely technical' concept of patent law, as well as, more broadly, of intellectual property in general, through the lens of incentive for innovation, in reality, today appears inadequate to fully meet the multiple and constantly changing characteristics of the biotechnological innovation 'prism'. Contrarily, the central importance on the subject regarding the issue of balancing the interests at play, in light of the fundamental values at the foundation of the legal systems under examination, seems clear, as outlined at a constitutional level.[30]

---

[27] In this regard, besides referring above, to Part III, see Siew-Kuan, 'Immoral Inventions: Interaction between Ethics and Biotechnology Patent Law', cit., p. 940 et seq.; O. Mills, *Biotechnological Inventions. Moral Restraints and Patent Law*, cit., p. 45 et seq.; M.A. Bagley, 'A Global Controversy: The Role of Morality in Biotechnology Patent Law', in P. Yu (ed.), *Intellectual Property and Information Wealth: Issues and Practices in the Digital Age*, Westport (USA): Praeger Press, 2006, p. 317 et seq. (318 et seq.).

[28] 561 F.3d 1351 (Fed. Cir. 2009).

[29] Therefore, for instance, it is held that under the Kubin standard many of the single gene patents that Myriad has foreclosed would have gone down on obviousness grounds anyway. What is more, such a standard is likely having a similar effect on cDNA patents as well, undercutting the significance of the Court allowing them to survive as patentable subject matter (see J. Conley, 'Myriad, Finally: Supreme Court Surprises by not Surprising', in 2013 *Genomics Law Report*, accessed 19 December at http://www.genomicslawreport.com/index.php/2013/06/18/myriad-finally-supreme-court-surprises-by-not-surprising, 18 June 2013, p. 2).

[30] See G. Ghidini and A. Stazi, 'Freedom to Conduct a Business, Competition and IP', in C. Geiger (ed.), *Research Handbook on Human Rights and Intellectual Property*, Cheltenham, UK and Northampton, MA, USA, Edward Elgar, 2015 (forthcoming); D. Wielsch, *Zugangsregeln. Die Rechtsverfassung der Wissensteilung*, Tübingen (Germany): Mohr Siebeck, 2008, pp. 8 and 66 et seq.; C. Geiger, '"Constitutionalising" Intellectual Property Law? The Influence of

In this sense, it is worthwhile to highlight the relationship – always growing, with procedures of scientific and economic development – between intellectual property and the system of fundamental rights.[31] In this regard, there are always more frequently cases in which the courts are called to evaluate the legitimacy of the granting or the use of intellectual property rights with regard to a fundamental principle of constitutional status considered to be antagonistic to them, such as human dignity, freedom of information, freedom of scientific research, and so on.[32]

In particular, in the biotechnology field, the decisions taken by the EU Court of Justice on the validity of Directive 98/44/EC in light of the principles of human dignity and the non-patentability of discoveries come to mind, as well as the recent *Brüstle* case regarding the compatibility of patenting of embryonic stem cells with the fundamental principles of human dignity and integrity, in addition to the U.S. Supreme

---

Fundamental Rights on Intellectual Property in the European Union' (2006) 37 *IIC* 371 et seq.

[31] See, in particular, C. Geiger, 'Fundamental Rights as Common Principles of European (and International) Intellectual Property Law', in A. Ohly (ed.), *Common Principles of European Intellectual Property Law*, cit., p. 223 et seq.; L.R. Helfer and G.W. Austin, *Intellectual Property: Mapping the Global Interface*, Cambridge: Cambridge University Press, 2011, p. 1 et seq.; W. Grosheide (ed.), *Intellectual Property and Human Rights: A Paradox*, Cheltenham, UK and Northampton, MA, USA, Edward Elgar, 2010, *passim*; P.L.C. Torremans, *Intellectual Property and Human Rights*, Alphen aan den Rijn (Netherlands): Kluwer Law International, 2008, *passim*; P. Yu, 'Reconceptualizing Intellectual Property Interests in a Human Rights Framework' (2007) 40 *UC Davis Law Review* 1039 et seq.; L.R. Helfer, 'Toward a Human Rights Framework for Intellectual Property', therein, 971 et seq., and 'Human Rights and Intellectual Property: Conflict or Coexistence?' (2003) 5 *Minnesota Intellectual Property Review* 47 et seq.; A.R. Chapman, 'A Human Rights Perspective on Intellectual Property, Scientific Progress, and Access to the Benefits of Science', in WIPO, *Intellectual Property and Human Rights*, WIPO Publication No. 762(E), Geneva, WIPO, 1999, *passim*.

[32] See C. Geiger, '"Constitutionalising" Intellectual Property Law? The Influence of Fundamental Rights on Intellectual Property in the European Union', cit., p. 371 et seq.; T. Braegelmann, 'The Constitutional Scope and Limits to Copyright Law in the United States, in Comparison with the Scope and Limits Imposed by Constitutional and European Law on Copyright Law in Germany' (2009) 27 *Cardozo Arts & Entertainment Law Journal* 99 et seq.

Court rulings on the *Myriad* and *Mayo* cases regarding the relationship between patents on genes, freedom of research and the right to health.[33]

The benchmark in regard to the discipline of biotechnological inventions,[34] then, appears to be found in the interpretation and application of the same in light of the fundamental rights provided at a constitutional level in the legal model of reference, according, in fact, to a 'constitutionally oriented' reading.

This is in view of the pursuit of two essential and closely connected aims, that are: (a) the reconciliation of the interests at stake with fundamental rights which are specifically relevant to the case at hand, in order to avoid that damage may arise from the granting or use of exclusive patents: in biotechnology this perspective, as has been seen, is particularly important with respect to the bioethical issues it brings up (in practice, essentially in the European model); and (b) the use of fundamental rights as tools of 'limitation' in the context of ownership and 'opening' of the sphere of the exclusive, in order to rebalance a patent system which, on both sides of the Atlantic, is always more unbalanced in favour of rights owners.

---

[33] After such Supreme Court decisions, in fact, potential competitors results enable them to enter the market with less risk of being rightfully sued for infringement, patients seem to be allowed to obtain the availability of a second-opinion testing, and basic researchers are free to do research with isolated genomic DNA (while as hinted in the United States there is no effective research exemption to patent infringement, even in the nonprofit sector); see J. Conley, 'Myriad, Finally: Supreme Court Surprises by not Surprising', cit., p. 1. In regard to all the cited U.S. and European decisions, besides referring to the analyses carried out above in Parts III and IV, refer also to that which will be said again below.

[34] As, further, for intellectual property *tout court*. In this sense, among others, see G. Ghidini and A. Stazi, 'Freedom to Conduct a Business, Competition and IP', cit.; C. Geiger, 'The Constitutional Dimension of Intellectual Property', in Torremans, *Intellectual Property and Human Rights*, cit., p. 101 et seq.; A. Ohly, *Geistiges Eigentum und Gemeinfreiheit: Forschrungsperspektiven*, in A. Ohly and D. Klippel, (eds), *Geistiges Eigentum und Gemeinfreiheit*, Tübingen (Germany): Mohr, 2007, p. 2 et seq.

## V.2 THE UTILITY OF THE COMPARATIVE METHOD FOR AN 'INTEGRATED' VISION, IN LIGHT OF RELEVANT INSTITUTIONAL ROLES FOR THE RECONCILIATION OF INTERESTS AT STAKE AND RELATIONSHIPS WITH OTHER SCIENCES AND HUMANISTIC DISCIPLINES

Patent law in general, and that of biotechnology patents in particular (although especially in the United States model this characteristic tends to be hidden behind the 'veil' of technicality), is in effect an instrument of public policy.[35]

Patent law is aimed at operating, actually, as a sort of 'law for the regulation of innovation markets', in which the established authorities are called to 'dole out' the granting and limits of exclusive rights and the uses derived from them, in particular, in light of the competitive antibodies and bioethical counterlimits referenced previously.

In this view, it is necessary to consider two essential elements as fundamental for the proper classification and adequate comprehension of the discipline on the subject, which are: (a) the process of formation and application of the rules, according to the so-called 'legal process' approach,[36] and in particular the role of the legislature and/or government in the identification of the most appropriate discipline in terms of public policy (being that they are bodies which are institutionally responsible and democratically legitimized to do so), but also, as was seen, the connected relationship between primary legislative sources and other relevant sources on the subject of the legal system or systems under

---

[35] In this regard, as far as is relevant in this case, see, among others: O. Mills, *Biotechnological Inventions. Moral Restraints and Patent Law*, cit., p. 1 et seq. On the theme of public policy, *ex multis*, see L.M. Surhone, M.T. Timpledon and S.F. Marseken, *Public Policy*, Saarbrücken (Germany): VDM Verlag Dr. Mueller e.K., 2010, *passim*; Wikipedia.org, entries *Public policy* and *Public policy doctrine*, available at http://en.wikipedia.org/wiki/Public_policy and http://en.wikipedia.org/wiki/Public_policy_doctrine.

[36] In regard to which, see especially the fundamental work of H.M. Hart Jr. and A.M. Sacks, *The Legal Process. Basic Problems in the Making and Application of Law* (W.N. Eskridge Jr. and P.P. Frickey (eds)), New York: Foundation Press, 1994, *passim* (in particular 1 et seq. and 102 et seq.); see also H.M. Hart Jr. and H. Wechsler, *The Federal Courts and the Federal System* (R.H. Fallon Jr., J.F. Manning, D.J. Meltzer and D.L. Shapiro (eds)), New York: Foundation Press, 2009 (6th edn.), *passim*.

examination, among which in particular that of case law;[37] and (b) the relationship between patent law for biotechnology and other sciences and humanistic disciplines, which in particular, besides the natural sciences, are those of politics and economics, as well as philosophy and bioethics.

For these purposes, in light of the analyses carried out, it appears clear that the comparative method is a significantly useful instrument. Despite the difficulty in obtaining exhaustive information with respect to the actors and the dynamics of the process for the formation of rules in various systems, in fact, the use of the comparative method gives rise to at least two extremely significant benefits, that are: '1) the tendency to push analytic categories to higher levels of abstraction in order to bridge differences between legal systems; 2) the tendency to force the researcher to expand the analysis to include the whole legal system and its relationship with the rest of human culture and its material and spiritual context in order to understand the differences and similarities observed'.[38]

As far as an analysis of the legal process is concerned, it is first necessary to specify that, referencing the adoption of this perspective, it is intended in the sense – which is still completely current[39] – of comparative analysis of the roles taken on by competent institutions on the subject under discussion here, as well as the related operative dynamics and spheres of influence.

---

[37] Including in the definition of case law, as said, also decisions of patent offices.

[38] J.C. Reitz, 'How To Do Comparative Law' (1998) 46 *American Journal of Comparative Law* 617 et seq.; see also G. Mousourakis, 'How Comparative Law Can Contribute to the Development of a General Theory of Legal Evolution' (2007) 14 *Tilburg Law Review: Journal of International and Comparative Law* 277 et seq., who, further, in regard to the issue of the definition of comparative law as a method or as a science, affirms that: 'At its simplest level, the description of differences and similarities between legal systems, the comparative method allows us to acquire a better understanding of the characteristic features of particular institutions or rules. But as the comparative method becomes more sophisticated, for instance, when the socio-economic and political structures, historical background and cultural patterns that underpin the institutions or rules are taken into account, the comparative method begins to produce explanations based on interrelated variables – explanations which become progressively more scientific in nature'.

[39] As it was opportunely shown by E.L. Rubin, 'The New Legal Process, the Synthesis of Discourse, and the Microanalysis of Institutions' (1996) 109 *Harvard Law Review* 1393, 1403 et seq.

What is more, the same perspective of the meaning accepted here is, on one hand, 'oriented' in light of subsequent teaching coming from approaches of 'law and economics' and 'critical legal studies', as revisited by the 'post-Chicagoan' and 'outsider scholarship' schools.[40] This is, particularly, in regard to (a) the decisive role of the political choices and relationships of power in the formation of the rules, given that patent law, as said, is essentially an instrument of policy, and (b) the correlated mechanisms often of 'second best' functioning of the process: think of the centrality, demonstrated on the subject, of the role of case law, in place of the traditional criteria of reserving to the legislature – often 'slow' in intervening or 'captured' by lobbies – based on the *numerus clausus* concept.

On the other hand, with respect to the law and economics and critical legal studies approaches, the comparative method adopted leads to the belief that the instrumental reconstruction of rights is not shared, evaluating that and in the measure in which they are actually capable of

---

[40] In regard to such teachings, as far as critical principles specifically indicated by them are concerned towards the legal process (recalled immediately below), see again, E.L. Rubin, 'The New Legal Process, the Synthesis of Discourse, and the Microanalysis of Institutions', cit., p. 1411 et seq. In regard to the law and economics approach, see among others, R.A. Posner, *The Economic Analysis of Law*, New York: Aspen Publishers, 2011 (8th edn), *passim*; D.C. Mueller, *Public Choice III*, Cambridge: Cambridge University Press, 2008 (3rd ed.) p. 1 et seq.; D.R. Mayhew, *Congress: The Electoral Connection*, New Haven (USA)/London: Yale University Press, 2004 (2nd edn), p. 13 et seq.; W.M. Landes and R.A. Posner, *The Economic Structure of Intellectual Property Law*, cit., p. 1 ss.; M.P. Fiorina, *Congress: Keystone of the Washington Establishment*, New Haven (USA)/London: Yale University Press, 1989 (2nd edn), p. 39 et seq.; W.A. Niskanen Jr., *Bureaucracy and Representative Government*, Chicago/New York: Aldine-Altherton, 1971, p. 38 et seq.; G. Tullock, *The Politics of Bureaucracy*, Washington DC: Public Affairs Press, 1965, p. 134 et seq.; J.M. Buchanan and G. Tullock, *The Calculus of Consent: Logical Foundations of Constitutional Democracy*, Ann Arbor (USA): University of Michigan Press, 1962, p. 3 et seq. In regard to the approach of *critical legal studies*, see among others, D. Kennedy, *Legal Education and the Reproduction of Hierarchy: A Polemic Against the System – A Critical Edition*, New York: New York University Press, 2004, *passim*; M.J. Horwitz, *The Transformation of American Law 1780–1860*, Cambridge (USA): Harvard University Press, 1977, p. 30 et seq.; R.M. Unger, *The Critical Legal Studies Movement*, Cambridge (USA): Harvard University Press, 1986, p. 5 et seq.; M. Horkheimer and T.W. Adorno, *Dialectic of Enlightenment*, New York: Continuum, 1972, p. 20 et seq.

pursuing their underlying social aims;[41] a reconstruction which may be acceptable perhaps in a utilitarian model such as the American one is, but not in one based on fundamental rights such as the European one is.

Rather, on the theme of the patentability of biotechnological inventions, given the peculiarities which set it apart, it is possible to show how the reconstruction according to which – inspired by the European model – it is held appropriate to frame the regulation of intellectual property rights and their limits in the context of fundamental rights specifically relevant to the case in question appears particularly convincing.[42]

Once the range of the analytical approach taken has been clarified, regarding the role of the legislature and/or government, it appears proper to believe that – in general, except for the *caveats* reported above – their intervention may lead to significant advantages, in view of the framework of biotechnological patent law as an instrument of policy.

In this sense, first of all there are advantages coming from the democratic legitimizing of such bodies, which allow them to respond to the needs of the collective in a more direct and informed manner than might be possible through case law, reflecting adequately the social values relevant in those circumstances. There is, then, the advantage of having general rules established prior to the problem appearing, which may allow it to be prevented. Still, there are advantages from the absence of the obligation to justify the issuing of rules in light of fundamental principles (especially reasonability) or of precedents, as happens instead for case law. What is more, the legislature and government have the resources and capability to synthesize vast amounts of data and opinions coming from various sectors, and to formulate general guidelines which can be applied to a broad range of situations.[43]

---

[41] On the contrary, E.L. Rubin, 'The New Legal Process, the Synthesis of Discourse, and the Microanalysis of Institutions', cit., p. 1397, highlighted that: 'The Constitution's role is … noteworthy for legal process scholars, particularly given their frequent emphasis on public law. It guarantees, in their view, a series of basic human rights, perhaps even a system of human rights, as a matter of positive law, and authorizes the courts to enforce these rights'. In this regard, see also, especially, J.H. Choper, *Judicial Review and the National Political Process: A Functional Reconsideration of the Role of the Supreme Court*, Chicago: University of Chicago Press, 1980, p. 4 et seq.; J.H. Ely, *Democracy and Distrust: A Theory of Judicial Review*, Cambridge (USA): Harvard University Press, 1980, p. 73 et seq.

[42] In this regard, refer also, extensively, below.

[43] See, in particular, C.H. Baron, 'Bioethics and Law in the United States: A Legal Process Perspective', (2007) *Diritto pubblico comparato ed europeo* 1653 et seq., 1658 et seq.

However, in the face of new challenges arising from the development of biotechnological inventions, in practice, case law finds itself overwhelmed by a significant 'para-regulatory' task, often being called – even in the essentially civil law model such as that which exists in Europe – to function as a true creator of law (in place of a legislature which as noted above is often 'slow' to react or 'captured' by lobbies).[44] This is through a work which from the expansive interpretation often leads to it crossing over to performing true actions in order to close gaps or fallacies in the system, or rather to make 'technology-specific' rules compatible with the general principles or with other disciplines in force in the same system. As a consequence, on this matter the recent criticism made regarding the reserve solely to the legislature seems to have found its true reason for being, on the basis of the principle of the obligatory nature of intellectual property rights, of the determination of their category and the powers they grant, as well as relative contexts and limits of extension, and thus of the balancing mechanisms between exclusive control and access to natural resources and/or information.

Coming to the relationship between biotechnological patent law and other sciences and humanistic disciplines, in general, it must be noted that the law, being involved in every social relationship, is in a certain sense an element of each of these, as they are, in some sense, an element of law. The legal framework regulates and influences every single action of the people who live in an organized society; thus, the aspects of this which are specifically relevant must be taken into consideration in every study on social behaviour.

On the other hand, given that the law is only one of the sciences and humanistic disciplines, jurists should always keep the knowledge coming from the others in mind in consideration of the circumstance under examination.[45]

---

[44] In this regard, see, among others, N. Thompson, 'Gene Blues', in *Washington Monthly*, April 2001, p. 9 et seq. (14).

[45] In this sense, H.N. Butler and L.E. Ribstein, *Legal Process and the Discovery of Better Policies for Fostering Innovation and Growth*, in The Kauffman Task Force on Law, Innovation, and Growth, '*Rules for Growth: Promoting Innovation and Growth Through Legal Reform*', Kansas City (USA), Ewing Marion Kauffman Foundation, 2011, p. 464, have correctly noted – on the basis of the vision expressed by F. von Hayek, 'The Use of Knowledge in Society' (1945) 35 *American Economic Review* 519 et seq. – that 'experts, academics, and lawmakers lack the necessary knowledge to make decisions about which legal and regulatory structures maximize welfare. Even the most brilliant and thoughtful policymakers cannot take account of the enormous number of political, social, technological, and economic variables that determine

The analysis of these profiles would be in and of itself easier within a specific legal system, given that the institutional and socio-economic procedures and dynamics differ within each system, and the differences may be seen in the distinct operative and/or professional duties and responsibilities.[46]

The comparative method, however, allows for the adoption of an analytical view which is, on the one hand, 'integral' through the evaluation of the elements of knowledge of the situation coming from various sciences and humanistic disciplines; on the other 'enriched' by the study of the peculiarities of the specific attitude of such elements in the various legal systems.[47]

## V.3 THE RELEVANCE OF PERSPECTIVES IN THE LEGAL PROCESS FOR THE PURPOSE OF TRANSPARENCY OF RELEVANT PLAYERS, INTERESTS AND TOOLS

Biotechnology patent law, in particular in the United States model, but also in the European one, is essentially a case law matrix.[48] This characteristic, further, considering the constant evolution of biotechnological inventions and the need for the decisions on the matter to constantly 'keep up' with the developments of science (and thus the market), seem in effect almost inevitable.[49]

---

how their plans will operate in the unknowable real world of the future. In order to deal with this knowledge problem, Hayek championed the market system as a vast network that can process the necessary information'; getting from this the consideration according to which: 'Consistent with this approach, consideration of how to alter laws and regulations to increase innovation and economic growth must include an analysis of how to harness market-like mechanisms to create an information network that leads to the discovery of better policies over time'.

[46] See H.M. Hart and A.M. Sacks, *The Legal Process. Basic Problems in the Making and Application of Law*, cit., p. 174 et seq.

[47] This in particular, as noted, through a lens of the legal process which is particularly relevant to the subject.

[48] *Lato sensu*, as said.

[49] In this regard, among others, see C.H. Baron, 'Bioethics and Law in the United States: A Legal Process Perspective' cit., who highlights, in particular, that in the United States judges 'are appointed to the bench and do not stand for popular election. As a result, they are not as vulnerable as legislators to the threat of political retribution. Moreover, the courts offered other advantages over the legislature as institutions for modifying the law to accommodate the pressures on

In this sense, it is easy to understand, then, how it was case law which first granted patent protection for biotechnological inventions, followed later by legislation. The adoption of the TRIPS Agreement tended to unify biotechnological patent law at a global level. Today, member countries of the Agreement provide the patentability of various living organisms, like microorganisms, plants, animals and human genetic material, produced using biotechnology. On the other hand, as has been seen, the prevailing opinion is that the production of transgenic human beings and related research should not be encouraged; thus, for instance, a cloned human embryo or the processes which alter the germinal line of human beings are not patentable.

The consideration of the central role of case law, however, must be contextualized in the broadest perspective of the legal process which gives rise to the rules on the matter. The construction and analysis of a system of institutionalized procedures of composition of the various social circumstances, in fact, imply necessarily the inclusion of various essential elements, that are in particular: (a) what are the relevant public institutions, or the private bodies, on the matter; (b) what type of circumstances is each of the structures and procedures of the system designed to address; (c) what are the instruments of intervention at their disposal in order for them to carry out their duties; and (d) who are the subjects or stakeholders who will promote such circumstances.[50]

---

society created by progress in medicine and biotechnology. Although, common law judges clearly "make law", the tradition is for them to do so by incorporating values from, and building incrementally upon, judicial precedents established in earlier cases. And, in deciding the cases before them, they are expected to write opinions that explain their decisions in a fashion that limits, as much as possible, the rule justifying the result to only what is necessary to reach the decision on the facts of the particular case. Unlike legislators, they need not draft laws in broad, sweeping form and in a fashion which anticipates and accommodates all possible exceptions. They can make law tentatively. They can establish a rule in one case; broaden the rule in a later case if it is determined to have been fashioned too narrowly, limit it if it was stated too broadly, or overrule it altogether if the court concludes it to have been a mistake'.

[50] See, among others, G. Mousourakis, 'How Comparative Law Can Contribute to the Development of a General Theory of Legal Evolution' cit., 272 et seq. (287 et seq.); A. Watson, 'Comparative Law and Legal Change' (1978) 37 *Cambridge Law Journal* 313 et seq.; as far as is relevant in this case, see R.S. Crespi, 'Ethico-Legal Issues in Biomedicine Patenting: A Patent Professional Viewpoint', cit., p. 117 et seq.

Clearly, the sources of the rules which are significant for the discipline of biotechnological inventions, as for other sectors, are extremely varied, and their analysis, as has been seen, is complex.

Also in this context, the issues which arise are derived from the dynamics of socio-economic relations,[51] and in particular from the situation of constant scientific and applied developments from research on the matter.

In any legal system, at a public level, the responsibility to put in place the measures needed for the settlement of these instances is divided, as is known, among the institutions which hold various powers (legislative, executive and judicial). In a federal model such as in the United States, this is further divided between the federal system and that of each single state.[52] In the European model, the institutions of the European Union can adopt more or less binding measures for the single Member states.[53]

On the subject of biotechnological inventions, it is necessary to note that in the United States intellectual property is protected by an *ad hoc* provision inserted in the Constitution, which grants Congress the power to adopt regulations in relation to patents. In Europe, even in the presence of a specific provision within the Charter of Fundamental Rights of the European Union[54] and of important harmonizing directives, in its current state, intellectual property remains essentially a national discipline, lacking directly binding, uniform legislation or centralized mechanisms of enforcement.[55]

The division of duties among various institutions occurs only on the basis of the type of issue raised, depending on whether it originated from

---

[51] On this point, see H.M. Hart and A.M. Sacks, *The Legal Process. Basic Problems in the Making and Application of Law*, cit., p. 159, where it is specified that 'the shape of the problems as they present themselves for official settlement is the shape which has been given them by the successes or failures of the manifold processes of private adjustment'.

[52] In this regard, see, *ex multis*, H.M. Hart and H. Wechsler, *The Federal Courts and the Federal System*, cit., *passim*; Hart and Sacks, *The Legal Process. Basic Problems in the Making and Application of Law*, cit., p. 160 et seq.

[53] See, *ex multis*, P. Craig, *EU Administrative Law*, Oxford: Oxford University Press, 2012 (2nd edn), p. 3 et seq.; P. Craig and G. de Burca, *EU Law: Text, Cases, and Materials*, Oxford: Oxford University Press, 2011 (6th edn), *passim*.

[54] Proclaimed on 7 December 2000 in Nice, and subsequently, in an adapted version, on 12 December 2007 in Strasbourg.

[55] At least until the entry into force of the 'European Union patent', approved in December 2012, which would constitute a single patent right valid for the entire territory of the EU (in this regard, refer to that which was said above, as well as to Part IV).

a dispute for which there is a type of compensation requested, or that – even eventually originating from past disputes – the reason for the request for action is to obtain a change in the rules for the future. The most important institutions which deal with the negotiation of issues in the first case are the courts, jointly with administrative agencies which may be established to deal with specific types of issues (for instance the USPTO and the EPO);[56] both of these models of institutions has been given special importance in the field of biotechnological inventions, on both sides of the Atlantic.

What is more, as has been extensively noted, the adoption of a decision and/or the granting of a remedy in relation to a single issue or controversy may involve a determination which produces, also significant, effects for the future. In this sense, especially in the United States model, but also in the European one, the courts or administrative agencies function as the 'primary interface' for the resolution of issues with respect to which no solution has been found privately. The legislature or government, then, may not need to intervene with regulatory changes until it is clear that they are necessary, being in these cases able to refer to previous solutions identified by the courts or agencies (which remain in this way the substratum of the adopted legislation).

Should, as happens – especially in the U.S. model, but partially also in the European one – in regard to biotechnological inventions, and in particular concerning bioethical profiles raised by them, the primary regulation be absent or be constituted by provisions of a general nature which are not exhaustive regarding particular issues, and the legislature or government do not intervene with *ad hoc* regulations, then the decisions of the courts or agencies may be in effect properly binding sources for similar cases – in a 'formal' sense in the American common law model, 'material' in the European civil law model[57] – and an essential point of reference for the discipline on the matter.

In this scenario, bodies which are of significant interest for the analysis of the United States and European models are those of the advisory commissions on bioethics, whose recommendations can provide – to political decision makers and/or the collective – clarifications regarding

---

[56] Agencies responsible, respectively in the United States and Europe, for granting patents.

[57] On the role of case law in the two models, see, *ex multis*, D. Nelken and E. Orücü, *Comparative Law: A Handbook*, Portland (USA): Hart Publishing, 2007, p. 411 et seq.

controversial facts or issues, counsel on the best interpretation of the provisions in force or to offer insights regarding future rules.[58]

In the United States model, also in view of fulfilling obligations which would be derived for the U.S. from the UNESCO Universal Declaration on the Human Genome and Human Rights, in 1995 President Clinton established the National Bioethics Advisory Commission, whose main task was to present reports on bioethical profiles concerning research in the biological and medical fields, especially regarding the genome and human cloning.[59]

In 2001 President Bush, by virtue of the subsequent developments in biotechnological research, deemed it appropriate to establish the President's Council on Bioethics, in order that it might provide opinions on the bioethics issues connected to progress in biomedical science and technology.

Most recently, in 2009 President Obama, stating that he held it opportune to replace the Council with an agency which provided practical policy options, rather than philosophical lines, established the Presidential Commission for the Study of Bioethical Issues.

The function of the advisory commission for bioethics, further, has often been the subject of criticism, given that it is held to be the expression of politically oriented evaluations, in consideration of the direct dependence of the commissions on political decision makers, and

---

[58] In this regard, see, among others, W. Rei and J. Yeh, 'Promises and Pitfalls of Using National Bioethics Commissions as an Institution to Facilitate Deliberative Democracy – Lessons from the Policy Making of Human Embryonic Stem Cell Research' (2009) 4 *National Taiwan University Law Review* 69 et seq.; S. Dodds and C. Thomson, 'Bioethics and Democracy: Competing Roles of National Bioethics Organizations' (2006) 20 *Bioethics* 326 et seq.; M.F. Riley and R.A. Merrill, 'Regulating Reproductive Genetics: A Review of American Bioethics Commissions and Comparison to the British Human Fertilisation and Embryology Authority' (2005) 6 *Columbia Science and Technology Law Review* 1 et seq.; M. Fuchs, *National ethics councils. Their backgrounds, functions and modes of operation compared*, Berlin: Nationaler Ethikrat, 2005; B.H. Gray, 'Bioethics Commissions: What Can We Learn from Past Successes and Failures?', in R.E. Bulger, E.M. Bobby and H.V. Fineberg (U.S. Institute of Medicine; eds), *Society's Choices: Social and Ethical Decision Making in Biomedicine*, Washington DC: National Academies Press, 1995, p. 261 et seq.

[59] In regard to the other national commissions on bioethics, also prior to those cited, see A. McCoy, 'Biotechnology and Embryonic Stem Cells: A Comparative Analysis of the Laws and Policies of the United States and Other Nations' (2009) 8 *Loyola Law and Technology Annual* 63 et seq. (68 et seq.) (as well as information reported at http://www.bioethics.gov, accessed 19 December 2014).

in particular of the spoil system created by them; in particular, the President's Council on Bioethics – recently, as noted, replaced by the Presidential Commission for the Study of Bioethical Issues – has been criticized for the fact that its evaluations appear to be in line with the political affiliation of the President in office at the time.[60]

In the European model, the most significant advisory body on the subject is currently the European Group on Ethics in Science and New

---

60 See, for instance, W. Rei and J. Yeh, 'Promises and Pitfalls of Using National Bioethics Commissions as an Institution to Facilitate Deliberative Democracy – Lessons from the Policy Making of Human Embryonic Stem Cell Research', cit., p. 73 and 85 et seq., where it is also highlighted that in 2004 a declaration of 62 known scientists accused President Bush of manipulating the objectivity of scientific knowledge for his political agenda, substituting expert scientists with other unqualified and biased ones, or censuring reports of governmental scientists which were not in line with his agenda. Similar criticism, further, was made regarding President Obama's actions; see C.C. Campbell, 'Dissolution of bioethics council is a loss for America', in *St. Louis Post Dispatch*, 2 July 2009, accessed 19 December 2014 at http://www.colleen-campbell.com/P-D_Columns/PD090702Bioethics.htm. See also U. Schuklenk, 'National Bioethics Commissions and Partisan Politics' (2008) 22 *Bioethics* ii et seq.; M.F. Riley and R. Merrill, 'Regulating Reproductive Genetics: A Review of American Bioethics Commissions and Comparison to the British Human Fertilisation and Embryology Authority', cit., p. 8 et seq.; R. Steinbrook, 'Science, Politics and Federal Advisory Committees' (2004) 350 *New England Journal of Medicine* 1454 et seq.; G.J. Annas, 'Will the Real Bioethics (Commission) Please Stand Up?' (1994) 24 *Hastings Center Report* 19 et seq., who showed, in particular that 'bioethics has been called on primarily by politicians to help them neutralize contentious issues, or to provide ethical cover for policy decisions that have already been made … and when called on has usually been called on late and treated like a second-class citizen'. In order to remedy such issues in the functioning of advisory commissions on bioethics, various criteria was proposed which must be satisfied by such commissions, in particular: pluralism in its composition, the declaration of conflicts of interest, independence of funding of the components, the clarity of the legislative mandate, the accuracy and integrity in the analysis of facts, transparency and public participation in its operation, the manifestation of the degree of consensus in decision-making and space for dissenting views, the duty of the government to respond to the recommendations; see W. Rei and J. Yeh, 'Promises and Pitfalls of Using National Bioethics Commissions as an Institution to Facilitate Deliberative Democracy – Lessons from the Policy Making of Human Embryonic Stem Cell Research', cit., p. 88 et seq.; R.E. Bulger, E.M. Bobby and H.V. Fineberg, *Criteria for Success*, in Bulger, Bobby and Fineberg *Society's Choices: Social and Ethical Decision Making in Biomedicine*, cit., p. 150 et seq.

Technologies,[61] which declares itself neutral, independent, pluralistic and multidisciplinary, and counsels the European Commission on ethical issues connected to science and new technology, by request of the Commission, which may be solicited by the European Parliament or Council, or on its own initiative.

Besides being chosen on the basis of a differentiated geographic distribution among the various Member states of the European Union, the members provide a vast range of skills in various disciplines, among which are biology, genetics, medicine, pharmacology, law, ethics, philosophy, theology, and so on, in order to be able to face the multiple ethical issues arising from the constant progress of science and technology. They are requested to provide their own opinions to the Commission personally and with full independence. Their selection is carried out on the basis of the publication of a notice of request for an expression of interest published on the internet.

The Group adopts its Opinions following consultation and debate with the main experts in the various fields of interest, including a public round table to which representatives of European institutions, sector experts, individual stakeholders, such as NGOs, patient and consumer organizations and industrial stakeholders are invited.[62]

Considering this group of characteristics, then, the reasons for which the function of the European Group on Ethics is not the subject of significant criticism is clear, as happens instead for the corresponding body in the U.S. system.[63]

For the purposes of a proper categorization of the roles of each of the various public institutions, of their instruments for action and their relative efficacy, as well as the reasoning at the foundation of the discipline and the single rules in force, finally, it is necessary to hold the individual stakeholders' interests on the subject and the reasons for the respective circumstances promoted by them in consideration.

---

[61] Established in December 1997 to replace the previous Group of Advisers on the Ethical Implications of Biotechnology.

[62] In this regard, see Bureau of European Policy Advisers (European Commission), '*The role of ethics in international biomedical research – Report of the 2nd meeting of the European Commission's International Dialogue on Bioethics – Madrid, 04-05 March 2010*', Luxembourg, Publications Office of the European Union, 2010, p. 6 et seq.

[63] What is more, the EGE itself, as said, suggested the establishment of an independent advisory board responsible for ethical evaluations during the examination of patent applications, where they involve specific issues of this type ('Opinion n°16 – 07/05/2002 – Ethical aspects of patenting inventions involving human stem cells', cit., p. 18).

In this sense, first of all, the rules on the matter may also be dictated or inspired by different sources and/or individuals, within or outside the traditional regulatory instruments (legislation, regulations, case law), in particular guidelines or documents of professional organizations or scientific/social associations for civil rights or representatives of researchers, patients or relatives of patients; codes of self-regulation or professional ethics; petitions of environmental activists, and so on.[64]

The public choice theory, which is based on the application of economic methods to the study of the behaviour of actors working in the political arena,[65] then, tends to analyze, further – through the interest group theory[66] – the role of organized interest groups in the process of adopting public choices. This approach highlights how it is not the majority of individual citizens, but rather concentrated interest groups, who are the main force capable of influencing political, and in particular legislative, decisions. Lobbies, in fact – being composed by a discreet number of members who share similar objectives – have greater benefits and lower costs than single individuals.[67]

---

64 See S. Penasa, 'The Need for a Procedural Approach to Human Embryonic Stem Cell Research: An Emerging Regulatory Model within EU' (2011) 3 *Dilemata: International Journal of Applied Ethics* 39 et seq.

65 In regard to the 'public choice theory', see, among others, C.K. Rowley and F. Schneider (eds), *The Encyclopedia of Public Choice*, Boston: Springer, 2003, *passim*; D.A. Farber and P.P. Frickey, *Law and Public Policy*, Chicago: University of Chicago Press, 1991, p. 1; as well as the work which is held to be the foundation of this school, that is, M. Buchanan and G. Tullock, *The Calculus of Consent: Logical Foundations of Constitutional Democracy*, cit., *passim.*

66 In regard to which, see, among others, M.M. Golden, 'Interest Groups in the Rule-Making Process: Who participates? Whose Voices Get Heard?' (1998) 8 *Journal of Public Administration Research and Theory* 245 et seq.; C.S. Thomas (ed.), *First World Interest Groups: A Comparative Perspective*, Westport (USA): Greenwood Press, 1993, *passim*; G.S. Becker, 'A Theory of Competition Among Pressure Groups for Political Influence' (1983) 98 *Quarterly Journal of Economics* 371 et seq.; R.E. McCormick and R.D. Tollison, *Politicians, Legislation, and the Economy: An Inquiry into the Interest-Group Theory of Government*, Boston: Martinus-Nijhoff, 1981, *passim*; W.M. Landes and R.A. Posner, 'The Independent Judiciary in an Interest Group Perspective' (1975) 18 *Journal of Law and Economics* 875 et seq.

67 For a single individual, in fact, the costs of an activist behaviour with respect to a discipline or decision which will often also benefit all others or rather various other citizens would exceed the individual expected benefits. In particular, as far as legal actions against patents, it has been shown that they are particularly costly: for instance, in the United States, a researcher who undertakes this action with respect to a patent could expect to pay over $500,000 just

Interest groups – such as associations of biotechnological or pharmaceutical firms, associations for civil rights, environmental activist movements, animal rights organizations, and associations or groups of researchers or patients, and so on – are mainly active, but not only, in regard to legislative and executive bodies, attempting also to influence judicial decisions. In this last context, in the U.S. model, the instrument used is that of observations which are expressed in documents presented throughout the trial as *amici curiae*;[68] in this regard, it is clear that the lobbies have greater interests and better ways to finance these biased technical documents than do private citizens, who are subject instead to the aforementioned problem of costs and benefits and to that of free riding.

With respect to solicitation arriving from interest groups, as seen, United States and European patent systems have responded in rather different ways.[69] The European legislature introduced a rule to Directive 98/44/EC which includes some relevant concessions with regard to their situation; the EPO has acknowledged their arguments in the context of the examination of single patent applications, through minor changes to its approach, while the Court of Justice recently demonstrated that it would strongly characterize its action on the matter following the requests made by some of them. In the American system, instead, these instances have more rarely appeared capable of producing significant effects, though recent decisions of the Supreme Court on the *Mayo* and *Myriad* cases seem potentially likely to show manifestations of a possible inversion of this tendency.

In particular, the organization of the Patent Office in the United States provides generally that the examiners have a scientific background and are trained from the legal point of view once they have entered into

---

in legal fees (see R. Portman, 'Legislative Restriction on Medical and Surgical Procedure Patents Removes Impediment to Medical Progress' (1996) 4 *University of Baltimore Intellectual Property Law Journal* 91 et seq.); for a laboratory, then, it would be cheaper to pay a fee for the user's license for a gene – passing that cost on to patients – rather than to file a lawsuit (see L.B. Andrews, 'Genes and Patent Policy: Rethinking Intellectual Property Rights', cit., p. 806).

68 See in particular W.M. Landes and R.A. Posner, *The Economic Structure of Intellectual Property Law*, cit., p. 411, where in fact it is shown that the *amicus curiae brief* constitutes proof of activity of interest groups.

69 In this regard, see S. Parthasarathy, 'Whose Knowledge? What Values? The Comparative Politics of Patenting Life Forms in the United States and Europe' (2011) 44 *Policy Sciences* 267 et seq.

service, in order to evaluate patent applications and patentability requirements based on their technical skill[70] and with the help of the specific Manual of Patent Examination Practice, which includes the relevant legislation and precedents.

Interest groups, for their part, participate in the regulatory and decision making processes through lawyers and representatives of firms and universities which specialize in the subject, with testimonies before Congress or with *amicus briefs* to the courts.[71]

Congress, though it has provided a significant contribution to the work of the Patent Office,[72] has had, in effect, a limited influence on it, through committees such as the former House Committee on Patents and now the House Judiciary Committee (naturally subject to the possible influence of lobbies), which concentrate on verifying that the USPTO operates in the correct manner, and rarely examine the foundation of the system.

Beyond this option, the opportunities for interest groups to influence the decisions of the EPO are limited, given the mechanism of random review it has established. However, although it has recently been modified with the provision of the possibility for third parties to continue to present arguments and evidence after the review,[73] it has remained characterized by a context which is confined to the verification of the existence of new elements compared to the prior art with respect to the patent.[74]

Finally, jurisdictional control has recently become more focused, with the establishment of the single court – the Court of Appeals for the

[70] On this point, see in particular, K.W. Swanson, 'The Emergence of the Professional Patent Practitioner' (2009) 50 *Technology and Culture* 519 et seq.

[71] For an accurate diachronic reconstruction of the main expressions of such participation on the subject of biotechnological inventions, as well as for specific considerations on the relative incidence of these, see Parthasarathy, 'Whose Knowledge? What Values? The Comparative Politics of Patenting Life Forms in the United States and Europe', cit., p. 272 et seq.

[72] See J. Duffy, 'The FCC and the Patent System: Progressive Ideals, Jacksonian Realism, and the Technology of Regulation' (2000) 71 *University of Colorado Law Review* 1071 et seq.

[73] While such possibilities in the past were precluded, because from that moment the USPTO communicated only with the owner of the patent.

[74] See S. Parthasarathy, 'Whose Knowledge? What Values? The Comparative Politics of Patenting Life Forms in the United States and Europe', cit., p. 269 et seq.

Federal Circuit ('CAFC') – to deal with all appeals on the matter[75] (except naturally for the further appeal to the Supreme Court).

The unification of duties under the responsibility of the CAFC, on the one hand has allowed for the formation of a highly specialized body, but on the other has reduced the range of the legal perspectives which may emerge during the examination of patent issues.[76]

In Europe, the scenario of patent policy has rather similar connotations to that of the United States system, with similar institutions, bureaucrats and interest groups, and with the same context of relevant knowledge and skill. Here as well, generally, the personnel of the EPO has a scientific background and acquires legal training in service, basing the examination of the application on legislation, on precedents and on the manual of Guidelines for Examination.

The EPO is subject to legislative and legal controls and, in the field of biotechnology, those of a large number of public and private interest groups – such as in particular non-governmental organizations, large firms and patent lawyers, as well as, to a lesser extent, those of universities, groups of patients and researchers – who attempt to influence the practices of the EPO through the presentation of briefs opposed to the granting of certain patents.[77]

On the other hand, with respect to the United States system there are some cited differences, which have led to the formation of a patent policy which is more susceptible to instances coming from such interest groups, especially when it comes to the theme of the patentability of living beings.

The EPO manages a centralized system of granting European patents, which are, as is known, valid for several Member states of the EPC, but leave space for the various jurisdictions of the single systems, providing the option – preferred by many – of appealing directly through their own

[75] In that regard, besides referring above to Part III, see M.J. Adelman, 'The New World of Patents Created by the Court of Appeals for the Federal Circuit' (1986) 20 *University of Michigan Journal of Law Reform* 979 et seq.

[76] Adopting, further, a markedly 'pro-patent' approach. With respect to such consequences of the establishment of the CAFC, see in particular, J.F. Merz and N.M. Pace, 'Trends in Patent Litigation: The Apparent Influence of Strengthened Patents Attributable to the Court of Appeals for the Federal Circuit' (1994) 76 *Journal of the Patent and Trademark Office Society* 579 et seq.

[77] See D. Harhoff and M. Reitzig, 'Determinants of Opposition against EPO Patent Grants – The Case of Biotechnology and Pharmaceuticals' (2004) 22 *International Journal of Industrial Organization* 443 et seq.

domestic system of appeal at two levels (Technical Board of Appeal and Enlarged Board of Appeal).

The respect of national differences is ensured, also, through the provision of the exclusion from patentability of innovations which are contrary to public order or to morality, parameters which in fact allow interested third parties to introduce new forms of knowledge and skill to the European patent system.

Moreover, the opposition procedure provided with respect to the decisions of the EPO allows greater opportunities of intervention by third parties, compared to the mechanism of review in the U.S. system: the EPC, in fact, allows a patent to be brought under discussion within nine months from its granting, with respect to any one of its requirements; in this manner, in fact there is space given to the presentation before the EPO of even non-traditional evidence, such as bioethical articles, policy reports, public opinion surveys, and so on.[78]

Finally, perhaps the greatest difference between the two systems appears to be in relation to the policies respectively adopted concerning scientific and technological innovations,[79] given that at the European level not only is the principle of precaution applied, which places the burden of proof that a technology is safe in the hands of the manufacturers, but more broadly, for the purposes of examination of the applications for patent protection – in virtue of the aforementioned structure of the discipline on the matter – alongside 'quantitative' data of scientific knowledge and economic impact, 'qualitative' parameters of a social, bioethical and ecological nature, that introduce differentiated, and often prevalent, criteria of evaluation that are taken more and more into consideration with respect to traditional ones.

---

[78] See S. Parthasarathy, 'Whose Knowledge? What Values? The Comparative Politics of Patenting Life Forms in the United States and Europe', cit., p. 271.

[79] For interesting considerations in regard to the respective policy in the United States and European models, see, among others, S. Jasanoff, *Designs on Nature: Science and Democracy in Europe and the United States*, Princeton (USA): Princeton University Press, 2007, *passim*; L. Levidow, S. Carr and D. Wield, 'European Union regulation of agri-biotechnology: Precautionary links between science, expertise, and policy' (2005) 32 *Science and Public Policy* 261 et seq.; D. Vogel, 'The Hare and the Tortoise Revisited: The New Politics of Risk Regulation in Europe and the United States', in M. Levin and M. Shapiro (eds), *Transatlantic Policymaking in an Age of Austerity*, Washington DC: Georgetown University Press, 2004, p. 177 et seq.; J.A. Tickner and S. Wright, 'The Precautionary Principle and Democratizing Expertise: A US Perspective' (2003) 30 *Science and Public Policy* 213 et seq.

## V.4 RELATIONSHIPS OF PATENT LAW IN BIOTECHNOLOGY WITH OTHER DISCIPLINES AND HUMANITIES

### V.4.1 Relations with Political and Economic Sciences and the Sphere of Extension of Patent Protection between Applied Research and Access to Information

In the view of the comparative method, further, the examination of the discipline on biotechnological inventions in the United States and European models has been seen to take into account its relation, on the one hand, to political science and economics, and on the other to philosophy and bioethics.

In this view, as far as the political, economic framework of the subject, as said, both in the U.S. and European models, the essential nature of the biotechnological patent discipline emerged as an instrument of policy, aimed at encouraging innovation as a fundamental element of economic growth and development. The patentability of findings, in fact, allows innovators a period of exclusivity over the invention (in exchange for public disclosure of the way it works), which is sufficient to recover the significant investments necessary for the research and development of biotechnological products.[80]

While at the beginning of the 20th century the theme of innovation was framed in terms of 'creative destruction', the most recent economic doctrine shows the role played by various factors in the promotion of growth and productivity in the long term. These include not only proper investments for innovation, but also learning-by-doing, human capital, as well as institutions.[81]

---

[80] In this regard, besides referring to that which is highlighted above in Part I, see, among others, J.H. Graham *et al.*, 'High Technology Entrepreneurs and the Patent System: Results of the 2008 Berkeley Patent Survey' (2010) 24 *Berkeley Technology Law Journal* 1255 et seq.; C.M. Holman, 'Learning from Litigation: What Can Lawsuits Teach Us about the Role of Human Gene Patents in Research and Innovation?' (2009) 18 *Kansas Journal of Law and Public Policy* 215 et seq.

[81] In this regard, see WIPO, '2011 World Intellectual Property Report. The Changing Face of Innovation', cit., p. 23 et seq.; P.M. Romer, 'Which Part of Globalization Matters for Catch-Up Growth?' (2010) 100 *American Economic Review* 94 et seq.; G.M. Grossman and E. Helpman, 'Endogenous Innovation in the Theory of Growth' (1994) 8 *Journal of Economic Perspectives* 23 et seq.; P. Aghion and P. Howitt, 'A Model of Growth Through Creative Destruction'

The discipline of biotechnological inventions, as noted, has been and still is at the centre of constant debate regarding the forms and limitations of recognition of patent protection for them.

The discipline concerning the patentability of such inventions has its roots in the United States case law. In 1980, with the *Chakrabarty* decision, the Supreme Court laid the foundation for the granting of protection and for the evolution of biotechnology patent law. In subsequent years, the path beaten by such decisions regarding the patentability of microorganisms was followed for plants, animals and human genetic material.

Following the *Chakrabarty* case, other legal systems, and in particular the European system, also began to recognize the patentability of living organisms. In this way the European legislature, considering the economic and financial impact of modern biotechnological research, aligned itself with the U.S. model, with the idea of bridging the gap between European firms and those overseas.[82]

In the *Chakrabarty* case, the central issue regarding the patentability of biotechnological inventions was faced in relation to the distinction between living organisms existing in nature and those produced by human ingenuity, recognizing patentability only in the latter case. As a consequence, the subsequent decisions of the courts and patent offices, on both sides of the Atlantic, held that DNA in its natural state cannot be

---

(1992) 60 *Econometrica* 323 et seq.; J.A. Schumpeter, *Capitalism, Socialism and Democracy*, New York: Harper, 1942, *passim*.

[82] In the EU, interventions on the matter which followed the U.S. interventions could identify themselves as an example of the so-called phenomena of 'competition between legal systems', that is the race of countries or organizations of regional integration aimed at providing more favorable disciplines for private individuals operating within their own jurisdictions, as well as to orient in its favour the cross border flow of resources (so-called 'race to the top'), also possibly through actual barriers to entry in a protectionist sense. In this regard, besides referring below, see P. Drahos, 'The TRIPS Review and the CBD: A Dress Rehearsal?', in P. Drahos and M. Blakeney (eds), *IP in Biodiversity and Agriculture: Regulating the Biosphere*, London: Sweet & Maxwell, 2001, p. 53 et seq. (58). In regard to the risk that such a phenomena may lead rather to a 'race to the bottom', that is rules which are more permissive and which answer only to biased interests (that indeed in this case does not appear to take place, given that in Europe the rules governing the matter appear to have preserved in any case – although according to some improperly, as we shall recall below – its original principles), see Butler and Ribstein, 'Legal Process and the Discovery of Better Policies for Fostering Innovation and Growth' cit., p. 463 et seq.

patented, while DNA which has been isolated and purified or significantly modified may be patented, given that it is the result of human ingenuity.[83]

The view taken in the *Chakrabarty* ruling, then, was accepted in various international agreements and declarations, as well as in European legislation. Thus, the UNESCO Universal Declaration on the Human Genome and Human Rights of 1997 prohibits profiting from the human genome in its natural state, the European Patent Convention states that biological materials and processes may be patented if they are the result of an inventive step, and Directive 98/44/EC provides that natural plant varieties, animals and processes may not be patented, while biological material which has been isolated from nature and purified may be.

On the theme of the patentability of biological materials, further, still today the fundamental question that is asked is in relation to the criteria on the basis of which the subject of protection may consist in a product of human ingenuity, and not instead of nature.[84]

In this regard, it has been shown that DNA produced in a laboratory is not structurally and functionally identical to that existing in nature. According to the supporters of this position, during the process of isolation and purification of natural DNA, organic mutations take place such that make it not identical to its natural form. Researchers, further, may introduce specific mutations, for instance by removing or adding DNA sequences.[85] What is more, even where researchers do not make changes to isolated and purified DNA, it is not in its purified form, identical to natural DNA, given that in nature it exists in an impure form. On the basis of such an assumption, according to which isolated and purified DNA constitutes a patentable invention while that occurring in nature is not, clearly, it must be guaranteed that exclusive patent right claims cannot exist for DNA existing in nature, so that anyone can study it. Thus, a patent concerning DNA should be interpreted in order to

---

[83] See D.B. Resnik, 'DNA Patents and Scientific Discovery and Innovation: Assessing Benefits and Risks' (2001) 7 *Science and Engineering Ethics* 29 et seq.; M. Sagoff, 'DNA Patents: Making Ends Meet', in A.R. Chapman (ed.), *Perspectives on Gene Patenting. Religion, Science, and Industry in Dialogue*, Washington DC: American Association for the Advancement of Science, 1999, p. 245 et seq.; J.J. Doll, 'The Patenting of DNA' (1998) 280 *Science* 689 et seq.

[84] In this sense, see D.B. Resnik, *Owning the Genome. A Moral Analysis of DNA Patenting*, New York: State University of New York Press, 2004, p. 83.

[85] To produce cDNA, for instance, sequences which do not codify for protein are removed. Researchers can add sequences of nucleotides to the DNA in order to induce modifications in the proteins produced.

guarantee the inventor exclusive rights for its isolated and purified form, but no right over the natural form.[86]

This reconstruction has been recently brought under discussion in the U.S. system itself,[87] by recent Supreme Court interventions regarding the *Mayo* and *Myriad* cases, in March 2012 and in June 2013. In the decision on the first,[88] a patent concerning a method aimed at optimizing the therapeutic efficacy of a medicine for the treatment of a gastrointestinal illness was declared invalid, holding that the claims in effect concerned the 'laws of nature' underlying the method itself, regarding in particular a 'well-understood, routine, conventional activity previously engaged in by scientists in the field'. This mutation of approach with respect to previous ones on the subject appeared immediately to give rise to significant implications for the patentability not only of isolated genes, but more generally of purified natural products and other innovations based on biological matter existing in nature.

As a consequence, the Supreme Court itself intervened a few days later in the *Myriad* controversy through the granting of a *certiorari* and by referring the case to the Federal Circuit, so that it might reconsider its decision in light of the aforementioned *Mayo* ruling. In fact, as was shown also in an *amicus brief* filed by the Department of Justice in June 2012,[89] the *Mayo* decision provided important indications with respect to the issue of whether the difference between isolated and natural DNA is significant enough to make the first considerably different from the second for the purposes of 35 U.S.C. § 101.

According to the criteria outlined in *Mayo*, in the *Myriad* case the Federal Circuit could have held that the information codified in DNA is a

---

86 In this regard, besides referring to the doctrine cited above in Part III, see again, D.B. Resnik, *Owning the Genome. A Moral Analysis of DNA Patenting*, cit., p. 89.

87 Contrarily, in Europe 'it has generally been the case that if the methods to isolate a DNA sequence are routine and the starting materials are available, there will be no inventive step': S.A. Jameson, 'A Comparison of the Patentability and Patent Scope of Biotechnological Inventions in the United States and European Union' (2007) 35 *AIPLA Quarterly Journal* 193 et seq. (222 et seq.). This, further, is with the possibility of acknowledging the patentability should isolation be particularly difficult; see A. McCoy, 'Biotechnology and Embryonic Stem Cells: A Comparative Analysis of the Laws and Policies of the United States and Other Nations', cit., p. 80.

88 Supreme Court, *Mayo Collaborative Services, DBA Mayo Medical Laboratories, et al. v. Prometheus Laboratories, Inc.*, 566 U.S. (2012).

89 'Brief for the United States as amicus curiae in support of neither party', No. 2010-1406, 15 June 2012.

'law of nature', that DNA is a 'product of nature', and that the isolation of DNA consists in a process which was already well known and predictable at the moment of the application; thus, isolated DNA would not be patentable, given that the claims would concern, in effect, a 'law of nature' and a 'product of nature'.

On the contrary, in August 2012, the Federal Circuit Court of Appeals confirmed the validity of Myriad's patents on 'isolated' genes related to breast and ovarian cancer.[90]

Thus, in June 2013 the Supreme Court confirmed that Myriad's DNA main claim falls within the law of nature exception, and so is not patentable.[91] According to the Court's decision, in fact, Myriad did not create or alter either the genetic information encoded in the BCRA1 and BCRA2 genes or the genetic structure of the DNA. It found important and useful genes,[92] but innovative or brilliant discovery alone does not satisfy the U.S.C. § 101 requirement.

However, the Court specified that the decision does not include, on one hand, cDNA, which is not a 'product of nature'. Thus, it is eligible for a patent under § 101, as its creation results in a not naturally occurring exon-only molecule.[93] On the other hand, the decision itself does not

---

[90] More specifically, the Federal Circuit upheld Myriad's composition claims to isolated DNA (reversing the District court), invalidated Myriad's method claims for comparing or analyzing gene sequences (affirming the District court), and upheld Myriad's method claims for screening potential cancer therapeutics (reversing the District court); see M. Textor, *Gene Patents at Home and Abroad: Should the WTO Take Action in Light of Myriad?*, May 2013, accessed 19 December 2014 at http://ssrn.com/abstract=2282676, p. 9.

[91] In this sense, the Court recalled the wording used lastly in the *Mayo* decision, according to which: 'laws of nature, natural phenomena, and abstract ideas … are basic tools of scientific and technological work', that lie beyond the domain of patent protection. As the Court had held there, in fact, without this exception, there would be considerable danger that the grant of patents would 'tie up' the use of such tools and thereby 'inhibit future innovation premised upon them'; this consequence, evidently, would be at odds with the very point of patents, which, as known, exist to promote creation. In this respect, it is worth noting how a recent study affirmed that sequence patents would already cover the entire human genome; see J. Rosenfeld and C.E. Mason, 'Pervasive Sequence Patents Cover the Entire Human Genome' (2013) 5 *Genome Medicine* 27 et seq.

[92] In fact, Myriad's principal contribution was uncovering the precise location and genetic sequence of the BRCA1 and BRCA2 genes within chromosomes 17 and 13.

[93] This distinction has its roots mainly in the decision of the case *Amgen v. Chugai* (927 F.2d 1200, 18 USPQ 2d 1016, 1991), where the Federal Circuit focused the subject matter of the claim on a purified and isolated DNA sequence

cover method claims, patents on new applications of knowledge about the BRCA1 and BRCA2 genes, or the patentability of DNA in which the order of the naturally occurring nucleotides has been altered.[94]

So, every patent drafted similarly to Myriad's broadest claim – an isolated DNA coding for a specific protein – is now invalid. Vice versa, claims related to cDNA versions of genes continue to pass the threshold test, though they are still subject to scrutiny under all the other patentability requirements.[95]

Finally, in March 2014, the USPTO issued a memorandum titled 'Guidance For Determining Subject Matter Eligibility Of Claims Reciting Or Involving Laws of Nature, Natural Phenomena, & Natural

---

encoding human erythropoietin, interpreting the term 'purified' as meaning essentially only the coding regions, that is, only the novel purified and isolated sequence which coded for EPO (see, also for the other relevant precedents, M. Textor, *Gene Patents at Home and Abroad: Should the WTO Take Action in Light of Myriad?*', cit., pp. 23–26; D.M. Gitter, 'International Conflicts over Patenting Human DNA Sequences in the United States and the European Union: An Argument for Compulsory Licensing and a Fair-Use Exemption' (2001) 76 *New York University Law Review* 1623 et seq.).

[94] U.S. Supreme Court, *Association for Molecular Pathology et al. v. Myriad Genetics Inc. et al.*, 569 U.S. (2013).

[95] As far as concerns Myriad, in particular, about three-quarters of its BRCA-related patents haven't been invalidated, including claims to cDNA and some of its methods. Moreover, the invalidated patents would have begun to expire in the next year. Last but not least, Myriad's most valuable asset may be the proprietary database it has built up through its testing monopoly: this database consists of test results, i.e. DNA sequences, and associated health outcomes, and gives Myriad a relevant advantage in interpreting BRCA gene mutations, especially the lesser-known one (see J. Conley, 'Myriad, Finally: Supreme Court Surprises by not Surprising', cit., p. 1; and, also for an interesting analysis of the previous European decisions and related developments concerning Myriad's patents, G. Matthijs, I. Huys, G. Van Overwalle and D. Stoppa-Lyonnet, 'The European BRCA Patent Oppositions and Appeals: Coloring Inside the Lines' (2013) 31 *Nature Biotechnology* 704 et seq., especially 709). In this respect, then, a new episode of the 'Myriad saga' before antitrust authorities could be projected (regarding the profiles involved in the relationship between access to databases and antitrust, and the relative case law, see, among others, D. Lim Tze Wei, 'Regulating Access to Databases Through Antitrust Law: A Missing Perspective in the Database Debate' (2006) *Stanford Technology Law Review* 7 et seq.; E. Derclaye, 'The IMS Health Decision: A Triple Victory' (2004) 27 *World Competition* 397 et seq.; D.M. Gitter, 'Strong Medicine for Competition Ills: The Judgment of the European Court of Justice in the IMS Health Action and Its Implications for Microsoft Corporation' (2004) 15 *Duke Journal of Comparative & International Law* 153 et seq.).

Products'.[96] The Guidance implements a new procedure to address changes in the law relating to subject matter eligibility under 35 U.S.C. § 101 in view of the above-mentioned *Myriad* and *Mayo* decisions.[97]

Another interesting profile for the purposes of policy regarding biotechnological inventions, then, concerns the issues – inherent to the collective well-being, as well as of useful pursuit for competition – raised by them concerning the profile of access to findings by third parties, subsequent innovators or patients, consumers, and so on.[98]

As far as the profile in relation to the context of the extension of patent protection for biotechnological inventions is concerned, in the first place, both in the United States and European models, the traditional distinction between patents for inventions of products and processes seems relevant.

The patent for a product is held to provide protection of the product regardless of how it is obtained and for all its possible uses. With reference to biotechnological inventions, on the other hand, under the first profile, the opinion of those who state that the patent grants exclusive production rights for the finding only when it is produced using the process described in the patent application seems agreeable.[99] Under the second profile, in the field of genetics the intentions are linked to the identification of the function of the gene, which usually is not the only function it may carry out: in this context, extending protection to all possible functions of the gene would be excessive, given that the contribution to scientific and social progress by the inventor concerns exclusively the specific function of the identified gene; subsequent inventions regarding the same gene may be considered dependent on the first only when they are to some degree logically connected to its technical teaching.

In the patent for a process, the exclusive covers in fact the process, even if it does not lead directly to the product but satisfies in any case

---

[96] See USPTO, '2014 Procedure For Subject Matter Eligibility Analysis of Claims Reciting or Involving Laws of Nature/Natural Principles, Natural Phenomena, and/or Natural Products (March 2014)', accessed 19 December 2014 at http://www.uspto.gov/patents/law/exam/examguide.jsp.

[97] Regarding the industry's criticism and the ongoing debate on the Guidance, see *Life sciences special* (2014) 240 *Managing Intellectual Property* (June) 30 et seq.

[98] In this regard, besides referring to the doctrine cited previously, see for all, G. Ghidini, *Innovation, Competition and Consumer Welfare in Intellectual Property Law*, cit., *passim*.

[99] In this sense, see among others V. Di Cataldo, 'Fra tutela assoluta del prodotto brevettato e limitazione ai procedimenti descritti ed agli usi rivendicati' (2004) *Rivista di diritto industriale* 111 et seq. (117).

industrial interests, or should it lead to the creation of a product, in this latter case with the possibility of extension also to the product created through that process, where it is the direct and necessary result of it. The extension of protection for the process to the product, within the limits indicated, allows third parties to be granted patent protection for an identical product as long as it was obtained through a different process, or with instruments coming from the first process unless the product is directly descendent from it: this is, in this sense, a product-by-process claim.

These approaches in regard to biotechnological inventions of products or procedures are inspired, evidently, by the desire to avoid needlessly hindering 'subsequent innovation', allowing for the free evolution of technological innovation. As a consequence, the invention which improves the execution of the same type of use with a solution which perfects the previous invention shall be dependent on it, while the invention which, though it uses elements which are the subject of other exclusive rights, combines them in such a manner as to give rise to a new useful result which could not be carried out with the first, shall not be dependent on it. Similarly, the invention which transfers previous ideas to a different and distant sector of use, producing a new useful result shall not be so dependent.[100]

In the United States, these positions are facing some peculiar aspects of patent law, along with, as noted, specific characteristics of the institutional structure with which it finds itself interacting. In particular, among such peculiarities, it is necessary to remember those regarding: (a) the narrowness of exceptions for experimental use;[101] (b) the failure

---

[100] See again G. Ghidini, *Innovation, Competition and Consumer Welfare in Intellectual Property Law*, cit., p. 69 et seq. Thus, the patent of a certain segment of DNA aimed at producing a protein covers its commercial uses which are aimed at the production of the protein, but if it later emerges that the protein can carry out other functions, or that its functions may increase when it is associated with other structures, the subsequent inventor can patent the different use or the different characteristics of the protein without the first being able to oppose it. Still, the owner of the patent on the protein obtained through a certain procedure may not claim an exclusive right for the production of the protein through different methods, which should be the subject of an independent patent.

[101] Excluding commercial uses and those 'solely for business reasons and not for amusement, to satisfy idle curiosity or for strictly philosophical inquiry' (J.M. Mueller, 'No "Dilettante Affair": Rethinking the Experimental Use Exception to Patent Infringement for Biomedical Research Tools' (2001) 76 *Washington Law Review* 1 et seq., 24, citing the ruling of the *Federal Circuit* on the *Roche Products v. Bolar Pharmaceutical* case, 733 F.2d 858, 1984), but also the

to use the tool of the compulsory license (except for the hypothesis, till now only 'on the books', of the so called 'march-in right');[102] and (c) the broadness with which patent protection has traditionally been granted, independently from the specific indication of the function concretely carried out by the molecule of DNA in the patent application (a principle which, rather, is specifically provided for in more recent European legislation on the subject, such as in German and Italian legislation).[103]

---

'research that utilized a patented invention at a university if there is a remote commercial purpose consistent with a university's legitimate business objectives', including 'educating and enlightening students and faculty participating in (noncommercial research projects)' (A. McCoy, 'Biotechnology and Embryonic Stem Cells: A Comparative Analysis of the Laws and Policies of the United States and Other Nations', cit., p. 82 et seq., citing the decision of the Federal Circuit in the *Madey v. Duke University* case, 307 F.3d 1351, 2002; see M.S. Mireles, 'Adoption of the Bayh-Dole Act in Developed Countries: Added Pressure for a Broad Research Exemption in the United States?' (2007) 59 *Medical Law Review* 259 et seq., 278). In particular, a mechanism of 'reach-through royalty' was suggested for non-consensual uses of 'research tools not readily available for licensing on reasonable terms or via anonymous marketplace purchase': J.M. Mueller, 'No "Dilettante Affair": Rethinking the Experimental Use Exception to Patent Infringement for Biomedical Research Tools', cit., p. 58; see also R. Eisenberg, 'Patents and the Progress of Science: Exclusive Rights and Experimental Use' (1989) 56 *University of Chicago Law Review* 1017 et seq.

[102] Which, according to the 'Bayh-Dole Act', would permit the federal government to require a federally-funded patentee to grant licenses under his patent to third-party applicants where the patent holder has failed to achieve sufficient practical application of the invention, or to alleviate health or safety needs which are not reasonably satisfied by the contractor, assignee or their licensees (35 U.S.C. § 203(1)(a)-(b), 1994). See A. McCoy, 'Biotechnology and Embryonic Stem Cells: A Comparative Analysis of the Laws and Policies of the United States and Other Nations', cit., p. 86 et seq.; D. Beardsley, 'A Two-Front Assault on the Stem Cell Patents' (2007) 6 *John Marshall Review of Intellectual Property Law* 501 et seq. (523); C.R. Carroll, 'Selling the Stem Cell: The Licensing of the Stem Cell Patent and Possible Antitrust Consequences' (2002) *University of Illinois Journal of Law, Technology & Policy* 435 et seq. (463 et seq.).

[103] According to the United States 'Utility Examination Guidelines', in fact, '(a) patent on a composition gives exclusive rights to the composition for a limited time, even if the inventor disclosed only a single use for the composition' ('Guidelines', January 2001, 66 Fed. Reg. 1092, 1095). However, U.S. law allows an inventor who develops a new use for a patented compound to get a process patent for that new use, notwithstanding that the DNA is itself patented; see M. Textor, *Gene Patents at Home and Abroad: Should the WTO Take Action in Light of Myriad?*, cit., p. 28–29; D.M. Gitter, 'International Conflicts over Patenting Human DNA Sequences in the United States and the European Union:

Under a different profile, then, granting a large number of industrial exclusives on primary research results, the so-called 'upstream discoveries', threatens to hinder subsequent research aimed at the development of the product, so-called 'downstream research', increasing transaction costs as well. In this regard, the concept of 'tragedy of anticommons' has been developed. While in general the patent may consist in a significant incentive for investments for high risk research projects and contribute to an equal distribution of profits, the granting of exclusive rights for basic research risks hindering subsequent research.[104] What is more, where there are different patent owners for various basic elements needed for the development of a product, giving rise to the so-called 'patent thicket', the price of the product will be higher than if a single individual was the owner of all the patents, or rather produces the product himself, with the result of 'obstruct[ing] entry to some markets and so imped[ing] innovation'.[105] Thus, it is necessary to make the (difficult) distinction between

---

An Argument for Compulsory Licensing and a Fair-Use Exemption', cit., p. 1666. Regarding these issues, see also W. Lesser, 'Myriad & Prometheus, Laws & Products of Nature: Are the Courts Considering an Economic Non-Statutory Subject Matter Exclusion?' (2013) 53 *IDEA – The Intellectual Property Law Review* 173 et seq.; H.C. Wegner, *Mayo v. Prometheus Patent-Eligibility: Whither Myriad and Gsk v. Classen*, 2012, accessed 19 December 2014 at www.grayonclaims.com/hal, *passim*; C.A. Fowler, 'Ending Genetic Monopolies: How the TRIPS Agreement's Failure to Exclude Gene Patents Thwarts Innovation and Hurts Consumers Worldwide' (2010) 25 *American University International Law Review* 1073 et seq. (esp. 1091 and 1097 et seq.).

104 The main instrument through which an anticommon may arise is constituted by the Reach Through License Agreement (RTLA), clauses which are inserted into the licensing contracts through which the owners of patents on research tools are ensured rights for subsequent innovation, rights which may consist in royalties on the profits of the final product, a license on the product or in an option to buy of such a license. On the theme of the 'tragedy of anticommons', and the 'gridlock economy' which arises from it, see in particular: M.A. Heller, *The Gridlock Economy. How Too Much Ownership Wrecks Markets, Stops Innovation, and Costs Lives*, cit., *passim*, and 'The Tragedy of Anticommons', cit., p. 621 et seq.; M.A. Heller and R.S. Eisenberg, 'Can Patent Deter Innovation? The Anticommons in Biomedical Research', cit., p. 698 et seq.

105 Thus I. Hargreaves, *Digital Opportunity. A Review of Intellectual Property and Growth. An Independent Report by Professor Ian Hargreaves*, cit., p. 5. On the theme of the 'patent thicket', see especially, C. Shapiro, 'Navigating the Patent Thicket: Cross Licenses, Patent Pools, and Standard-Setting', in A.B. Jaffe, J. Lerner and S. Stern (eds), *Innovation Policy and the Economy, Volume 1*, Cambridge (USA): MIT Press, 2001, p. 119 et seq.; see also, with respect to the implications in terms of antitrust: M. Ganslandt, 'Intellectual Property Rights and Competition Policy', in K.E. Maskus (ed.), *Intellectual Property, Growth and*

'applied research', aimed at the development of industrial applications and as such worthy of patent incentive, and 'pure research', which is free from a specific purpose, and in any case from results of a practical nature, and therefore not patentable.[106]

In a sector like that of biotechnology, characterized by the cumulative effect of the innovative process, often based on the use of well-known techniques or materials, alongside the problem of anticommons, is that of the high transaction costs needed for negotiations with multiple owners of exclusive rights, as well as the relative strategic behaviours which arise from the latter, with effects which clearly fall on, upon final analysis, not only the individuals who are the manufacturers of the innovation, but also – indirectly – the licensors of the invention or on the consumers of the final product.[107]

In order to remedy the problematic issues referred to above, then, several possible solutions have been suggested.

In general, it has been suggested allowing exclusive patents to coexist with a mechanism of access to biotechnological findings by third parties,

---

*Trade* (Frontiers of Economics and Globalization, Volume 2), Bradford (UK): Emerald Group, 2009, p. 233 et seq.; D.L. Rubinfeld and R. Maness, 'The Strategic Use of Patents: Implications for Antitrust', in F. Lévêque and H.A. Shelanski (eds), *Antitrust, Patents and Copyright: EU and US Perspectives*, Cheltenham, UK and Northampton, MA, USA, Edward Elgar, 2005, p. 85 et seq.; L.B. Andrews, 'Genes and Patent Policy: Rethinking Intellectual Property Rights', cit., p. 805, who on this point also showed that 'the combined profits of the producers are lower in the presence of complementary monopolies. So, if there are several patent holders whose permission is needed to create a gene therapy (and any one of them could block the production of the gene therapy), inefficiencies in the market are created, potentially harming both the patent holder and the patent users'.

[106] In regard to which, see in particular G. Ghidini, *Innovation, Competition and Consumer Welfare in Intellectual Property Law*, cit., p. 37 et seq.

[107] In this regard, besides referring below, see, among others, V. Falce, 'Innovation in the New Technological Industries: Looking for a Consistent Cooperative Model', in E. Arezzo and G. Ghidini (eds), *Biotechnology and Software Patent Law: A Comparative Review of New Developments*, Cheltenham, UK and Northampton, MA, USA, Edward Elgar, 2011, p. 40 et seq. (42 et seq.); F.M. Scherer, 'The Economics of Human Gene Patents' (2002) 77 *Academic Medicine* 1348 et seq.; R.S. Eisenberg, 'Bargaining over the Transfer of Proprietary Research Tools: Is This Market Failing or Emerging?', in R. Dreyfuss, D.L. First, and H. Zimmerman, *Expanding the Bounds of Intellectual Property: Innovation Policy for the Knowledge Society*, cit., p. 209 et seq.; M.A. Heller and R.S. Eisenberg, 'Can Patent Deter Innovation? The Anticommons in Biomedical Research', cit., p. 700.

where such use appears reasonable. The introduction of such a 'fair use' would be aimed at resolving the problem of market failure occurring due to the presence of anticommons and related transaction costs.[108] From this perspective, article 30 of the TRIPS Agreement, which allows for exceptions to exclusive rights granted by a patent, on the condition that it does not affect the normal use of it and the legitimate interests of its owner,[109] could be interpreted in a manner so as to allow access to genetic resources based on fair use criteria, in order to develop innovation and the promotion of scientific research and education.

A primary instrument that is held to be useful, in this sense, is that of the exception of experimental use – which allows scientific research 'on' the invention and not further research 'through' the invention – with respect to which the extension to the context of application was suggested in the American system, where it is found to be rather limited,[110] and a clearer definition and harmonization of them in the European system.[111]

The TRIPS Agreement, in article 30, has been interpreted to permit an experimental-use exception for patent law that allows exceptions for infringements granted for: (a) private, noncommercial purposes; (b) research; (c) experimentation for testing or improvement; and (d) educational purposes. However, an express statement regarding an experimental-use defense against a claim of patent infringement does not currently exist in the TRIPS Agreement. Vice versa, a specific statement in the Agreement allowing for a genetic experimental use exception would likely facilitate the diminishing of the human gene patent controversy. Such an exception would be also in line with the provision of article 8 of the Agreement, which permits Member countries to 'adopt

---

108 In this sense, see M.A. O'Rourke, 'Toward a Doctrine of Fair Use in Patent Law' (2000) 100 *Columbia Law Review* 1177 et seq.

109 In particular, according to such provision: 'Members may provide limited exceptions to the exclusive rights conferred by a patent, provided that such exceptions do not unreasonably conflict with a normal exploitation of the patent and do not unreasonably prejudice the legitimate interests of the patent owner, taking account of the legitimate interests of third parties'.

110 Now strictly limited, as noted above, to use 'for amusement, to satisfy idle curiosity, or for strictly philosophical inquiry' (Federal Circuit, *Roche Products, Inc. v. Bolar Pharmaceutical Co.*, 733 F.2d 858, 1984, and *Madey v. Duke University*, 307 F.3d 1351, 2002).

111 See S. Aymé, G. Matthijs and S. Soini, 'Patenting and Licensing in Genetic Testing – Recommendations of the European Society of Human Genetics', cit., p. S7, and 'Patenting and Licensing in Genetic Testing: Ethical, Legal and Social Issues', cit., p. S27.

measures necessary to protect public health and nutrition, and to promote the public interest in sectors of vital importance to their socio-economic and technological development', as well as 'measures … to prevent the abuse of intellectual property rights by rights holders'.[112]

Another possible way would be that of the compulsory license of dependence, which would help to avoid the effect of removing incentive from research and block exploitation of its results which occurs in the case of multiple related inventions.[113] Further, the solution of the compulsory license has a limited sphere of application, in that it assumes the existence of a dependent patent and thus is applicable only to the cases in which the biotechnological product makes the violation of the patent of another necessary. In this case, the compulsory license would be an instrument in the hands of legislators and courts in order to keep the invention available to the public. In the case of failure to use the invention by the patent owner, or of improper use, or when the invention is not available to the public at reasonable prices, every interested person may request a compulsory license. On the other hand, there are countries, like the United States, in which compulsory licenses for biotechnological inventions are considered unfavourably (except for the cited hypothesis of the 'march-in right').[114]

The TRIPS Agreement, in articles 8 and 31, allows Member countries to introduce compulsory licenses, for reasons of public interest, providing that a series of specific criteria are respected;[115] the provision, then, is not

---

[112] See M. Textor, *Gene Patents at Home and Abroad: Should the WTO Take Action in Light of Myriad?*, cit., p. 34; M. Blakeney, *Trade Related Aspects of Intellectual Property: A Concise Guide to the Trips Agreement*, London: Sweet & Maxwell, 1996, p. 87.

[113] On this theme, see among others, S. Aymé, G. Matthijs and S. Soini, 'Patenting and Licensing in Genetic Testing: Ethical, Legal and Social Issues', cit., p. S27 et seq.

[114] In regard to which, see above.

[115] Among which, in particular: '(a) authorization of such use shall be considered on its individual merits; (b) such use may only be permitted if, prior to such use, the proposed user has made efforts to obtain authorization from the right holder on reasonable commercial terms and conditions and that such efforts have not been successful within a reasonable period of time. … ; (c) the scope and duration of such use shall be limited to the purpose for which it was authorized. … ; (d) such use shall be non-exclusive; … (h) the right holder shall be paid adequate remuneration in the circumstances of each case, taking into account the economic value of the authorization', etc. In this framework, therefore, subsequent scientists would require the consent of the initial patent owner before starting research. A different option could be an amendment of art.

compulsory and leaves the states free to introduce these licenses or not.[116] This possibility, in any case, is consistent with the objectives established in article 7 of the Agreement,[117] as well as with that stated subsequently in the 'Doha Declaration' of 2001.[118]

---

31 (b) to relax the authorization requirement to a lower standard of notice of intent to use gene patents for private, noncommercial purposes; research; experimentation for testing or improvement; and educational purposes (M.Textor, *Gene Patents at Home and Abroad: Should the WTO Take Action in Light of Myriad?*, cit., pp. 33–34).

[116] In this regard, see especially, S. Aymé, G. Matthijs and S. Soini, 'Patenting and Licensing in Genetic Testing: Ethical, Legal and Social Issues', cit., p. S27 et seq.: C. Correa, 'Can the TRIPs Agreement Foster Technology Transfer to Developing Countries?', in K.E. Maskus and J.H. Reichman (eds), *International Public Goods and Transfer of Technology under a Globalized Intellectual Property Regime*, Cambridge: Cambridge University Press, 2005, p. 227 et seq.; J. Watal, *Intellectual Property Rights in the WTO and Developing Countries*, The Hague: Kluwer Law International, 2001, p. 317 et seq.

[117] On the basis of which: 'The protection and enforcement of intellectual property rights should contribute to the promotion of technological innovation and to the transfer and dissemination of technology, to the mutual advantage of producers and users of technological knowledge and in a manner conducive to social and economic welfare, and to a balance of rights and obligations'. In this regard, see especially: D. Matthews, *Globalising Intellectual Property Rights: The TRIPs Agreement*, London: Routledge, 2002, p. 49.

[118] WTO, 'Ministerial Declaration of 14 November 2001', WT/MIN(01)/DEC/2, 41 I.L.M. 746 (2002), so-called 'Doha Declaration', which in particular, in para. 4, provides: 'The TRIPs Agreement does not and should not prevent Members from taking measures to protect public health. Accordingly, while reiterating our commitment to the TRIPs Agreement, we affirm that the Agreement can and should be interpreted and implemented in a manner supportive of WTO Members' right to protect public health and, in particular, to promote access to medicines for all. In this connection, we reaffirm the right of WTO Members to use, to the full, the provisions in the TRIPs Agreement, which provide flexibility for this purpose'. In this regard, see Gitter, 'International Conflicts Over Patenting Human DNA Sequences in the United States and the European Union: An Argument for Compulsory Licensing and a Fair-Use Exemption' cit., p. 1623 et seq., who proposes a reform of the international framework of the matter based on two criteria, that are: (a) a system of compulsory licenses for which genetic patent owners are held to grant licenses to third party researchers who intend to develop derivative inventions from it for profit, in exchange for a royalty proportionate to the profits earned from the product developed by the licensee; and (b) an exemption from experimental use with respect to this system for public or non-profit researchers. On these themes, see also Fowler, 'Ending Genetic Monopolies: How the TRIPs Agreement's Failure to Exclude Gene Patents Thwarts Innovation and Hurts Consumers

In this scenario, it appears clear that for the purposes of progress in research on the subject of biotechnology the theme of cooperation mechanisms among stakeholders and relative transaction costs is central. In this regard, agreements on research and development between firms and universities are particularly interesting, through which the former tend to retain ownership over rights for the results of research they finance, while university researchers are allowed access to findings which are protected by the intellectual property rights of the firm, or sometimes also the possibility of publishing or for external funding; this, also, with the risk of negative effects on research – with regard to traditional public

---

Worldwide', cit., p. 1099 et seq., who affirms in particular, contrasting the opinion of those who hold that patents do not exclude that owners may find themselves in competition with similar products which may carry out the same function and satisfy the same need (see N. Pires De Carvalho, *The Trips Regime of Antitrust and Undisclosed Information*, The Hague: Kluwer Law International, 2008, p. 34), that 'the very process of information gathering and innovation would cease to exist as we know it if researchers were entirely unable to share knowledge and build upon it in hopes of creating new inventions'; G. Ghidini, *Innovation, Competition and Consumer Welfare in Intellectual Property Law*, cit., p. 71 et seq.; C.M. Ho, 'Patent Breaking or Balancing?: Separating Strands of Fact from Fiction Under TRIPS' (2009) 34 *North Carolina Journal of International Law and Commercial Regulation* 371 et seq.; N.S. Sreenivasulu and C.B. Raju, *Biotechnology and patent law. Patenting living beings*, Noida (India); Manupatra, 2008, p. 236; M. Boldrin and D.K. Levine, *Against Intellectual Monopoly*, cit., p. 73 et seq.; R. Cooper Dreyfuss, 'TRIPS Round II: Should Users Strike Back?' (2004) 71 *University of Chicago Law Review* 21 et seq. (22); D.B. Resnik, *Owning the Genome. A Moral Analysis of DNA Patenting*, cit., p. 80 s., who highlights that: 'The high price of medications does not itself demonstrate that the patent system is unjust because these medications might not even be available, unless a private company had invested money to develop them. … In the short-term, it seems highly unfair that people do not have access to these medications due to their prohibitive cost. However, in the long run, the cost will go down and people will have greater access. Of course, governments could eliminate these initial high costs by sponsoring biotechnology research and development so that patients would not need to depend on private investment to obtain access to new medications. But does any government, including the U.S. government, have $ 70 billion to spare on more biomedical R & D?'; in the same 'conservative' sense as patents, C.M. Holman, *Gene Patents Under Fire: Weighing the Costs and Benefits*, in E. Arezzo and G. Ghidini (eds), *Biotechnology and Software Patent Law: A Comparative Review of New Developments*, cit., p. 260 et seq. (267 et seq.), who highlights the excesses in the interpretations of these as property rights which hinder innovation and the importance of limits in the sphere of the exclusive which are implicit in the claims.

funding – such as for instance the concentration of funding on more profitable products or processes, or a delay in the publishing of results.[119]

In the same sense, it was suggested that cooperation be created through the use of so-called 'patent pools' – agreements among patent owners in order to grant licenses of one or more of their patents together to third parties – which may be inspired by different motives, among which are the overlapping of patent claims which hinder the marketing and adoption of technology, or the perspective of elevated organization costs, but also philanthropic reasons or to encourage research. However, the diffusion of these types of agreements, besides possibly giving rise to violations of antitrust laws,[120] on the subject of biotechnology is obstructed by the fact that various patents generally concern competitive products or procedures; thus the owner of a patent upstream has an interest in making his right valid in order to prevent the use of patents downstream, and not to enter into an agreement with owners of these patents and be subject to their competition, with the relative lack of converging interest of various related patent owners to use them jointly.[121]

---

[119] In this regard, see among others: WIPO, '2011 World Intellectual Property Report. The Changing Face of Innovation', cit., p. 139 et seq.; S. Aymé, G. Matthijs and S. Soini, 'Patenting and Licensing in Genetic Testing: Ethical, Legal and Social Issues', cit., p. S33.

[120] In this regard, see R.J. Gilbert, 'Antitrust for Patent Pools: A Century of Policy Evolution' (2004) 3 *Stanford Technology Law Review* 3 et seq.; S.C. Carlson, 'Patent Pools and the Antitrust Dilemma' (1999) 16 *Yale Journal of Regulation* 359 et seq.

[121] Unlike that which occurs in the sphere of information technology, where the impulses of standardization and the need for interoperability push towards the spread of patent pools. In argument, see WIPO, '2011 World Intellectual Property Report. The Changing Face of Innovation', cit., pp. 67 et seq. and 109 et seq.; V. Falce, 'Innovation in the New Technological Industries: Looking for a Consistent Cooperative Model', cit., p. 45 et seq.; G. van Overwalle (ed.), *Gene Patents and Collaborative Licensing Models: Patent Pools, Clearinghouses, Open Source Models and Liability Regimes*, Cambridge: Cambridge University Press, 2009, p. 1 et seq.; H. Ullrich, 'Patent Pools – Policy and Problems', in J. Drexl (ed.), *Research Handbook on Intellectual Property and Competition Law*, Cheltenham, UK and Northampton, MA, USA, Edward Elgar, 2008, p. 139 et seq.; S. Aymé, G. Matthijs and S. Soini, 'Patenting and Licensing in Genetic Testing: Ethical, Legal and Social Issues', cit., p. S29; A.D. Melamed and D. Lerch, 'Uncertain Patents, Antitrust, and Patent Pools', in C.D. Ehlermann and I. Atanasiu (eds), *European Competition Law Annual 2005: The Interaction Between Competition Law And Intellectual Property Law*, Oxford/Portland (USA): Hart Publishing, 2006, p. 275 et seq., where, in particular, it is stated that

Basically similar considerations, then, are relevant in regard to the further cooperative mechanism of so-called 'clearing houses', aimed at the creation of a platform, or database, for the exchange of information – whether it be raw data or technical knowledge, whether it is patentable or not – regarding a certain innovative path, which is shared freely (according to an 'open source' model) or on the basis of standardized licensing agreements.[122]

Most recently, a further proposal to limit the sphere of patent protection on biotechnology, concerns, in fact, the application of the open source model – used in the software field – also for the circulation and use of biotechnological inventions.[123] However, a similar model is

---

a system of 'patent pools can, in other words, promote the dissemination of technology both by lowering transaction costs associated with such dissemination and by lowering the prices for such technology', keeping in mind that such systems may be just as restrictive of competition; A. Krattiger, S. Kowalski, R. Eiss and A. Taubman, 'Intellectual Property Management Strategies to Accelerate the Development and Access of Vaccines and Diagnostics: Case Studies on Pandemic Influenza, Malaria and SARS' (2006) 2 *Innovation Strategy Today* 67 et seq.; B. Verbeure, E. van Zimmeren, G. Matthijs and G. van Overwalle, 'Patent Pools and Diagnostic Testing' (2006) 24 *Trends in Biotechnology* 115 et seq.; G. van Overwalle, E. van Zimmeren, B. Verbeure and G. Matthijs, 'Models for facilitating access to patents on genetic inventions' (2006) *Nature Review Genetics* 143 et seq.; J. Lerner and J. Tirole, *Efficient Patent Pools*, NBER Working Paper 9175, 2002, accessed 19 December 2014 at http://www.nber.org/papers/w9175.pdf; J. Clark, *et al.*, *Patent Pools: A Solution to the Problem of Access in Biotechnology Patents?*, 2000, accessed 19 December 2014 at http://www.uspto.gov/web/offices/pac/dapp/opla/patentpool.pdf.

[122] In this regard, see especially V. Falce, 'Innovation in the New Technological Industries: Looking for a Consistent Cooperative Model', cit., p. 50 et seq.; G. van Overwalle (ed.), *Gene Patents and Collaborative Licensing Models: Patent Pools, Clearinghouses, Open Source Models and Liability Regimes*, cit., p. 63 et seq.; S. Aymé, G. Matthijs and S. Soini, 'Patenting and Licensing in Genetic Testing: Ethical, Legal and Social Issues', cit., p. S29; E. van Zimmeren *et al.*, 'A Clearinghouse for Diagnostic Testing: The Solution to Ensure Access to and Use of Patented Genetic Inventions?' (2006) *Bulletin of the World Health Organization* 352 et seq.; E.R. Gold, 'Biotechnology Patents: Strategies for Meeting Economic and Ethical Concerns' (2002) 30 *Nature Genetics* 359 et seq.

[123] See, among others, A. McCoy, 'Biotechnology and Embryonic Stem Cells: A Comparative Analysis of the Laws and Policies of the United States and Other Nations', cit., p. 88 et seq.; K.M. Nolan-Stevaux, 'Open Source Biology: A Means to Address the Access and Research Gaps?' (2007) 23 *Santa Clara Computer & High Technology Law Journal*, 2007, p. 271 et seq.

considered hardly practical with regard to these innovations,[124] for which it is the tendency to suggest that it share space with the proprietary model, occupying a different sphere and giving rise to competition between the two models aimed at increasing the overall well-being of society.[125]

In any case, according to the framework of the discipline on the matter which appears preferable, the underlying objective of it is to ensure a proper balance between the protection of investments and access to essential information for research and innovation.

In this sense, where the view of considering the patent as an essential element for stimulating research in biotechnology and thus for scientific progress and social and economic development is shared, from the considerations proposed up to this point one may deduce the importance – both in the U.S. and European models – of patentability requirements as important tools in the hands of legislators and case law, in order to provide adequate solutions for the various issues which have arisen. Among others, the requirement of industriality seems particularly significant, which requires the identification of specific methods of use for the finding, allowing the limitation of the exclusive patent granting it for a specific application of the invention.

---

124 Considering, in particular, the complexity of inventive and strategic behaviour which characterizes the scientific community compared to the interests on the subject of software, and the amount of funding which is necessary for biotechnological research, often with venture capital which requires adequate returns. In this sense, see S. Aymé, G. Matthijs and S. Soini, 'Patenting and Licensing in Genetic Testing: Ethical, Legal and Social Issues', cit., p. S30; G. van Overwalle, E. van Zimmeren, B. Verbeure and G. Matthijs, 'Models for facilitating access to patents on genetic inventions', cit., p. 143 et seq.

125 In this regard, see G. van Overwalle (ed.), *Gene Patents and Collaborative Licensing Models: Patent Pools, Clearinghouses, Open Source Models and Liability Regimes*, cit., p. 169 et seq.; A.W. Torrance, 'Open Source Human Evolution' (2009) 30 *Washington University Journal of Law & Policy* 93 et seq.; J. Eng Jr, 'From Software to Life Sciences: The Spreading of the Open Source Production to New Technological Areas' (2005) 24 *Temple Journal of Science, Technology & Environmental Law* 419 et seq.; A. Pollack, 'Open-Source Practices for Biotechnology', *New York Times*, 10 February 2005, p. C8; R. Feldman, 'The Open Source Biotechnology Movement: Is It Patent Misuse?' (2004) 6 *Minnesota Journal of Law, Science and Technology* 117 et seq.; D.W. Opderback, 'The Penguin's Genome, or Coase and Open Source Biotechnology' (2004) 18 *Harvard Journal of Law and Technology* 167 et seq.; M.S. McBride, 'Bioinformatics and Intellectual Property Rights' (2002) 17 *Berkeley Technology Law Journal* 1331 et seq.

### V.4.2 Relations with Bioethical Profiles between their Central Role in the European Model and the Utilitarian Approach in the United States: Two Cultures or a 'Common Core'?

As far as philosophical profiles, and in particular bioethics, of biotechnological inventions, then, are concerned, bioethics highlights how these innovations give rise to new problems for human existence, involving the risk of objectifying life and genetic manipulation, with unpredictable results. Thus, bioethicists note the necessity to use instruments of liberalism, and in particular the adoption of policy, rules and procedures to resolve such problems, as well as any controversies arising from them.[126]

On the other hand, the role of bioethical profiles on the theme of patentability for biotechnological inventions is the subject of extremely contrasting opinions, with strong opposition concerning their emergence in the patent discipline and the decisions made in this regard; that is, as has been seen, especially in the U.S. system – in which there are no regulations based on morality, nor has it found almost any space as such in case law – but also in the European system, in which, instead, both the EPC and Directive 98/44/EC provide causes for exclusion from the patentability inherent to it.[127]

---

126 See B. Bryan, 'Biotechnology, Bioethics and Liberalism: Problematizing Risk, Consent and Law' (2003) 11 *Health Law Journal* 119 et seq. (128 et seq.); in regard to liberalism see, *ex multis*, A. Wolfe, *The Future of Liberalism*, New York/London; Vintage, 2010, *passim*; J. Rawls, *Political Liberalism: Expanded Edition*, New York: Columbia University Press, 2005 (2nd edn), *passim*; P. Manent, *An Intellectual History of Liberalism*, Princeton: Princeton University Press, 1996, *passim*; Wikipedia.org, entry *Liberalism*, accessed 19 December 2014 at http://en.wikipedia.org/wiki/Liberalism (with the bibliography cited therein).

127 In this sense, in American doctrine, it has been observed that '(i)n modern times ... issues of morality and of public policy are irrelevant in the determination of whether or not an invention has patentable utility' (J.P. Daniel, 'Of Mice and "Manimal": The Patent & Trademark Office's Latest Stance Against Patent Protection for Human-Based Inventions' (1999) 7 *Journal of Intellectual Property Law* 99 et seq., 121 et seq.); in European doctrine, others have noted that patent law is in 'startling and marked isolation from matters cultural, political and ethical', in part because it is '*continually* presented as a neutral, inert system, which is above or beyond ethics', putting this down to the fact that 'the image of patent law (has been) understood and explained in positivist terms and thus is logically premised on the absence of morality' (L. Bently and B. Sherman, 'The Ethics of Patenting: Toward a Transgenic Patent System' (1995) 3 *Medical Law Review* 275 et seq.).

The distinction between law and ethics is moved, in particular, according to the principle of legal positivism,[128] which implies that a rule or decision is the result of certain procedures, and thus must be respected regardless of the fact of whether it is correct or not.[129]

What is more, it has been shown that this distinction is not exactly between law and ethics, but rather between one aspect of ethics in the creation of law and another. The statement that the regulation or decision must be respected until it has been amended according to relative procedures – or that it is in some way correct because it has been adopted in the correct manner – represents itself as an ethical concept, based on the recognition of an assumption according to which the lack of respect for institutional structures may affect the foundation of civil society, in the absence of which ethics and justice in any sense would be impossible to enforce. Further, it is a significantly complex ethical concept, involving many issues which are not easily solved when considering which rules must be considered established and the limits of the title of obedience even in undisputed frameworks. Still, according to this reconstruction, from the moment in which rules are adopted, ethics may carry out a fundamental role in supporting the decision maker in the evaluation of the aims and possible means for implementing them.[130]

In the same sense, it has been shown that one of the objectives of the patent system is to enhance research and development, encouraging public disclosure of new knowledge: thus, it would be strange if the systems provided incentives for immoral findings or those which are

---

[128] In virtue of which, as known: 'In any legal system, whether a given norm is legally valid, and hence whether it forms part of the law of that system, depends on its sources, not its merits' (see J. Gardner, 'Legal Positivism: 5½ Myths' (2001) 46 *American Journal of Jurisprudence* 199 et seq.; see also, among others: L. Greene, entry *Legal Positivism*, in *Stanford Encyclopedia of Philosophy*, 2003, accessed 19 December 2014 at http://plato.stanford.edu/entries/legal-positivism; H.L.A. Hart, *The Concept of Law*, Oxford: Clarendon Press, 1994 (2nd edn), *passim*; J. Raz, *The Authority of Law: Essays on Law and Morality*, Oxford: Clarendon Press, 1979, *passim*).

[129] See, among others, M.D. Howe, 'The Positivism of Mr. Justice Holmes' (1951) 64 *Harvard Law Review* 529 et seq., and M.D. Howe, 'Holmes' Positivism – A Brief Rejoinder', therein, 937 et seq.; H.M. Hart Jr., 'Holmes' Positivism – An Addendum', therein, p. 929 et seq.; O.W. Holmes, 'The Path of the Law' (1897) 10 *Harvard Law Review* 457 et seq.

[130] See H.M. Hart and A.M. Sacks, *The Legal Process. Basic Problems in the Making and Application of Law*, cit., p. 109 et seq.

contrary to public order. On the contrary, the granting of patents for such innovations would damage the reputation of the patent system.[131]

With reference to the issue of the patentability of biotechnological inventions, then, various opponents have stated that it would give rise to immoral patents over nature, holding that all DNA, created in nature or produced in a laboratory, is not a product of human ingenuity, and so must not be patented in any case.[132]

Several arguments, in particular, have been given against the patentability of biotechnological inventions.[133]

One argument, which may be defined as utilitarian, states that the patenting of products of nature should not be allowed, given that this would suffocate innovation and scientific discovery. In this sense, take for instance the negative consequences which would arise should it be allowed to patent newly discovered species, mathematical formulas or natural phenomena. In general, this argument against patents for products of nature appears consistent with the consideration on the basis of which the regulations on the subject of intellectual property should be balanced with the promotion of social objectives.[134]

---

[131] See E. Siew-Kuan, 'Immoral Inventions: Interaction between Ethics and Biotechnology Patent Law', cit., p. 945 et seq., who adds that: 'Arguably, an adapted version of the European model based on general immorality exclusion-cum-specific examples/guidance customized to accommodate the public interest of each nation has much to be commended in this respect. Hopefully, common principles of moral values would serve to chart the course where difficulties prove intractable'; A. McCoy, 'Biotechnology and Embryonic Stem Cells: A Comparative Analysis of the Laws and Policies of the United States and Other Nations', cit., p. 93.

[132] In this sense, see L. Andrews and D. Nelkin, *Body Bazaar: The Market for Human Tissue in the Biotechnology Age*, New York: Crown Publishers, 2001, p. 42 et seq.; J. Rifkin, *The Biotech Century: Harnessing the Gene and Remaking the World*, New York: J.P. Tarcher/Putnam, 1998, p. 37 et seq.; A. Kimbrell, *The Human Body Shop. The Cloning, Engineering and Marketing of Life*, Washington DC: Gateway, 1997 (2nd edn), *passim*.

[133] In this regard, besides referring above to Part I, see among others, for a critical analysis of such arguments, E.R. Gold, 'The Ethics of Biotechnological Intellectual Property', cit., p. 47 et seq.; F. Vandenabeele, *Patentability of Living Organisms: Legal and Ethical Aspects of the Question*, Montreal: McGill University, 2005, p. 64 et seq.; D.B. Resnik, *Owning the Genome. A Moral Analysis of DNA Patenting*, cit., p. 74 et seq.

[134] Besides referring to that which is shown above, see among others: G. Ghidini, *Innovation, Competition and Consumer Welfare in Intellectual Property Law*, cit., p. 50 et seq.; D.B. Resnik, *DNA Patents and Scientific Discovery and Innovation: Assessing Benefits and Risks*, cit., p. 29 et seq.; one may also refer to

Other arguments against the patentability of biotechnological inventions can be defined as having a moral character.

Among these, one argument has a theological nature and concerns the fact that products of nature were created by God and not by human beings.[135] In this sense, the patenting of nature could be considered a form of indignity, given that it would give rise to the exclusive right for something that was not created by man, but rather by God. While this argument is without a doubt true from a theological perspective, its applicability to a legal context is debatable.[136]

Another widespread argument considers it immoral to patent human DNA in that the human genome represents our common heritage, which belongs to all and cannot be privately owned or marketed.[137] The UNESCO Universal Declaration on the Human Genome and Human Rights of 1997, in this regard, states that the human genome is the 'common heritage of mankind', but '(i)n a symbolic sense', with the intention of minimizing the meaning of the notion to a guiding principle. This notion, in reality, essentially concerns the equal use of some resources and the respect of peoples' right to development. It is not opposed to the granting of exclusive rights and provides the alienation of resources based on existing property rights, not allowing, however, for their ownership and thus voiding the function of property rights in certain areas. In this, then, the central concept is not that of property, but rather that of equal use of certain resources. What is more, it cannot even be

---

A. Stazi, 'Intellectual Property and Consumer Law in the Knowledge Economy' (2010) *Il diritto di autore* 342 et seq.

[135] For this argument, inspired at its origin by the Book of Genesis 1:1, according to which: 'In the beginning God created the heavens and the Earth', besides referring above to Part I, see for instance, C.B. Mitchell, 'A Southern Baptist looks at Patenting Life', in A.R. Chapman, *Perspectives on Gene Patenting. Religion, Science, and Industry in Dialogue*, cit., p. 167 et seq. (176), according to which 'by claiming exclusionary property rights in a genetically-altered composition of matter, the human manipulator is assuming a place which belongs alone to God, the divine Artificer'; M.J. Hanson, 'Religious Voices in Biotechnology. The Case of Gene Patenting' (1997) 27 *Hastings Center Law Report* 1 et seq.

[136] On this point, see D.B. Resnik, *Owning the Genome. A Moral Analysis of DNA Patenting*, cit., p. 75 et seq.

[137] In this sense, besides referring to doctrine cited in Part I, see Y. Joly, 'Genetic Research Tools: Recent Trends and Future Outlook', in Gold and Knoppers, *Biotechnology IP & Ethics*, cit., p. 39 et seq. (47 et seq.); M. Sturges, 'Who Should Hold Property Rights to the Human Genome? An Application of the Common Heritage of Mankind' (1997) 13 *American University International Review* 219 et seq.

asserted that the human genome is really the heritage of humanity. In fact, 98 per cent of it is identical to that of chimpanzees and a significant percentage to that of other species, while only two per cent is exclusively human. Thus, it is not correct to state that the notion excludes the possibility for granting patents on genes.[138]

Another argument holds, then, that the patent for products of nature, and in particular for DNA, would give rise to a form of commoditization of it, which would violate the dignity and inviolable value of the natural world.[139] However, with respect to this argument, it has been shown that it also implies the immorality of other types of proprietary control over nature, beyond the patent, such as for its sale, purchase, loan, and so on. From this point of view, the problem of the patent would appear to be essentially that of allowing a human being to be considered the inventor or creator of that which already exists in nature. In this manner, further, one must return to the aforementioned issue regarding the criteria according to which the object of protection can be considered a product of human ingenuity and not of nature.[140]

In the European model, the bioethical implications of biotechnological inventions are held to be an obstacle to the modification and patentability of living organisms. In this sense, the traditional approach is based on the assumption under which human beings cannot create living organisms and cannot own them through patents, in that the life of a living organism cannot be monopolized, nor can it be treated as a commodity. The patentability of living organisms, according to this reconstruction, would be equivalent to enslaving them, violating their dignity.[141]

---

138 See in particular, C. Kuppuswamy, *The International Legal Governance of the Human Genome*, London: Routledge, 2009, p. 140 et seq.; D.B. Resnik, *Owning the Genome. A Moral Analysis of DNA Patenting*, cit., p. 13 et seq. and 77 et seq.

139 See M.J. Hanson, 'Biotechnology and Commodification Within Health Care' (1999) 24 *Journal of Medicine and Philosophy* 267 et seq.; J. Rifkin, *The Biotech Century: Harnessing the Gene and Remaking the World*, cit., *passim*; A. Kimbrell, *The Human Body Shop. The Cloning, Engineering and Marketing of Life*, cit., *passim*; A. Shiva, *Biopiracy. The Plunder of Nature and Knowledge*, Boston: South End Press, 1996, p. 19 et seq.

140 See again D.B. Resnik, *Owning the Genome. A Moral Analysis of DNA Patenting*, cit., p. 83.

141 In this regard, besides referring above to Chapter 4 and below, see among others, S. Aymé, G. Matthijs and S. Soini, 'Patenting and Licensing in Genetic Testing: Ethical, Legal and Social Issues', cit., p. S16 et seq.

On the other hand, over time the progress of biotechnology and its enormous utility has outweighed several of these bioethical considerations, leading some commentators to believe that they could be sacrificed in view of the benefits of biotechnological inventions. In the European context, however, the importance of an approach to the subject which is inspired by the balancing of benefits coming from biotechnology with bioethical instances is clear: biotechnology is capable of producing transgenic human beings, or using human embryos, but for such cases patentability is prohibited, in that such practices are considered immoral. In view of the aforementioned balance of values, in fact, it is held that in these cases the bioethical instances outweigh the potential benefits.[142]

In the U.S. system, starting from the application regarding the human-animal chimera presented by Newman and Rifkin, despite the limitation of the context and the involvement of third parties, the review mechanism before the USPTO has been used at various times to present cases of a moral character. However, as has been seen, in general these issues have not been accepted as such, while they were eventually accepted 'indirectly', focusing the attention on the relative implications in terms of the fulfilment of patentability requirements.[143]

The USPTO, on the other hand, has acknowledged in part the importance of such cases implementing an internal 'Sensitive Application Warning System', aimed at the early identification by workers and then office managers of those patent applications which may appear likely to giving rise to debate and criticism, so that they may give particular attention to controversial aspects throughout the examination, as well as allowing them to prepare for any publicity which may arise from it. This system, in any case, is internal and informal, is not codified and lacks operative tools, beyond the judgment and skill of the examiners, to ensure its function.[144]

At a European level, however, these issues have received greater attention from both the European Parliament and Commission and the EPO, who, in their decisions on the subject of biotechnological patents, have taken their bioethical, social, economic and ecological implications

---

142 See N.S. Sreenivasulu and C.B. Raju, *Biotechnology and Patent Law. Patenting Living Beings*, cit., p. 232 et seq.

143 In this regard, refer to that which is highlighted above, and in particular to Part III, as well as to the doctrine referred to therein.

144 See S. Parthasarathy, 'Whose Knowledge? What Values? The Comparative Politics of Patenting Life Forms in the United States and Europe', cit., p. 279 et seq.

into consideration. Directive 98/44/EC accepted such instances, in particular, through the specification of the clause for public order and morality already present in the EPC, and the ban of inventions which involve the use of the human embryo for industrial or commercial purposes, of procedures for the modification of germinal genetic makeup or cloning of human beings, and patents for plant varieties and animal breeds, as well as through the introduction of a test for patents regarding animals based on the measuring of benefits/risks, in order to ensure that the benefits for humanity outweigh the risks for the animals involved.[145] As a consequence, as has been seen, in the decisions of the EPO and the Court of Justice various patents have been denied, limited or revoked on the basis of opposition on appeal inherent to the aforementioned profiles.

Within the EPO, further, various initiatives have been developed for the training of personnel regarding the issues under discussion, and similarly to the USPTO a system for 'Sensitive Cases' has been developed, which in this case is initiated at the classifying stage of the application, and formalized through preprinted labels and a specific procedure by the protocol office and technical departments. More broadly, then, at the moment of issuing a patent which may be susceptible to creating disputes or controversy, or when they come about, or with respect to problematic issues in general, the EPO often uses communication tools such as press releases or informative charts, which describe the issues of public interest and the reasons at the foundation of its decision.[146]

Bioethical limits to biotechnological patents, however, do not appear in and of themselves capable of preventing the diffusion of morally reprehensible inventions, in that there is not a 'necessary' relation between

---

[145] On these aspects, besides referring to Part IV, see *ex multis*, E. Siew-Kuan, 'Immoral Inventions: Interaction between Ethics and Biotechnology Patent Law', cit., p. 937 et seq.; I. Schneider, 'The Evolution of a "European Mode" of Patent Governance: Legislation on the Patenting of Biotechnology in the European Union and Germany', in J.P. Gaudillière, D.J. Kevles and H.J. Rheinberger (eds), *Living Properties: Making Knowledge and Controlling Ownership in the History of Biology*, Berlin: Max-Planck Institute for the History of Science, 2009, p. 187 et seq.; M.A. Bagley, *A Global Controversy: The Role of Morality in Biotechnology Patent Law*, cit., p. 330 et seq.; P. Drahos, 'Biotechnology Patents, Markets, and Morality' (1999) 21 *European Intellectual Property Review* 441 et seq.

[146] In this regard, see again S. Parthasarathy, 'Whose Knowledge? What Values? The Comparative Politics of Patenting Life Forms in the United States and Europe', cit., p. 280 et seq.

bioethical issues and patentability.[147] The function of hindering the production of a certain invention may – or better, as a rule, should – be pursued through regulations beyond patent regulations, through which states regulate research on ethically sensitive materials. Moreover, where it is not patentable, the finding may eventually be imported[148] or protected using the regulations on trade secrets.[149]

Thus, regulations regarding bioethical profiles in patent law seem to lay down essentially 'connecting links' between policy pursued through patent law and the fundamental principles of the legal system, leaving the process by which choices on bioethical themes are carried out by the Member states within other regulations connected to patents unchanged.[150] In this regard, within the European model the provisions contained in the regulation of the Charter of Fundamental Rights of the European Union are particularly relevant, including, in particular, those relative to the principle of human dignity and the rights to life and integrity of the individual, to be reconciled with those, equally provided for, of freedom of scientific research and protection of intellectual property.[151] Such rights, in the sphere of biomedicine, are in particular:

---

[147] See F. Vandenabeele, *Patentability of Living Organisms: Legal and Ethical Aspects of the Question*, cit., p. 84 et seq.

[148] Think for instance of the protection of the embryo in the German system, in which as noted the production of embryos for research is prohibited, but stem cell lines obtained in compliance with the legal situation in the country of origin prior to May 2007 may be used for research (see '*Gesetz zur Sicherstellung des Embryonenschutzes im Zusammenhang mit Einfuhr und Verwendung menschlicher embryonaler Stammzellen*', 28 May 2002, as renewed in 2008, art. 4).

[149] With regard to the limits of this protection, further, see, among others, G. van Overwalle, *Study on the Patenting of Inventions Related to Human Stem Cell Research*, Darby (USA): Diane Publishing, 2003, p. 81 et seq.

[150] In this regard, see, among others: A. Ottolia, 'Moral Limits to Biotech Patents in Europe: A Quest for Higher Harmonization', in E. Arezzo and G. Ghidini (eds), *Biotechnology and Software Patent Law: A Comparative Review of New Developments*, cit., p. 309 et seq. (326 et seq.); M. Ricolfi, 'Biotechnology, Patents and Epistemic Approaches' (2002) *Journal of Biolaw & Business, Special Supplement* 77 et seq.

[151] Charter of Fundamental Rights of the European Union, arts 1–3, 13 and 17(2). In regard to the principle of human dignity, further, it must be shown how it is laid out in the Treaty of the European Union, in the current art. 2, as first among the values on which the Union is based, alongside, among others, the principle of freedom, equality and solidarity, which as said above in Part IV, are also indicated in the Charter of Nice as 'indivisible and universal' values; thus, in art. 10A, it is among the objectives which the Union intends to promote in the framework of international relations. In the principle of dignity, according to the

(a) free and informed consent of the individual involved, according to the methods defined by law; (b) the ban on eugenic practices, in particular those whose aim is the selection of people; (c) the ban on rendering the human body and its parts, as such, a source of profit; and (d) the ban on the reproductive cloning of human beings.[152]

Considering that the Charter of Nice, following the Treaty of Lisbon, took on the same legal value as treaties, it currently seems necessary for the regulations of the single Member states to take on the fundamental rights proclaimed within it as the basic criteria on which to build their own regulatory acts and legal decisions.[153] At the same time, however,

---

preferred reconstruction, two dimensions coexist, that are the 'subjective' (or of 'empowerment') and the 'objective' (or of 'constraint'), which characterize the operative dynamic. On this theme, besides referring to Part IV, see R. Andorno, 'Human Dignity and Human Rights as a Common Ground for a Global Bioethics' (2009) 34 *Journal of Medicine and Philosophy* 223 et seq.; P. Capps, *Human Dignity and the Foundations of International Law*, Oxford/Portland (USA): Hart Publishing, 2009, p. 106 et seq.; R. van der Graaf and J.J. van Delden, 'Clarifying Appeals to Dignity in Medical Ethics from an Historical Perspective' (2009) 23 *Bioethics* 151 et seq.; C. McCrudden, 'Human Dignity and Judicial Interpretation of Human Rights' (2008) 19 *European Journal of International Law* 655 et seq.; G. Bognetti, 'The Concept of Human Dignity in European and U.S. Constitutionalism', in G. Nolte (ed.), *European and U.S. Constitutionalism*, Cambridge: Cambridge University Press, 2005, p. 85 et seq.; J.Q. Whitman, '"Human Dignity" in Europe and the United States: The Social Foundations', therein, p. 108 et seq.; P. Häberle, 'Die Menschenwürde als Grundlage der staatlichen Gemeinschaft', in J. Isensee and P. Kirchhof (eds), *Handbuch des Staatsrechts der Bundesrepublick Deutschland, II, Verfassungsstaat*, Heidelberg: Müller Verlag, 2004 (3rd edn), p. 317 et seq.; J. Habermas, *The Future of Human Nature*, Cambridge: Polity Press, 2003, p. 1 et seq.; D. Beyleveld and R. Brownsword, *Human Dignity in Bioethics and Biolaw*, Oxford: Oxford University Press, 2002, p. 11 et seq.; M.C. Nussbaum, 'Objectification' (1995) 24 *Philosophy and Public Affairs* 249 et seq.; J.P. Fletcher, 'Human Dignity as a Constitutional Value (1984) 22 *University of Western Ontario Law Review* 171 et seq.

[152] Charter of Fundamental Rights of the European Union, art. 3. With regard to the significance of the principles of the Charter of Nice on the subject of biotechnology, see P. Šturma, 'The EU Charter of Fundamental Rights and the Governance of Biotechnology in the European Union', in F. Francioni and T. Scovazzi (eds), *Biotechnology and International Law*, Oxford/Portland (USA): Hart Publishing, 2006, p. 369 et seq.; N. Lenoir, 'Biotechnology, Bioethics and Law: Europe 21st Century Challenge' (2006) 69 *Modern Law Review* 1 et seq.

[153] In this view it may be understood, for instance, why Great Britain, Poland and the Czech Republic requested and received the so-called opt-out clause for the application of the regulations of the Charter.

such rights appear to be essential interpretative keys to the application of the patent discipline, through the 'general open clause' of public order, which in the current EU system seems to be stated in fact basically in the principles of the Charter of Nice.

The provision of moral clauses in European patent law, on the other hand, is held by doctrine as aimed at essentially producing legal certainty in the field of biotechnology, through interpretive instruments which contribute to the greater predictability of evaluation by patent offices and the courts, and thus to the development of the internal market. Such clauses, further, should not be considered where bioethical profiles of biotechnological inventions should be faced. They, rather, should be dealt with in further regulations which specifically discipline highly sensitive themes from a bioethical point of view, on the basis of the cultural and religious traditions of the single Member state.[154]

Such a reconstruction would be supported by the existence of so-called 'mixed' systems, in which, on the one hand, patentability and research on certain materials for bioethical reasons is prohibited; on the other, the importation of them from third party countries is permitted.[155] In the same sense, further, a high level of patent protection within a foreign system would push domestic firms to deviate from their research and development activities, as well as the patenting of the consequent results, through the same: from this, clearly, a phenomena of 'competition among legal systems' would arise, which could also, hypothetically, produce economic benefits, but which is certainly inconsistent with the effective achievement of bioethical purposes, in that such asymmetry could give rise to the attempt to 'externalize immorality'.[156]

---

[154] In this regard, see for instance: EU Court of Justice, *Association Eglise de scientologie de Paris and Scientology International Reserves Trust v The Prime Minister*, Case C-54/99, [2000] ECR I-1335. In doctrine, see, among others, A. Ottolia, *Moral Limits to Biotech Patents in Europe: A Quest for Higher Harmonization*, cit., p. 328; A. Plomer, *Stem Cell Patents: European Patent Law and Ethics Report*, 2006, accessed 19 December 2014 at http://www.nottingham.ac.uk/~llzwww/StemCellProject/project.report.pdf, p. 43; P. Spada, 'Etica dell'innovazione tecnologica ed etica del brevetto' (1996) *Rivista di diritto privato* 217 et seq.

[155] On this point, refer to the example of embryo protection in the German system, referred to above.

[156] See again, A. Ottolia, *Moral Limits to Biotech Patents in Europe: A Quest for Higher Harmonization*, cit., p. 331. On the theme of competition between legal systems, besides referring above, see, among others, H.N. Butler and L.E. Ribstein, 'Legal Process and the Discovery of Better Policies for Fostering Innovation and Growth', cit., p. 463 et seq.; E.A. O'Hara and L.E. Ribstein, *The*

In this regard, in the first place Directive 98/44/EC intervened to provide a series of minimum legal conditions needed for patenting; this is so regardless of the existence of other regulations specifically relevant to the discipline of bioethical profiles for scientific research. What is more, recently, the decision of the Court of Justice in the European Union in *Oliver Brüstle v. Greenpeace eV* seems to bring the framework outlined above regarding the relation between patent law and regulations specifically concerning bioethical themes even more clearly under discussion, giving rise to a broadening of the patent discipline's context of effectiveness with respect to bioethical issues.

This is, in particular, through the statement of some principles – regarding especially the notion of 'human embryo' – which, although they are outlined in the context of a decision adopted pursuant to patent rights, seem likely to exercise influence also over external regulations which concern bioethical issues. The same ruling, further, 'complains' about its own formal limits of efficacy in the sphere of patents, where it ends up leaving the task of verifying whether stem cells obtained from a human embryo in the blastocyst stage, in light of scientific developments, may give rise to the development of a human being and, thus, fall under the notion of human embryo, to national courts (in this perspective, see also the following *International Stem Cell* decision). Such a margin of discretion, in fact, appears aimed at not interfering with national provisions (for instance, of the Spanish system) which are more permissive regarding research on so-called pre-embryos.[157]

The *Brüstle* decision was highly criticized, in that it could reduce interest in obtaining patents for research results on the subject at a European level: which could lead to, according to some, a reduction in the interest of developing such research; according to others – in an observation which appears more likely – a push to patent such results in countries with disciplines which are less oriented towards bioethical profiles, such as for instance in the United States.[158]

---

*Law Market*, Oxford/New York: Oxford University Press, 2009, *passim*; X. Benavides, 'Competition Among Laws' (2007) 77 *Revista Jurídica de la Universidad de Puerto Rico*, p. 373 et seq., accessed 19 December 2014 at http://www.law.yale.edu/documents/pdf/sela/XimenaBenavides__English_.pdf; A. Ogus, 'Competition Between National Legal Systems: A Contribution of Economic Analysis to Comparative Law' (1999) 48 *International and Comparative Law Quarterly* 405 et seq.

[157] On these points see also, *amplius*, Parts II and IV.

[158] As shown in Part IV, however, a predictable consequence of the ruling could be to concentrate research on instruments which are more respectful of

The U.S. legal system, which led the way to the recognition of biotechnological patents with the *Chakrabarty* ruling, up to now has in general avoided taking moral issues into consideration in the context of the biotechnological patent discipline.

Congress has not introduced the evaluation of bioethical profiles in the patent system, and the Patent Office prefers to face moral instances indirectly, focusing its decisions regarding findings which raise such issues on the fact that the invention concerns non-patentable subject matter,[159] or does not meet the requirements for patentability,[160] and/or is

---

bioethics, in order to obtain patents regarding stem cells obtained from 'induced pluripotency' adult stem cells, or from embryonic stem cells which do not involve the destruction of the embryo (see European Centre for Law and Justice, '*Brüstle case: The ECLJ welcomes the decision of the Court of Justice of the European Union on the status of the human embryo*', 18 October 2011, accessed 19 December 2014 at http://eclj.org/Releases/Read.aspx?GUID=dbe0cfa9-b5b6-4c48-a419-60115f434ad4&s=eur).

[159] In this sense, think of the decision in which the Patent Office rejected the patent application presented by Professor Stuart Newman and the Sociologist Jeremy Rifkin, regarding a method for producing a creature which was part human and part animal. Initially, the PTO declared in a press release that: 'It is the position of the PTO that inventions directed to human/non human chimera could, under certain circumstances, not be patentable because, among other things, they would fail to meet the public policy and morality aspects of the utility requirement', giving the impression of being oriented toward the revival of the application of 'moral utility doctrine'. In the decision, however, the patent application was not rejected for this reason, but rather in that the hybrid included a human being, and thus did not fall under patentable subject matter. In regard to the decision, besides referring above, to Part III, see R. Hagglund, 'Patentability of Human-Animal Chimeras' (2008) 25 *Santa Clara Computer & High Technology Law Journal* 51 et seq.; R.P. Merges and J.F. Duffy, *Patent Law and Policy: Cases and Materials*, Newark (USA): LexisNexis, 2007 (4th edn), p. 131; S.M. Coughlin, 'The Newman's Application and the USPTO's Unnecessary Response' (2006) 5 *Chicago-Kent Journal of Intellectual Property* p. 90 et seq.; J. Wilson, 'Patenting Organisms: Intellectual Property Law Meets Biology', in D. Magnus, A. Caplan and G. McGee (eds), *Who Owns Life?*, New York: Prometheus Books, 2002, p. 25 et seq. (54).

[160] In this regard, consider the decision with which Board of Patent Appeals and Interferences of the PTO declared a patent of the Wisconsin Alumni Research Foundation on a technique which allowed the obtainment of embryonic stem cells invalid: the reason was not of a moral nature, but rather focused on the application of criteria of novelty and non-obviousness, in that the claims of the patent were already the subject of a previous patent from 1992. With regard to the *WARF* case, besides referring above to Part III, see, among others, S.R.

contrary to the ban on property rights for human beings laid out in the XIIIth Amendment.[161]

According to a part of the doctrine, this appears in line with the utilitarian approach adopted in the American system, which focuses the regulation of intellectual property in general on the guiding criteria of progress, which, pursuant to Article I, Section 8 of the U.S. Constitution,[162] represents the foundation and at the same time the limit of Congress's power to legislate on the matter.[163]

On the contrary another part of doctrine, in order to lead the U.S. system towards an effective analysis of the various bioethical issues constantly raised by biotechnological findings, suggested that the U.S. model be inspired by the European one, through the establishment of criteria for an evaluation which excludes some findings from the sphere of patentability in that they are contrary to public order or morality.[164]

---

Donnelly, 'The Patentability of Human Embryonic Stem Cells: Is the Inconsistent Application of the European Union Biotechnology Directive's Moral Exclusion Clause Undermining Investor Confidence in Europe, Providing a Competitive Advantage to the United States?' (2011) 20 *Dalhousie Journal of Legal Studies* 106 et seq. (126); J.M. Golden, 'WARF's Stem Cell Patents and Tensions between Public and Private Sector Approaches to Research' (2010) 38 *Journal of Law, Medicine and Ethics* 314 et seq.; A.R. Chapman, 'The Ethics of Patenting Human Embryonic Stem Cells' (2009) 19 *Kennedy Institute of Ethics Journal* 261 et seq. (275 et seq.).

161 On the basis of which: 'Neither slavery nor involuntary servitude, except as a punishment for crime where of the party shall have been duly convicted, shall exist within the United States, or any place subject to their jurisdiction. Congress shall have the power to enforce this article by appropriate legislation'.

162 In which it is provided, as such, that: 'The Congress shall have Power … To promote the Progress of Science and useful Arts, by securing for limited Times to Authors and Inventors the exclusive Right to their respective Writings and Discoveries'.

163 See A. Ottolia, *Moral Limits to Biotech Patents in Europe: A Quest for Higher Harmonization*, cit., p. 335 et seq.; Y. Benkler, 'Constitutional Bounds of Database Protection: The Role of Judicial Review in the Creation and Definition of Private Rights in Information' (2000) 15 *Berkeley Law & Technology Journal* p. 535 et seq.

164 See, in particular E. Siew-Kuan, 'Immoral Inventions: Interaction between Ethics and Biotechnology Patent Law', cit., p. 946 et seq.; A.R. Chapman, 'The Ethics of Patenting Human Embryonic Stem Cells', cit., p. 281 et seq., who, in particular, affirms that: 'Continuing disagreements within Europe over the interpretation of the relevant rules about patent exclusions underscore the difficulties of achieving consensus within a pluralistic cultural and social environment. A participatory, inclusive, and transparent process to shape such an

In this sense, as hinted, the recent U.S. Supreme Court decisions in *Myriad* and *Mayo* (and the subsequent USPTO Patent Eligibility Guidance), along with the EU Court of Justice decision in *Brüstle* and that of the EPO in *WARF*, may lead toward a common 'Western approach' to the regulation of biotechnological inventions, with specific regard to the limits of patentability.

Such an approach could be based on the balance of fundamental rights and public and private interests which are relevant on a case-by-case basis, resorting to the criteria of hierarchy and proportionality in order to regulate value-based choices and functional interactions between them.[165]

Finally, the same approach would be particularly relevant to the current perspective, which is aimed at enhanced transatlantic cooperation on the matter of intellectual property, specifically within the framework of the 'Transatlantic Trade and Investment Partnership'(TTIP) that is presently being negotiated.

*De iure condendo*, then, it may be noted that the approach not in compliance with legislation and case law in the U.S. and Europe regarding the bioethical profiles of biotechnological inventions may offer the opportunity for greater coordination of regulations on the matter at an international level. This may occur, for instance, through a revisionary act of the TRIPS Agreement, or the issuance of applicative guidelines, with which some essential bioethical limits are identified, inspired by universally shared fundamental rights,[166] in order to derive greater certainty of the law in

---

approach would therefore be important. Ethical screens should be formulated through an ethical process'.

[165] In this regard, see *amplius*, G. Ghidini and A. Stazi, *Freedom to Conduct a Business, Competition and IP*, cit.; G. Ghidini, 'Exclusion and Access in Copyright Law: The Unbalanced Features of the European Directive "On Information Society"', in G.B. Dinwoodie (ed.), *Methods and Perspectives in Intellectual Property*, Cheltenham, UK and Northampton, MA, USA, Edward Elgar, 2013, p. 307 et seq.; A. Stazi and M. Marzetti, 'Synergetic Interaction Between Intellectual Property and Consumer Protection: A Pragmatic Proposal to Rebalance Incentives and Access', in D. Beldiman (ed.), *Access to Information and Knowledge: 21st Century Challenges in Intellectual Property and Knowledge Governance*, Cheltenham, UK and Northampton, MA, USA, Edward Elgar, 2013, p. 189 et seq.

[166] In this sense, see C. Geiger, 'Fundamental Rights as Common Principles of European (and International) Intellectual Property Law', cit., p. 236, who states that 'the moral and cultural values of fundamental rights (as included for instance in the UDHR) are undisputed and could even represent a basis for worldwide harmonisation. … One should not forget, however, that the TRIPS-Agreement itself should be interpreted in the light of the UDHR.

regulations without specific provisions on such aspects (such as in the United States), as well as, in general, a 'minimum bioethical common denominator' which approximates the disciplines on the matter of biotechnological inventions regarding certain fundamental profiles, in view of a unified concept on the evolution of the human species and scientific progress.[167]

The TRIPS Agreement, in article 27, further, allows signing countries to introduce the exclusion from patentability into their legislation for reasons of public order or morality, as well as to protect the life or health of man or of animals, and in particular with regard to diagnostic, therapeutic or surgical methods for their treatment.[168] In this manner, single regulations are left free to determine the patentability of genes, gene sequences and in general of genetic products or processes. The ambiguity resulting from this has given rise to significant disparity between the disciplines on the matter,[169] as well as to several disputes among patent owners and competent authorities.[170]

---

According to some scholars, this results in the primacy of international human rights acts over trade liberalization rules'; on this point, see also G. Marceau, 'The WTO Dispute Settlement and Human Rights', in F.M. Abbott, C. Breining-Kaufmann and T. Cottier (eds), *International Trade and Human Rights Foundations and Conceptual Issues*, Ann Arbor (USA): University of Michigan Press, 2006, p. 181 et seq.

167 This coordination, further, would be useful also to prevent the phenomena of possible competition between legal systems referred to previously mentioning its inconsistency with the effective achievement of moral purposes.

168 See TRIPS Agreement, art. 27(2) and (3). In general, the TRIPS Agreement binds adhering countries only to allow the introduction of a series of intellectual property rights aimed at reinforcing protection at a national level, without specific obligations regarding the reduction or elimination of this protection (see R. Cooper Dreyfuss, 'TRIPS Round II: Should Users Strike Back?', cit., p. 21 et seq.).

169 As seen, through the analysis of legislation and case law in the United States and European models; think only, for instance, of the various decisions on the cases of *WARF* and *Myriad* in the U.S. and European systems, or the recent European decision on the *Brüstle* case, an expression of a quite distant approach (at least in its current state) from that of the U.S. system, but that has also been the subject of criticism within single Member states of the EU. On the theme of the significant differences between the disciplines involved in the TRIPS Agreement, see, among others, C.A. Fowler, 'Ending Genetic Monopolies: How the TRIPs Agreement's Failure to Exclude Gene Patents Thwarts Innovation and Hurts Consumers Worldwide', cit., p. 1094; K.J. Strandburg, 'Evolving Innovation Paradigms and the Global Intellectual Property Regime' (2009) 41 *Connecticut Law Review* 861 et seq. (896).

170 See in this regard, WHO, 'Genetics, genomics and the patenting of DNA: Review of potential implications for health in developing countries', 2005,

The section of the Agreement relative to patents, then, could be modified, for instance, introducing the compulsory exclusion from patentability of materials existing in nature,[171] or of inventions which give rise to results which are universally considered abhorrent,[172] or still, of inventions which are based on genetic material whose extraction from a human being damages her/him. In particular, article 27 could include the provision of exclusion from patentability for such findings for reasons of public health,[173] and eventually of the need for patent offices, in facing investigations which would likely involve such issues, to work side by side with bioethics committees which shall support them in their evaluation.[174]

Finally, should the introduction of such revisions not be possible, a further instrument of minimal harmonization of the regulatory framework and market on the matter of biotechnological inventions in the United States and European models could be identified in the adoption of an agreement based on a sort of 'Safe Harbor Principles',[175] on the basis of

---

accessed 19 December 2014 at http://www.who.int/genomics/patentingDNA/en, p. 25 et seq. Further, on the basis of the TRIPS Agreement, the legal controversies occurred only following appeal by the Member states, without the possibility for a single patent owner to file an appeal for violation of the Agreement; on this point, see J. Gibson, *Intellectual Property, Medicine and Health: Current Debates*, Farnham (UK): Ashgate Publishing, 2009, p. 160.

[171] In this sense, see L. Palombi, 'The Impact of TRIPS on the Validity of the European Biotechnology Directive' (2005) 2 *Journal of International Business and Law* 62 et seq. (70).

[172] Such as, for instance, transgenic humans or those which originate from reproductive cloning; in this sense, see N.S. Sreenivasulu and C.B. Raju, *Biotechnology and Patent Law. Patenting Living Beings*, cit., p. 236.

[173] This, in fact, appears to be in jeopardy due to the repression of development and commercialization of genetic tests, medicines, etc. On this point, see C.A. Fowler, 'Ending Genetic Monopolies: How the TRIPs Agreement's Failure to Exclude Gene Patents Thwarts Innovation and Hurts Consumers Worldwide', cit., p. 1103; M. Swenarchuk, 'Of Harvard Mice and Prairie Farmers: Canadian Patents on Life', in R. Grinspun and Y. Shamsie (eds), *Whose Canada?: Continental Integration, Fortress North America, and the Corporate Agenda*, Montreal: Mcgill Queen's University Press, 2007, p. 481 et seq. (488).

[174] In regard to the opportunity of the establishment of specialized committees, where necessary, to work side by side with patent offices for the investigation of bioethical profiles (in an advisory position), besides referring above, see N.S. Sreenivasulu and C.B. Raju, *Biotechnology and Patent Law. Patenting Living Beings*, cit., p. 236; D.B. Resnik, *Owning the Genome. A Moral Analysis of DNA Patenting*, cit., p. 200 et seq.

[175] As is well known, such a framework is provided for privacy protection (accessed 19 December 2014 at http://export.gov/safeharbor).

which firms subject to another system – or, with specific regard to bioethical profiles, considering the more restrictive European system, American companies – may freely carry out their research and development activities, and in particular obtain relevant exclusive rights, after undertaking to respect essential principles dictated by the discipline in force on the matter within the other system.[176]

---

[176] That is, in fact, within the European system.

# Index